THE ROLE OF IoT AND BLOCKCHAIN

Techniques and Applications

THE ROLE OF IoT AND BLOCKCHAIN

Techniques and Applications

Edited by
Sanjay K. Kuanar, PhD
Brojo Kishore Mishra, PhD
Sheng-Lung Peng, PhD
Daniel D. Dasig, Jr., PhD

AAP | APPLE ACADEMIC PRESS

First edition published 2022

Apple Academic Press Inc.
1265 Goldenrod Circle, NE,
Palm Bay, FL 32905 USA

4164 Lakeshore Road, Burlington,
ON, L7L 1A4 Canada

CRC Press
6000 Broken Sound Parkway NW,
Suite 300, Boca Raton, FL 33487-2742 USA

2 Park Square, Milton Park,
Abingdon, Oxon, OX14 4RN UK

Library and Archives Canada Cataloguing in Publication

Title: The role of IoT and blockchain : techniques and applications / edited by Sanjay K. Kuanar, PhD, Brojo Kishore Mishra, PhD, Sheng-Lung Peng, PhD, Daniel D. Dasig, Jr., PhD.

Names: Kuanar, Sanjay Kumar, editor. | Mishra, Brojo Kishore, 1979- editor. | Peng, Sheng-Lung, editor. | Dasig, Daniel D., Jr., editor.

Description: First edition. | The chapters are selected papers from the International Conference on Recent Trends in IoT and Blockchain, held at the GIET University, Gunupur, Odisha, India, 19-20 October 2019. | Includes bibliographical references and index.

Identifiers: Canadiana (print) 20210340142 | Canadiana (ebook) 20210340290 | ISBN 9781771889148 (hardcover) | ISBN 9781774638101 (softcover) | ISBN 9781003048367 (ebook)

Subjects: LCSH: Blockchains (Databases)—Congresses. | LCSH: Internet of things—Congresses. | LCGFT: Conference papers and proceedings.

Classification: LCC QA76.9.B56 R65 2022 | DDC 005.74—dc23

Library of Congress Cataloging-in-Publication Data

CIP data on file with US Library of Congress

ISBN: 978-1-77188-914-8 (hbk)
ISBN: 978-1-77463-810-1 (pbk)
ISBN: 978-1-00304-836-7 (ebk)

About the Editors

Sanjay K. Kuanar, PhD
Professor and Head, Computer Science and Engineering Department, GIET University, Odisha, India

Sanjay K. Kuanar, PhD, is Professor and Head of Computer Science and Engineering Department of GIET University, Odisha, India. He has (co)authored several publications in refereed SCI journals, including *IEEE Transactions on Multimedia, IEEE Transactions on Cybernetics*, and Elsevier's *Journal of Visual Communication and Image Representation*, for which he has also acted as a reviewer. He has also participated in many international conferences. His research interests include computer vision, pattern recognition, machine learning, and multimedia computing. His primary research is focused on exploring different machine learning techniques for solving several challenging problems in summarizing and segmenting videos. Dr. Kuanar earned his PhD from Jadavpur University, Kolkata, his ME (Computer Engineering) degree from Jadavpur University, and his BTech in Computer Science and Engineering from Utkal University.

Brojo Kishore Mishra, PhD
Professor, Computer Science and Engineering Department, GIET University, Gunupur, Odisha, India

Brojo Kishore Mishra, PhD, is a Professor in the Computer Science and Engineering Department at the GIET University, Gunupur, Odisha, India. He has published more than 32 research papers in national and international conference proceedings, over 27 research papers in peer-reviewed journals, over 24 book chapters, two authored books, and four edited books. His research interests include data mining and big data analysis, machine learning, and security. He received his PhD degree in Computer Science from the Berhampur University.

Sheng-Lung Peng, PhD

Professor, Department of Computer Science and Information Engineering, National Dong Hwa University (NDHU), Taiwan

Sheng-Lung Peng, PhD, is a Professor in the Department of Computer Science and Information Engineering at National Dong Hwa University (NDHU), Taiwan. Dr. Peng has edited several special issues for journals, such as *Soft Computing, Journal of Internet Technology, Journal of Computers,* and *MDPI Algorithms.* He is also a reviewer for many journals. He published more than 100 papers in international journals and conferences. Dr. Peng is now the Dean of the Library and Information Services Office of NDHU, an honorary Professor of the Beijing Information Science and Technology University of China, and a visiting Professor at the Ningxia Institute of Science and Technology of China. He is also the regional director of the ACM-ICPC Contest Council for Taiwan, a director of the Institute of Information and Computing Machinery (IICM), and a director of the Information Service Association of the Chinese Colleges and of the Taiwan Association of Cloud Computing (TACC). He is also a supervisor of the Chinese Information Literacy Association, the Association of Algorithms and Computation Theory (AACT), and the Interlibrary Cooperation Association in Taiwan. He has also served as secretary-general of TACC, AACT, IICM. His research interests are in designing and analyzing algorithms for bioinformatics, combinatorics, data mining, and networks.

Daniel D. Dasig, Jr., PhD

Associate Professor, Electrical and Computer Engineering, De La Salle University, Dasmariñas, Philippines

Daniel D. Dasig, Jr., PhD, is a Professorial Lecturer in electrical engineering, environmental engineering, information technology, computer science, and education in the Philippines and abroad. He is an Associate Professor 6 of the Graduate Studies of College of Science and Computer Studies of De La Salle University Dasmarinas, Philippines, and a Professorial Lecturer at the College of Continuing and Advanced Professional Studies of the University of Makati, where he also served as Research Coordinator in the College of Computer Science. He served as Chair of Computer Engineering Department of Jose Rizal University, Philippines. He also works as part of the Change Management and software development teams of TELUS, a Canadian telecommunication company. He is a Lean Six Sigma Certified, ITIL V3 Certified, and Certified Project Management Expert. Dr. Dasig has authored and co-authored more than 100 academic and technical papers extending over a wide spectrum of research areas, such as in engineering, information technology, education, computer science, and social sciences. He has authored and co-authored books for higher education and senior high school programs. He has organized and co-organized local and international conferences and was a recipient of several awards and fellowships. He has held executive roles and active membership in more than 40 professional, research, and policymaking organizations, such as ICANN-NCSG, Internet Society, ICpEP, and PSITE. He has received numerous global and international awards, including the World Class Filipino for Engineering and Information Technology. He is an IEEE member of the Republic of the Philippines Section. He received his BSc in Computer Engineering, MSc in Engineering in Computer Engineering, and PhD in Electrical and Computer Engineering degrees.

Contents

Contributors

Md. Iman Ali
Research Scholar, School of Computer Application, Lovely Professional University, Phagwara, Punjab, India, E-mail: mdimanali@gmail.com

Auropremi Aspruha
Integrated Test Range, Defence Research Development Organization (DRDO), Chandipur, Odisha – 756025, India

P. Sudheer Babu
Department of Computer Science, GIET University, Gunupur, Odisha, India, E-mail: sudheerpunuri@giet.edu

Bikash Ranjan Bag
Department of Computer Science, Berhampur University, Berhampur – 761006, India, E-mail: bikashbag@gmail.com

Debrup Banerjee
Department of Computer Science and Engineering, Koneru Lakshmaiah Education Foundation, Vaddeswaram, Guntur, Andhra Pradesh, India

Ranjit Kumar Barai
Department of Electrical Engineering, Jadavpur University, Kolkata, West Bengal, India, E-mail: ranjit.k.barai@gmail.com

Anand Bhaskar
Assistant Professor, Sir Padampat Singhania University, Rajasthan, India, E-mail: anand.bhaskar@spsu.ac.in

Abhishek Bhattacharjee
Department of IT, C.V. Raman College of Engineering, Bhubaneswar, India, E-mail: abhishekb496@gmail.com

Nilesh Bhosale
Principal Software Engineer, Credit Vidya, Hyderabad, Telangana, India, E-mail: nileshxbhosle@gmail.com

Yugandhar Bokka
Research Scholar, GIET University, Gunupur, Odisha, India, E-mail: yug.599@gmail.com

Tapas Chhualsingh
Assistant Professor, Department of EEE, GIETU, Gunupur – 765022, Odisha, India

Debabrata Dansana
Department of Computer Science and Engineering GIET University, Gunupur, Odisha, India, E-mail: debabratadansana07@gmail.com

Dipak Das
Integrated Test Range, Defence Research Development Organization (DRDO), Chandipur, Odisha – 756025, India

Sudhansu Bala Das
Integrated Test Range, Defence Research Development Organization (DRDO), Chandipur,
Odisha – 756025, India

Kailas Devadkar
Department of Information Technology, Sardar Patel Institute of Technology, Mumbai – 400058,
Maharashtra, India

Subham Divakar
Department of ETC, C.V. Raman College of Engineering, Bhubaneswar, India,
E-mail: shubham.divakar@gmail.com

Umashankar Ghugar
Department of Computer Science, Berhampur University, Berhampur – 761006, India,
E-mail: ughugar@gmail.com

Chandan Kumar Giri
Associate Professor, Computer Science and Engineering, CUTM, Paralakhemundi, Odisha, India,
E-mail: chandankumargiri@gmail.com

K. Murali Gopal
Department of Computer Science and Engineering, School of Engineering, GIET University,
Gunupur – 765022, Odisha, India, E-mail: kmgopal@giet.edu

Sandeep Singh Jagdev
Ellen Technology (P), LTD, Jalandhar, Punjab, India

Amit Jain
Assistant Professor, Sir Padampat Singhania University, Rajasthan, India, E-mail: amit.jain@spsu.ac.in

Jemarani Jaypuria
Department of Computer Science, GIET, University, Gunupur, Odisha, India

Anand Kumar Jha
Department of Information Technology, CVRCE, CVRCE, Bhubaneswar, India

Pranay Jha
Research Scholar, School of Computer Application, Lovely Professional University, Phagwara,
Punjab, India, E-mail: pranay1988jha@gmail.com

Ferdin Joe John Joseph
Faculty of Information Technology, Thai-Nichi Institute of Technology, Bangkok, Thailand,
E-mail: ferdin@tni.ac.th

Dileep Kumar Kadali
Research Scholar, Department of CSE, GIET University, GIET University, Gunupur – 765022,
Odisha, India, E-mail: dileepkumarkadali@gmail.com

Sukhkirandeep Kaur
Assistant Professor, School of Computer Science and Engineering, Lovely Professional University,
Phagwara, Punjab, India, E-mail: sukhkirandeep.23328@lpu.co.in

Chetana Khetmal
Department of Computer Science, K.C College of Engineering, Mumbai, Maharashtra, India,
E-mail: khetmalchetna@gmail.com

K. Siva Krishna
Research Scholar, School of Computer Engineering (CSE), GIET University Gunupur, Odisha, India,
E-mail: ksivakrishna@giet.edu

Sanjay Kumar Kuanar
Department of Computer Science, GIET University, Gunupur, Odisha, India,
E-mail: sanjay.kuanar@giet.edu

Tanay Kulkarni
Department of Information Technology, Sardar Patel Institute of Technology, Mumbai – 400058,
Maharashtra, India, E-mail: tanaykulkarni06@gmail.com

A. V. S. Pavan Kumar
Department of Computer Science and Engineering, GIET University, Gunpur, Odisha, India,
E-mail: avspavankumar@giet.edu

D. Anand Kumar
Research Scholar, GIET University, Gunupur, Odisha, India, E-mail: anand.k.deva@gmail.com

D. Anil Kumar
GIET University, Gunpur, Rayagada, Odisha, India

Raghvendra Kumar
Department of Computer Science and Engineering, LNCT Group of College, Jabalpur, India

Chung Van Le
Duy Tan University, Da Nang, Vietnam

Chandrakanta Mahanty
Department of Computer Science and Engineering, GIET University, Gunupur, Odisha, India

Venkata Naresh Mandhala
Department of Computer Science and Engineering, Koneru Lakshmaiah Education Foundation,
Vaddeswaram, Guntur, Andhra Pradesh, India, E-mail: mvnaresh.mca@gmail.com

Anjana Mishra
Department of Information Technology, CVRCE, CVRCE, Bhubaneswar, India,
E-mail: anjanamishra2184@gmail.com

Brojo Kishore Mishra
Department of Computer Science and Engineering GIET University, Gunupur, Odisha, India,
E-mail: brojomishra@gmail.com

Debasis Mishra
Department of Electronics and Telecommunication, Veer Surendra Sai University of Technology,
Burla, Odisha – 768018, India

Gitanjali Mishra
Assistant Professor, GIET University, Gunupur, Odisha, India

Tusar Kanti Mishra
Department of CSE, Gayatri Vidya Parishad College of Engineering for Women, Visakhapatnam,
Andhra Pradesh, India, E-mail: tusar.k.mishra@gvpcew.ac.in

R. N. V. Jagan Mohan
Assistant Professor, Department of CSE, Swarnandhra College of Engineering and Technology,
Narasapur – 534280, Andhra Pradesh, India, E-mail: mohanrnvj@gmail.com

Purnima Mokadam
Department of Information Technology, Sardar Patel Institute of Technology, Mumbai – 400058,
Maharashtra, India

Mrinal
Department of Information Technology, CVRCE, CVRCE, Bhubaneswar, India

S. Nagaprasad
Osmania University, Amberpet, Hyderabad, Telangana – 500007, India

V. B. Narasimha
Osmania University, Amberpet, Hyderabad, Telangana – 500007, India

K. Lakshmi Narayana
PhD Scholar, Department of CSE, Pondicherry Engineering College, Puducherry, India,
E-mail: kodavali.lakshmi@gmail.com

Geetanjali Nayak
Department of CSE, Gayatri Vidya Parishad College of Engineering, Visakhapatnam, Andhra Pradesh,
India, E-mail: geetanjalinayak2019vizag

Gouri Sankar Nayak
MTech Research Scholar, School of Computer Engineering (CSE), GIET University Gunupur, Odisha,
India, E-mail: mr.gouri4u@gmail.com

Neelamadhab Padhy
Associate Professor, School of Computer Engineering (CSE), GIET University Gunupur, Odisha, India,
E-mails: dr.neelamadhab@giet.edu; dr.neelamadhab@gmail.com

Bhavani Sankar Panda
Assistant Professor, Computer Science and Engineering, GIET University, Gunupur, Odisha, India,
E-mail: chintupanda@gmail.com

Devpriya Panda
Department of CSE and IT, GIET University, Gunupur, Odisha, India

P. K. Panda
Department of Computer Application, North Orissa University, Baripada – 757003, Odisha, India,
E-mail: panda3k32003@gmail.com

Ribhu Abhusan Panda
Department of Electronics and Telecommunication, Veer Surendra Sai University of Technology, Burla,
Odisha – 768018, India, E-mail: ribhupanda@gmail.com

Chhabi Rani Panigrahi
Rama Devi Women's University, Bhubaneswar, India

Ranjeet Kumar Panigrahi
Department of Computer Science, GIET, University, Gunupur, Odisha, India

Rasmita Panigrahi
PhD Research Scholar, School of Computer Engineering (CSE), GIET University Gunupur, Odisha,
India, E-mail: rasmita@giet.edu

S. Gopal Krishna Patro
Department of Computer Science and Engineering GIET University, Gunupur, Odisha, India,
E-mail: sgkpatro2008@gmail.com

Sibo Prasad Patro
Research Scholar, School of Computer Engineering (CSE), GIET University Gunupur, Odisha, India,
E-mail: sibofromgiet@giet.edu

Sonali Shirish Patwe
PhD Research Scholar, Symbiosis International University, Lavale, Pune, India

Vijaya Pinjarkar
PhD Scholar, Sir Padampat Singhania University, Rajasthan, India, E-mail: vijaya.pinjarkar@spsu.ac.in

Nihar Ranjan Pradhan
Department of Computer Science, National Institute of Technology Meghalaya, Shillong – 793003, Meghalaya, India, E-mail: niharpradhan@nitm.ac.in

Adarsh Pratik
Department of IT, C.V. Raman College of Engineering, Bhubaneswar, India, E-mail: adarshpratik.2010@gmail.com

Rojalina Priyadarshini
Department of IT, C.V. Raman College of Engineering, Bhubaneswar, India, E-mail: priyadarhini.rojalina@gmail.com

S. Priyanga
MTech, Department of CSE, Pondicherry Engineering College, Puducherry, India

Bhuvan Puri
D.A.V Institute of Engineering and Technology, Jalandhar, Punjab, India, E-mail: bhuvanpuri239@gmail.com

Vikram Puri
Duy Tan University, Da Nang, Vietnam

Ritika Raj
Department of Information Technology, CVRCE, CVRCE, Bhubaneswar, India

R. Rajasekhar
Assistant Professor, Department of CSE, GIET University, Gunupur, Odisha, India, E-mail: rrajasekhar@giet.edu

Kali Charan Rath
Associate Professor, Department of Mechanical Engineering, GIETU Gunupur – 765022, Odisha, India, E-mail: kalimtech@gmail.com

Abhishek Ray
Professor, KIIT University, Bhubaneswar, India

Rahul Deo Sah
Dr. Shyama Prasad Mukherjee University Ranchi, India, E-mail: rahuldeosah@gmail.com

Bandita Sahu
Department of Computer Science, GIET, University, Gunupur, Odisha, India

Rakesh Sahu
Assistant Professor, Department of EEE, GIETU, Gunupur – 765022, Odisha, India

K. Sathiyamurthy
Associate Professor, Department of CSE, Pondicherry Engineering College, Puducherry, India

Murali Krishna Senapaty
Assistant Professor, GIET University, Gunupur, Odisha, India

Ashok Sharma
Associate Professor, School of Computer Application, Lovely Professional University, Phagwara, Punjab, India, E-mail: drashoksharma@hotmail.co.in

Akhilendra Pratap Singh
Department of Computer Science, National Institute of Technology Meghalaya, Shillong – 793003, Meghalaya, India

Bhaskar Singh
Radharaman Institute of Science and Technology, Bhopal, Madhya Pradesh, India

Harsh Pratap Singh
Sri Satya Sai University of Technology and Medical Sciences, Sehore, Madhya Pradesh, India, E-mail: singhharshpratap@gmail.com

R. P. Singh
Sri Satya Sai University of Technology and Medical Sciences, Sehore, Madhya Pradesh, India

Rashmi Singh
Radharaman Institute of Science and Technology, Bhopal, Madhya Pradesh, India

M. Srikanth
Research Scholar, GIET University, Gunupur, Odisha, India, E-mail: srikanth.mandela@gmail.com

B. Sujatha
Osmania University, Amberpet, Hyderabad, Telangana – 500007, India

P. K. Swain
Department of Computer Application, North Orissa University, Baripada – 757003, Odisha, India, E-mail: prasantanou@gmail.com

Sabyasachi Tribedi
Department of Electrical Engineering, Jadavpur University, Kolkata, West Bengal, India, E-mail: stribedi15@gmail.com

Abbreviations

ABE	attribute-based-encryption
ADHD	attention deficit hyperactivity disorder
AI	artificial intelligence
ANFIS	adaptive neuro-fuzzy inference system
ANN	artificial neural network
ARM	architecture reference model
AWS	Amazon web services
BAN	body area network
BCG	ballistocardiogram
BDA	big data analytics
BETaaS	building the environment for the things as a service
BN	belief network
BPM	body pulse rate
BS	base station
BVS	beneath video system
CAD	coronary artery disease
CAMI	Caribbean agro meteorological initiative
CC	climate change
CC	cloud computing
CCI	cloud computing infrastructure
CCTV	closed camera television
CdS	cadmium sulfide
CH	cluster head
CHD	coronary heart disorder
CHMM	canny house monitor and manager
CII	critical information infrastructure
CNN	convolutional neural network
CO_2	carbon dioxide
CO_3^{2-}	carbonates
CP-ABE	ciphertext-policy attribute-based encryption
CPS	cyber-physical systems
CRM	customer relationship management
CSP	cloud service provider

DA	data analysis
DAaaS	data analytics-as-a-service
DAIaaS	data analytics infrastructure as a service
DAPaaS	data analytics platform as a service
DASaaS	data analytics software as a service
DB	database
DC	data center
DC-RBAC	data-centric RBAC
DDD	data-driven decision
DML	data manipulation language
DO	dissolved oxygen
DOB	date of birth
DPAE	demographic profiles based attribute encryption
DPBSV	dynamic prime number based security verification
DRI	data routing information
DRLO	deep, reinforcement, learning, and offloading
DT	DRI table
DTMF	dual tone multi frequency
EC	electrical conductivity
EDs	electronic devices
EESCDE	energy efficient SNR-based clustering in UWSN with data encryption
EH	energy harvesting
EHRs	electronic health records
EPs	emotional profiles
FCFS	first come first serve
FCL	fully connected layer
FER	facial emotion recognition
FTMH	fault tolerant multi-hop clustering
GAHA	green IoT agriculture and healthcare applications
GCP	Google cloud platform
GFLOPS	Giga FLOPS
GHQ	general health questionnaire
GIG	grayscale Image generation
GPIO	general-purpose input-output
GPS	global positioning system
GPUs	graphical processor units
HAN	home area network
HDFS	Hadoop distributed file system

HFSS	high frequency structure simulator
HGPCA	hierarchical GA and PSO-based computation algorithm
HMS	health monitoring system
HR	humanoid robots
IaaS	infrastructure as a service
IHD	ischemic heart disease
IIG	integral image generation
IIoE	industry internet of everything
IN	inspector node
IoT	internet of things
IP	interest points
IPD	interest point detection
IR	infra-red
ISP	internet service provider
KDD	knowledge discovery in databases
KNX	Konnex
LANs	local area networks
LDL	low density lipoprotein
LDR	light dependent resistor
M2M	machine to machine
MANET	mobile ad-hoc network
MBS	macro base station
MCC	mobile cloud computing
MCC	multi-class classifier
MDDSS	medical diagnostic choice aid structures
MEC	mobile edge computing
MECO	mobile edge computation offloading
MIC	many integrated core
MK	master-secret key
ML	machine learning
MLP	multi-layer perceptron
MVC	model-view-controller
NAN	neighborhood area network
NBDT	Naïve Bayesian and decision tree
NCH	non cluster head
NLP	natural language processing
NN	neural network
NO_2	nitrates
NRL	normalized routing load

NS2	network simulator 2
OCR	optical character recognition
OFDMA	orthogonal frequency-division multiple access
OKBC	open knowledge base connectivity
OpenCV	open source computer vision
OS	operating system
OT	operational technology
P2P	peer to peer
PaaS	platform as a service
PAN	personal area network
PCA	major component analysis
PCI	peripheral communication interface
PDD	pervasive developmental disorder
PDPR	packet dropping ratio
PDR	packet delivery ratio
PF	final prediction
PHM	prognostic health monitoring
PICO	population, intervention, comparison, and outcome
PIR	passive infra-red
PK	public-key
PM	particulate matter
PPDM	privacy preserving data mining
PWM	pulse width modulation
QoE	quality of experience
QoS	quality of service
RBAC	role-based access control
RCPSP	resource-constrained project scheduling problem
RDD	resilient distributed datasets
RF	random forest
RFID	radio frequency identification
RGM	response generation matrix
RL	reinforcement learning
RLO	reinforcement learning optimization
RMSE	root mean square error
RREP	route reply
RREQ	route request
RT	response time
SaaS	software as a service
SCH	sub cluster head

SDN	software-defined networking
SDT	selective data transmission
SDTO	software defined task offloading
SK	secret-key
SNAPP	social networks adapting pedagogical practice
SNR	signal-to noise ratio
SOA	service oriented architecture
SoC	system on chip
SOM	self-organizing map
SQL	structural query language
SSS	safe, scalable, and secure
SVM	support vector machine
SWOOP	semantic web ontology overview and perusal
TDMA	time-division multiple access
TDS	total dissolved solids
TDT	time space transmission
VM	virtual machine
VSWR	voltage standing wave ratio
WHO	World Health Organization
WPAN	wireless personal area network
WSN	wireless sensor network

Preface

Blockchain technology is receiving growing attention from various organizations and researchers as it provides magical solutions to the problems associated with the classical centralized architecture. Blockchain, whether public or private, is a distributed ledger with the capability of maintaining the integrity of transactions by decentralizing the ledger among participating users.

On the other hand, the Internet of Things (IoT) represents a revolution of the Internet that can connect nearly all environment devices over the Internet to share their data to create novel services and applications for improving our quality of life. Although the centralized IoT system provides countless benefits, it raises several challenges. Resolving these challenges can be done by integrating IoT with blockchain technology.

The Department of Computer Science and Engineering at the GIET University has organized the AICTE sponsored international conference on Recent Trends in IoT and Blockchain (ICRTIB-2019) during 19–20 October, 2019.

The current book provides an overview of IoT and blockchain concepts, IoT solutions with blockchain in the current business scenario, and much more.

Part I
IoT Techniques and Methodologies

CHAPTER 1

Data Mining Algorithms for IoT: A Succinct Study

CHANDRAKANTA MAHANTY,[1] BROJO KISHORE MISHRA,[1] and RAGHVENDRA KUMAR[2]

[1]*Department of Computer Science and Engineering, GIET University, Gunupur, Odisha, India*

[2]*Department of Computer Science and Engineering, LNCT College, Jabalpur, India*

ABSTRACT

IoT is a fresh idea which enables people to connect different sensors and intelligent machines to gather information from the surroundings in real time. It is anticipated that the number of effective IoT machines will increase to 0.01 trillion and 0.022 trillion by 2020 and 2025, respectively. Lots of analytical techniques are brought into IoT to render IoT smarter; data mining is among the most precious techniques. Data mining is the method of finding interesting knowledge and possibly relevant patterns from big data sets and using algorithms to extract crucial information. This article focuses on data mining framework for IoT, data mining functionalities and usage of data mining in IoT applications. After that, a survey on different data mining algorithms is presented. We also analyze the effectiveness and efficiency of different data mining algorithms (K-nearest neighbors, Naïve Bayes, support vector machine (SVM), C5.0, deep learning artificial neural networks (ANN), and ANNs). We reviewed the above mentioned algorithms and concluded that DLANNs, ANNs, C5.0 give relatively higher accuracy and memory-efficient as compared to other algorithms. To address IoT data mining problems such as managing large quantities of information, data analysis (DA), a big data mining system is suggested.

1.1   INTRODUCTION

Internet of things (IoT) relates to the next stage of the internet, containing hundreds of billions of nodes covering different elements from tiny ubiquitous sensor systems and portable equipments to big web servers and supercomputer clusters [1]. IoT operates with linking the entire world's stuff via the internet [2]. IoT incorporates fresh computing and communications systems (wireless communications systems, radio frequency identification (RFID) technologies, actuators, and sensors) and develops the next generation of internet [3]. Several scientists who work in multiple areas such as scholars, institutes, and departments of government gave a lot of importance in changing the today's internet by developing multiple technologies such as smart city, smart transportation, smart healthcare [4], smart home, and smart agriculture. IoT gets and stores a lot of information from sensors, wearables, smartphones, and other devices that are activated on the internet. In order to convert information into implementable information, it must be assessed utilizing suitable data mining techniques. Sensor information from a smart home, for example, is used for safety surveillance or home automation for elderly or disabled individuals, or traffic data is evaluated to determine an optimal ambulance path. The information gathered from IoT systems is being utilized for comprehend and supervise complicated environments within us, providing better decision-making, higher accuracy, effectiveness, and productivity. Analysis of the produced information must be recorded effectively in order to make this choice more precise and processing the same needs techniques. Data mining is among the most helpful techniques. Datasets store the information produced from IoT devices. The big volume of information processed using data mining technique estimates the model, generalizing it to fresh information as well. Data mining is therefore the method of finding precious data from bulk quantities of information stored in databases (DB) and data warehouses.

1.2   DATA MINING FOR IOT

Data mining in IoT is used to handle the big volume of information that IoT devices collect. Data mining includes discovering and analyzing information from the huge information set. Data mining's primary aim is to explore helpful patterns from big knowledge set obtained from machines, sensors, IoT devices [5]. Discovery of knowledge, analysis of patterns and collection of data are the words used in the IoT for data mining. Data mining's main goal is to create an effective and concise framework that is suited to the knowledge set. Thus, a number of researches focus on the use or development of efficient IoT data mining techniques. The findings outlined in [6] indicate that it is possible to utilize data mining algorithms to create IoT smarter which could provide better services. Based on prior information, a DM method could also be used to create a choice that let us tell the sale of a specific product, as example, cold drink sales higher in summer as compared to winter [7].

Data mining converts a set of data into a comprehensible framework and incorporates significant information that helps to gain insight into the raw data gathered from different IoT apps. IoT thus forms a network of devices that implanted with sensor, electronics, and network connectivity through which devices can gather and exchange information. The ideal combination of IoT and data mining leads to an innovative technology that benefits every part of the population. We were inspired to evaluate a data mining structure for IoT apps by the enormous quantity of data produced by IoT apps and knowledge discovery in databases (KDD). The IoT data mining framework is shown below by considering data mining for IoT.

Data mining is primarily divided into two procedures. First procedure is descriptive and the second procedure is predictive. Data mining is represented in a short and aggregated manner called descriptive data mining and it provides important overall information. In predictive information mining, information is evaluated in a series to build one or more information frameworks to estimate the actions of freshly created information sets [8] utilizing methods such as classification, clustering, regression, and trend analysis. In the course of knowledge discovery, data mining can be seen as an important technique. The stages which are required for this method depicted as below:

- **Stage 1: Cleaning of Data:** Noise and inconsistent information will be deleted in this process.
- **Stage 2: Data Integration:** It brings various sources of information.
- **Stage 3: Selection of Data:** In this level important information is recovered that is important to the method of evaluation.
- **Stage 4: Data Transformation of Data:** Aggregation activities is conducted in this stage in order to convert or enhance information into adequate knowledge for data extraction.
- **Stage 5: Data Mining:** Using smart techniques, the required information and patterns are obtained in this phase.

- **Stage 6: Evaluation of Patterns:** Exploration of interesting patterns which depict expertise based on certain pleasant procedures.
- **Stage 7: Knowledge Presentation:** Using visualization and information depiction methods, information is provided to the customer in this phase (Figure 1.1).

FIGURE 1.1   Data mining framework for IoT.

1.2.1   *DATA MINING FUNCTIONALITIES*

1. Classification allocates objects to target types or classes in a set. Its objective is to forecast the target class correctly in the information for each situation.
2. Clustering identifies clusters of comparable information objects and forms a cluster group.
3. Association assessment is the identification of membership rules that exhibit circumstances of object-value that often happen in a specified set of information.
4. Time series assessment includes techniques and tools for the assessment of time series data to obtain reliable data and other information features.
5. Outer assessment defines and models conceptual frameworks or patterns for objects whose conduct evolves over time.

1.2.2   *IOT APPLICATIONS BASED ON DATA MINING [9]*

1. **Health Care:** The development of the healthcare sector is apparent owing to the use and advances of IoT technologies. IoT technologies provide countless facilities to their consumers regarding their health like weight monitoring equipment, blood sugar levels, heart rate, and blood pressure information and stored it on a cloud-based system. To incorporate the above diverse information and provide precise patient information, a smart system must be evolved. The

information can be extracted from text using particular prescription medications and medical records of the patient doctor and from this; we can derive significant findings about the patient's current situation and the patient's survival chances. Clustering method is used to improve the patient's therapy and care. Outlier assessment may also be performed to recognize any uncommon trends that will be simple to detect any falsification.

2. **Home Automation:** Several IoT systems are used in the home automation scheme. The sensors used in IoT systems generate valuable information and patterns. The above crucial information and patterns are utilized to forecast forthcoming occurrences which will provide the consumer with effective automated communication. Data mining utilizes time series and classification analysis functionalities for smart home. Classification is being utilized by categorizing associated equipments which are strongly related and dependent on their use. Time series assessment is used for information that is produced by these instruments with their respective time stamps and can be forecast utilizing linear regression for a specific time using this future occurrence.

3. **In Controlling Traffic:** The traffic control systems based on IoT are global positioning system (GPS), smartphones, car sensors disposed throughout the town can produce information features like accident-prone zones, light car and heavy car occurrences and approximate driving time to a particular destination. The classification algorithm can be used in this situation to fix the issue of traffic jams. Depending on the elevated, medium, low chance of traffic jam occurrence in a specific region, selected regions can be grouped. After applying classification algorithm a novel classified framework is developed. This framework can be utilized to forecast the future time where more traffic jams occurred. According to the above forecast, the vehicle can pick the alternate path to reach the targeted station so that overcrowding issues will be prevented.

4. **Detection of Pipeline Leakage:** Maintaining leaks from water pipes for municipality is a very complicated task. Sensors will be utilized to analyze the noise of water that crosses via the tube using data mining methods, particularly by using the outdated tubes. To detect leaks in the tubes, the outlier detection algorithm is used. By applying this methodology in pipeline leakage, testing tasks can be simplified to detect water leaks and maintenance costs can be lowered to half as opposed to standard testing methods.

1.3   DATA MINING ALGORITHMS

1. **Support Vector Machines (SVMs):** It is a supervised learning paradigm focused on the concept of statistical learning. It is also related with learning algorithms that analyze information and acknowledge patterns. Usage of SVM is expanded from classification to regression. In market basket analysis, text classification [10] and acknowledgment of patterns in detecting diseases SVM method is used. This method is also suitable for complicated and noisy domains. Initially, SVM was intended to tackle the tradeoff, overfitting, and capability regulation of bias fluctuation.

2. **K-Nearest-Neighbor (KNN):** It is a voting algorithm as it uses sample points in terms of a given distance, measure to persuade the categories of information points being considered. KNN is also used efficiently for the identification of intrusion for multiple activities in wireless sensor networks (WSN) and IoT domains [11]. KNN is robust to noisy training data and is efficient when there are big training data.

3. **Naïve Bayes Classifier:** Purpose of Naïve Bayes is to build a policy that adds new items to the class. It provides Bayesian networks which are aligned with acyclic graphs and the nodes serve arbitrary variables. Edges are conditional dependencies; unconnected nodes are variables that are conditionally isolated from one another. These classifiers have a lot of power based on Bayesian networks, such as model interpretability and fitting to complicated contexts of information and classification issues. NB is a straightforward and robust classifier such as KNN. It derives the probability.

4. **C5.0:** It provides more precise and effective results among all these classifiers. C5 is a classifier that compares information to other classifiers with less time. Memory utilization is minimal in order to generate a decision tree and it also improves the precision. C5.0 algorithm offers selection of features, cross validation and decreased equipment for trimming errors. Quinlan subsequently created C5.0 which is the updated version of C4.5. C5.0 version is used to handle boosting and weighting [12].

5. **Artificial Neural Network (ANN):** These are focused on the imitation of the human brain neural system. These are highly effective in solving data mining duties with higher accuracy. ANNs are complicated and a large quantity of information processing is needed to solve a greater accuracy issue. Revolutionary learning algorithms which are focused on deep learning are further extensions of ANNs [13].

6. **Deep Learning Artificial Neural Networks (DLANNs):** These have an excessive capacity to learn and process huge data and deliver highly precise findings that other traditional machine and data mining algorithms cannot achieve. Deep learning can provide us with innovative ideas on IoT information that other data mining algorithms do not allow [14].

1.4   COMPARISON OF DIFFERENT DATA MINING ALGORITHMS

All the mentioned data mining algorithms are implemented in the R language. We use machine learning (ML) repository dataset [15]. We preprocess the data at first and make it appropriate for the classifier. We can understand the complete amount of instances correctly and incorrectly categorized with the aid of confusion matrix [16]. Comparisons of different data mining algorithms are depicted in Figure 1.2.

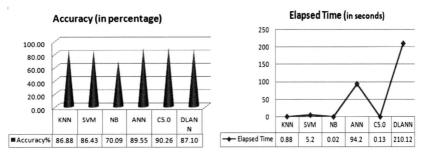

FIGURE 1.2   Classification accuracy and elapsed time for different algorithms (KNN, SVM, NB, ANN, C5.0, and DLANN).

ANN, C5.0, and DLANNs are having higher accuracy as compared to other algorithms. By rising the ages, hidden layers and neurons, classification accuracy could be accomplished. Because of complicated design, the DLANN algorithm has the largest runtime. A great deal of system resources are needed for DLANNs.

1.5   SUGGESTED BIG DATA FRAMEWORK FOR IOT

We recommend five layers of IoT and big data mining architecture which is illustrated in Figure 1.3:

1. **Devices:** IoT equipments (cameras, sensors, RFIDs, mobile phones, and smart devices) are coordinated into IoT frameworks.
2. **Raw Data:** IoT devices generate data, which can be integrated into the big data mining system, in different forms, such as structured, unstructured, and semi-structured data.
3. **Data Gather:** Data generated to produce clusters for storage and retrieval. Batch and real-time information can be endorsed and all information can be analyzed, combined, and reviewed.
4. **Processing of Data:** MapReduce, Hadoop, Storm, HDFS, and Oozie techniques are integrated for data processing.
5. **Service:** All data mining functions have come under this layer.
6. **Security and Privacy:** This safeguard information is carried against unauthorized access and confidentiality.

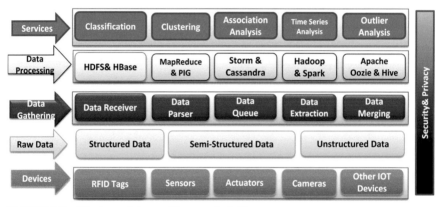

FIGURE 1.3   Big data mining architecture for IoT.

1.6   CONCLUSION

The idea of IoT emerges from a commitment to handle, automate, and discover all the world's smart devices, machines, actuators, auto-computing devices, and sensors. To make choices for both end-user and IoT devices, it is necessary to integrate data mining technique that supports strategy-making and system enhancement. Data mining includes the discovery of novel, exciting, and possibly helpful data patterns and the use of algorithms to obtain hidden information. It is a difficult job in information mining to collect this hidden insight from IoT data. In order to manage enormous IoT info, data mining algorithms are required. This article focuses on IoT data mining structure, data mining features, and data mining utilization in IoT apps. We also investigated the applicability of some data mining algorithms and found that, compared to other algorithms, DLANNS, ANN, and C5.0 can offer comparatively greater precision findings. This article discusses learning perspective implementation for IoT data mining, layers linked to IoT mining, and summary of IoT big data analytics (BDA). Big data mining mechanism is suggested to address the problems of IoT data mining.

KEYWORDS

- **artificial neural networks**
- **big data mining system**
- **data mining algorithms**
- **internet of things**
- **radio frequency identification**
- **raw data**

REFERENCES

1. Gubbi, J., Buyya, R., Marusic, S., & Palaniswami, M., (2013). Internet of things (IoT): A vision, architectural elements, and future directions. *Future Generation Computer Systems, 29*(7), 1645–1660.
2. Chen, F., Deng, P., Wan, J., Zhang, D., Vasilakos, A. V., & Rong, X., (2015). Data mining for the internet of things: Literature review and challenges. *International Journal of Distributed Sensor Networks, 11*(8), 431047.
3. Tsai, C. W., Lai, C. F., Chiang, M. C., & Yang, L. T., (2013). Data mining for internet of things: A survey. *IEEE Communications Surveys and Tutorials, 16*(1), 77–97.
4. Chen, H., Chung, W., Xu, J. J., Wang, G., Qin, Y., & Chau, M., (2004). Crime data mining: A general framework and some examples. *Computer, 4*, 50–56.
5. Shobanadevi, A., & Maragatham, G., (2017). Data mining techniques for IoT and big data: A survey. In: *2017 International Conference on Intelligent Sustainable Systems (ICISS)* (pp. 607–610). IEEE.
6. Ragavi, R., Srinithi, B., & Sofia, A., (2018). Data mining issues and challenges: A review. *IJARCCE, 7*, 118–121. doi: 10.17148/IJARCCE.2018.71125.
7. Batra, I., & Verma, S., (2018). Performance analysis of data mining techniques in IoT. In: *2018 4th International Conference on Computing Sciences (ICCS)* (pp. 194–199). IEEE.
8. Meghana, D., & Akarte, S. P., (2014). Predictive data mining: A generalized approach. *International Journal of Computer Science and Mobile Computing, 3*(1), 519–525.
9. Tapedia, K., & Wagh, A. M., (2016). Data mining for various internets of things applications. *International Journal of Research in Advent Technology.*

10. Burges, C. J., (1998). A tutorial on support vector machines for pattern recognition. *Data Mining and Knowledge Discovery*, *2*(2), 121–167.

11. Yang, X. S., (2019). *Introduction to Algorithms for Data Mining and Machine Learning.* Academic Press.

12. Hssina, B., Merbouha, A., Ezzikouri, H., & Erritali, M., (2014). A comparative study of decision tree ID3 and C4. 5. *International Journal of Advanced Computer Science and Applications*, *4*(2).

13. Mishra, M., & Srivastava, M., (2014). A view of artificial neural network. In: *2014 International Conference on Advances in Engineering and Technology Research (ICAETR-2014)* (pp. 1–3). Unnao. doi: 10.1109/ICAETR.2014.7012785.

14. Schmidhuber, J., (2015). Deep learning in neural networks: An overview. *Neural Networks*, *61*, 85–117.

15. Lichman, M., (2013). *UCI Machine Learning Repository*. Irvine, CA: University of California, School of Information, and Computer Science. http://archive.ics.uci.edu/ml (accessed 20 July 2020).

16. Alam, F., Mehmood, R., Katib, I., & Albeshri, A., (2016). Analysis of eight data mining algorithms for smarter internet of things (IoT). *Procedia Computer Science*, *98*, 437–442.

CHAPTER 2

Information Technology Evolutions in the Manufacturing Sectors: Industry 4.0, IoT, and Blockchain

KALI CHARAN RATH,[1] RAKESH SAHU,[2] and TAPAS CHHUALSINGH[2]

[1]Associate Professor, Department of Mechanical Engineering, GIETU Gunupur – 765022, Odisha, India, E-mail: kalimtech@gmail.com

[2]Assistant Professor, Department of EEE, GIETU, Gunupur – 765022, Odisha, India

ABSTRACT

Manufacturing is an important contributor to productivity, innovation, and trade. Technological revolution occurred with the change in time. After passing some revolution, today's industry handshake is with Industry 4.0. Industry 4.0 is a rising period of network and collaboration among parts, machines, and people that can possibly make huge creation and proficiency profits, upgrades in the personal satisfaction and maintainable natural results.

India is now rolling with digitalization and it has been seen its important and attractive growth market. Innovation is one of numerous components making makers reevaluate the manner in which they work. Computerization, apply autonomy, and the usage of cutting edge fabricating strategies have drastically changed the industrial facility floor and the assembling forms that happen there. This study provides an introduction to Industry 4.0 with the help of the cyber manufacturing system in association with IoT and blockchain.

2.1 INTRODUCTION

Manufacturing is the way of transforming various kinds of raw materials to final product and can be sold in the commercial center. Each physical item that one can purchase in a store or online is produced in some other place. Today, innovation is making the nation's economy move towards offering administrations rather than creating merchandise. Nonetheless, it is ending up clear to financial experts that a solid assembling industry is one of the trademark pointers of a sound, flourishing economy.

Producers' supply chains are modern, complex associations with various subtleties that can make straightforwardness and responsibility testing, particularly with regards to the co-ordinations of structure and sending new hardware and administration parts.

In an industry that computerizes things to help mankind, IT makes the assembling procedure not so much bulky but rather more mechanized. IT helps radically in conveying in the nick of time experiences, quick perceivability and consistent advancement for executing new age arrangements. Makers need to push their industry further and more distant as far as intricate schedules particularly, specially make and make-to-stock procedures so they can convey items on a design to-arrange necessity advertise.

Countering supply chain multifaceted nature is significant since the conventional production system is presently not being used. The profoundly mind-boggling supply chains are consequently loaded with issues, which provoked the requirement for effective administration and enhancement.

The fourth industrial revolution will be established on trusted networks that remove the need for middlemen. Simple processes that are currently tackled manually or in a segregated manner will be undertaken in an integrated way using digital, autonomous systems, underpinned in many cases by blockchain technology.

Extended number of model ventures with shorter between times drive the necessities for generation. The ability to reproduce manufacture growth thusly ends up being logically notable [1]. Potential zones for usage of IoT, Industry 4.0 in modern mechanization segment to satisfy the worldwide need is accessible in different businesses. The impact and the system of improvement is examined and proposed different strategies for the fourth era industrial revolution [2]. IoT based system can be best implemented and the data can be well assessed by user whereas the technology behind the blockchain can be implemented to provide the security and to keep away from the centralized system which avoids a common data sharing with others and improves security [3]. The applied structure takes into consideration multi move activity dependent on security encryption and information obtaining that gives a security certification, various activity, unwavering quality, and classification for execution upgrade in a framework [4]. Technological methodology supported by the horticultural and agricultural division on finding the issue, giving feasible, moderate, security, and adequate nourishment, feed, fiber, and fuel to purchasers, it is basic to guarantee these worth chains running easily and effectively by applying propelled web advances [5]. Multi-criteria examination is utilized by the blockchain innovation to assess and control the conceptual information for a synthetic industry and the models are altered for the business integrated with industry 4.0 [6]. Blockchains guarantee straightforward companion investigated works and activities in the vitality part in recognizing the specialized difficulties can illuminate for that application just as its potential downsides, and mechanical undertakings and new companies that are right now applying blockchain innovation to that zone [7].

Fundamental executions in assembling have likewise developed with regards to decentralizing 3D printing asset accessibility and in rethinking how information combination crosswise over assembling supply chains are actualized [8].

The expansion of IoT and selection of industrial internet undertakings has allowed security experts the chance to assess the new dangers and vulnerabilities that are being brought into associations around the globe. IoT and cybersecurity cross with various existing security disciplines. These orders include: (1) information technology (IT security) for ensuring data frameworks; (2) physical security to secure structures, workplaces, offices, and so forth; and (3) operational technology (OT) security to ensure operational frameworks for plant computerization and natural observing frameworks. Verifying IoT conditions will require drawing from every one of these three orders. The mix of the components can be depicted as computerized security.

The internet of things (IoT), particularly the modern IoT (IIoT), has quickly created and is getting a great deal of consideration in scholarly regions and industry, yet IoT protection dangers and security vulnerabilities are rising up out of absence of central security innovation. The blockchain procedure, because of its decentralization and data acknowledgment, was proposed as a decentralized and disseminated way to deal with assurance security prerequisites and rouse the improvement of the IoT and IIoT.

IoT based industrial research work is a hot cake in the research field. Cyber manufacturing is a challengeable area for software engineers and IT sectors to provide a challengeable security with smooth function of industry to put the customer satisfaction in an excellent manner.

This chapter provides a prologue to Industry 4.0 with the assistance of cyber manufacturing system in relationship with IoT and blockchain.

2.2   CYBER MANUFACTURING

Cyber manufacturing is an idea gotten from digital (cyber) physical frameworks (CPS) that alludes to a cutting edge fabricating framework that offers a data straightforward condition to encourage resource the executives, give reconfigurability, and look after efficiency.

The possibility of digital assembling originates from the way that internet-empowered administrations have included business esteem in financial parts, for example, retail, music, purchaser items, transportation, consumer products and medicinal services, etc. In any case, contrasted with existing internet-empowered parts, producing resources are less associated and less open progressively. In action the executives because of absence of linkage between manufacturing plants, conceivable overload in extra part stock, just as startling machine personal time. Such circumstance calls for network between machines as an establishment and examination over that as a need to make an interpretation of simple information into data that really encourages client basic leadership. Expected functionalities of cyber manufacturing frameworks incorporate machine network and information securing, machine wellbeing prognostics and assembling reconfigurability.

2.2.1   CYBER MANUFACTURING ASSOCIATED WITH INDUSTRY 4.0

Assembling industry has been in a consistent battle to deliver excellent products at the rate that customer's request. The mechanical time has enabled machines to determine the amount issue of assembling by IT framework through digital assembling. Digital assembling based on: simultaneous designing, disseminated fabricating, light-footed assembling, virtual assembling, virtual undertaking, digital/electronic business, web empowered CAD/CAM/PDM.

In cyber manufacturing, it is imagined that we will have fast appointing and decommissioning of framework components, which means critical time and cost will should be put into verifying trust in an arranged condition spreading over different business associations.

Cyber manufacturing (Figure 2.1) condition, the thought of arranged associations and machines makes an open door for item plans to be quickly created and afterward made with constrained human intercession through reuse of PC produced code or through programmed compilers that make an interpretation of structure information to machine guidelines.

FIGURE 2.1   Cyber-physical interface.

Quite a bit of this anticipated development is predicated on the utilization of IoT advancements, combined with distributed computing, information investigation, AI. In this, it is IoT that gives the extension between the computerized space, including new scientific techniques, and the physical area of the plant and inside the store network. This adjusts well to the Industry 4.0 vision of changing the supply chains into a keen system with good judgment and self-governing items that impart and interface with one another continuously. Especially as Industry 4.0 will involve the interlinkages of different specialized orders with minimal past association, for example, mechanical designing with data and interchanges innovation through information and communication technology.

System based correspondence for Industry 4.0 involves all innovations, systems, and conventions required technologies, networks, and protocols required to facilitate a communication relationship in the manufacturing network system.

2.2.2   CYBER MANUFACTURING PYRAMID

Cyber manufacturing is a noteworthy innovative technological part in the worldwide development on open manufacturing system. As digitalization have taken a significant and alluring development showcase, So, it is exceptionally fundamental requirement for the assembling ventures to make reconciliation of information technologies with items, frameworks, arrangements, and administrations over the total worth chain extending from design structure to production to the supportable maintenance.

In today's world, developments that incorporate IoT, 3D printing, man-made brainpower (AI), driverless vehicles, hereditary building, apply autonomy and savvy machines, drives us into the Fourth Industrial Revolution, or Industry 4.0. Another specialized development is bitcoins and its hidden blockchain innovation. Blockchains are viewed as the core of the Industry 4.0 where bitcoins enables a huge number of savvy gadgets to perform straightforward and frictionless budgetary exchanges, without human mediation however completely self-sufficient, in the IoT universe.

From the beginning of time, the assembling business has been in a consistent battle to create fantastic products at the rate that purchasers request them. While the mechanical time has enabled machines to determine the amount issue of assembling, it has prompted a flood of inferior low-quality merchandise. The new thought was found through cyber manufacturing. This idea exploits the progressions in data innovation, applying these plans to the assembling procedure. With a cyber manufacturing model, prescient investigation advancements will help organizations with operational choices by finding and featuring potential issues or changes in machine wellbeing and execution. By and large, this arrangement of cyber manufacturing will give ongoing linkage between the status of an organization's plants, current stock status, and conveyance framework through a progressive "cloud-based administration."

Cyber manufacturing has certain levels in its system, like: manufacturing resource, connection level, conversion level, cyber level, cognition level, and configuration level as shown in Figure 2.2.

The cyber manufacturing system uses ongoing improvements in IoT, distributed computing, haze figuring, administration arranged innovations, among others. Assembling assets and capacities can be exemplified, enrolled, and associated with one another legitimately or through the internet, in this manner empowering astute practices of assembling segments and frameworks, for example, self-awareness, self-expectation, self-advancement, and self-setup.

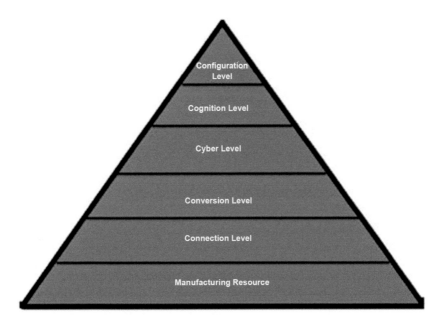

FIGURE 2.2 Cyber manufacturing pyramid.

2.3 IOT, BLOCKCHAIN, AND INDUSTRY-4.0

Today's world advancement is incorporate with IoT, 3D printing, computerized reasoning (AI), driverless vehicles, hereditary designing, mechanical autonomy and savvy machines, drives us into the Fourth Industrial Revolution, or Industry 4.0. Another specialized advancement is bitcoins and its hidden blockchain innovation. Blockchains are viewed as the core of the Industry 4.0 where bitcoins enables a huge number of shrewd gadgets to perform straightforward and frictionless money related exchanges, without human intercession yet completely self-sufficient, in the IoT universe.

IoT must be viewed as a series of advancements that have customarily not been associated and will presently be associated with an IP-based system. These advances are the most significant drivers of the computerized development. In the focal point of the institutionalization is the supposed "machine-to-machine" (M2M) correspondence.

IoT for corporate/mechanical use is an arrangement of incorporated PC organizes and associated modern offices with implicit sensors and programming for information accumulation and trade with the chance of computerized remote control and the executives without human intercession. The accompanying highlights describe the use of IoT are:

1. Generation of enormous information streams;
2. Multilevel information stockpiling;
3. Real-time preparing;
4. Big data analytics (BDA);
5. Forecasting;
6. Control and the executives.

In the event that all definitions are decreased to one shared factor, IoT just signifies: "the systems administration of articles which have the availability that empowers them to discuss freely with each other by means of the internet." The items are associated with a cloud-put together server with respect to which the information is accumulated, connected, and examined. For a model, being the "mind" of the machine, CNC assumes a key job in this: CNC gives significant creation information to improving generation groupings, just as machine status information for prescient upkeep of the machine.

The IoT is the thing that empowers articles and machines, for example, cell phones and sensors to "impart" with one another just as individuals to work out arrangements. The mix of such innovation enables items to work and take care of issues freely. The compelling advancement and usage of IoT will rely upon the quickened progress to the internet convention.

Blockchain innovation imitates a wedding function by utilizing the intensity of a conveyed record to keep confided in records of the "understanding/exchange" inside a welcoming network. The repeated records over various clients can approve the credibility of the synchronized substance of the understanding or exchange. Since 2008, the improvement of bitcoin turned into the principal case of a blockchain application. This across the board digital money gives a genuine answer for the test of trust in a decentralized framework. So, the blockchain is a developing rundown of records, called obstructs, that are connected utilizing cryptography. Each block contains a cryptographic hash of the past block, a timestamp, and exchange information. In other structure, blockchain can be characterized as decentralized, dispersed, and open advanced record that is utilized to record exchanges crosswise over numerous PCs so any included record can't be modified retroactively, without the change of every consequent block model.

In the course of the most recent decade, a few imaginative industrial facility computerization designs have been built up that influence the cloud computing (CC) worldview to improve the adaptability and adaptiveness of the shop floor. Be that as it may, their genuine reception by the assembling business is as yet restricted, due predominantly to execution and security requirements. The fourth modern insurgency incorporates IT frameworks with physical frameworks to get a digital physical framework (Figure 2.3) that gets this present reality a computer generated experience.

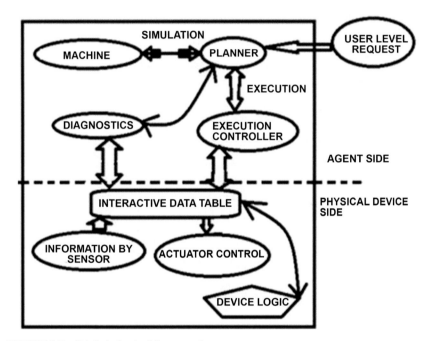

FIGURE 2.3   Digital physical framework.

Some central parts of Industry 4.0 are: CPF, digitization, integration of vertical and even worth chains, digitization of item and administration, innovative computerized plans of action, IoT or remote, automation or man-made reasoning or apply autonomy, and simulation/enlarged/augmented experience (Table 2.1).

The Fourth Industrial Revolution, authored by the World Economic Forum's Klaus Schwab a couple of years back, is no exemption. One of the additionally fascinating trendy expressions inside this is Industry 4.0, with supporters guaranteeing that the development of amazing registering and

AI-based advancements will change modern procedures. With the goal for organizations to change their modern procedures with problematic innovations, they first need to comprehend what those advancements are able to do, how they will affect the current biological system and what advantages they will bring.

TABLE 2.1 Advantages of Industry 4.0 Over Manual Manufacturing Industry

| Subjects | Data Source | General Manufacturing Industry | | Industry 4.0 | |
		Attributes	Technology	Attributes	Technology
Components	Sensor	Precision	Smart sensor and fault detection	Self-aware and self-predict	Useful life prediction by simulation softwares
Machine	Controller	Operation and quality control	Condition-based diagnostics	Self-aware, self-predict and self-compare	Health monitoring
Production system	Manufacturing system	Productivity and quality control	Work and waste reduction	Self-configuration, self-maintenance and self-organize for the system	Worry-free production system

Assembling organization can change to assembling 4.0 through certain substance, for example, the keen production line, digital physical observing frameworks, and extensively pursues these four standards:

- **Straightforward Data:** It is imperative that a virtual duplicate of the physical world can be made, and this must be done if information is uninhibitedly accessible, from crude sensor information appended to machines as far as possible up to a higher-esteem, logical data.
- **Interoperability of Parts:** The IoT supports Industry 4.0 and takes into consideration the different machines, gadgets, and sensors to discuss adequately with each other.
- **Specialized Help:** Innovation ought not to be put resources into only for it, yet rather on the grounds that it gives clear help to people, either in carrying out their responsibility or settling on educated choices.

- **Decentralized Choices:** By making data progressively straightforward and promptly accessible, it should move the basic leadership process nearer to the point of activity, with just remarkable choices raised up the chain of command.

Here the organization's recommendation to set up an assembling organization for Industry 4.0 with the following significant territories, are:

1. Create industry 4.0 methodology;
2. Select the correct group and the privilege computerized instruments;
3. Focus on improving procedures;
4. Implement new advances;
5. Improve data the board;
6. Appoint an individual or a group to lead the computerized change;
7. Understand shopper/customer needs.

As a modern innovation, blockchain offers a reachable reply for allowing a protected and decentralized open file that a significant mass of stimulating new innovation applications in a few regions such as IoT, cyber-physical systems (CPS), manufacturing, supply-chain, etc.

2.4 DIGITAL TECHNOLOGIES AND INDUSTRIAL TRANSFORMATIONS

Digitalization offers the assembling business fresh out of the box new chances and choices for successfully reacting to these market patterns. On the off chance that organizations need to misuse the maximum capacity of digitalization notwithstanding, they should have a correspondence arrange that is customized to their particular necessities. There is no other method to ensure far reaching access to every pertinent datum-nonstop and progressively. This thus makes the conditions for fundamentally quicker, progressively adaptable, and increasingly proficient generation forms.

1. **Level 1: Keen Association:** The capacity to oversee and secure information set aside a few minutes because of smart sensors and to move them with explicit correspondence conventions.
2. **Level 2: Information to-Data Change:** The capacity to total information and convert it to esteem included data.

3. **Level 3: Computerized Twin:** The capacity to speak to ongoing in an advanced reality.
4. **Level 4: Comprehension:** The capacity to recognize various situations and bolster a legitimate basic leadership process.
5. **Level 5: Setup:** Provides input on physical reality from computer generated reality and applies remedial activities to the past level.

Sensor systems in addition to IoT bolster all items in the generation procedure need to speak with the control framework remotely. Together with the sensor framework they make data information arrange. In view of this information, the focal administration framework can investigate creation methods and forms and improve them to accomplish considerably more noteworthy generation effectiveness.

Because of the sharp increment of arranged segments underway and their association with the organization's IT, security ideas must be reliable from start to finish. One promising arrangement is to execute a multilayered "safeguard top to bottom for industry." This standard makes the conditions essential for a planned methodology among item providers, framework integrators, and plant administrators. The usage of security forms, the preparation of staff and secure items are extremely significant segments while in transit to a safe creation plant.

The digitalized supply system is empowered by blockchain innovation going about as the system's center foundation. It is upheld and enabled by the IoT which gives expanded availability, advanced information social affair and superior systematic abilities. GPS beacons and keen sensors are additionally conveyed by the IoT condition which associates physical items with information, progressively proficient assembling forms, more judicious supply chains, and new business life systems. The reason for each computerized framework is a modern correspondence arranges which must be structured distinctively relying upon necessities.

Producers have focused on innovation, including blockchain, as a way to limit inventory network disturbance and suppress their store network-related uneasiness. By 2022, an expected 60% of real makers will be dependent on advanced stages, which will be in charge of supporting capacities that are justifiably in charge of 30% of their income. What's more, by 2025 the variety of new advancements will be implanted in the assembling area, with 20% of the top producers relying upon a mix of AI, IoT, intellectual frameworks, and the blockchain.

It has been found from different research surveys that the added substance assembling known as 3D printing, presently can't seem to accomplish standard appropriation; however, the business' force is going the correct way. A little more than 500,000 work area 3D printers have been sold so far in 2018, with these side interest level machines filling in as a sign that more individuals are trying things out of added substance fabricating. This speaks to 52% year-over-year development, a pace more noteworthy than the 38% development recorded in 2017. In light of current circumstances, 1.5 million 3D printers will be sold in 2021; however, these patterns will be in general fluctuation.

However, this innovation is still in its early stages, and in our view, there are many assembling use case situations set forward with blockchain innovation that could some way or another be similarly also executed with customary assembling data frameworks. Decentralized systems have been considered widely before and have their very own vulnerabilities. Be that as it may, late progressions in circulated figuring, web of things and advances in information investigation has given another stimulus to reexamining decentralized systems for assembling activities.

Prevailing in the following modern period requires assembling organiza-tions to characterize and shape their guiding principle drivers empowered by computerized innovations. Industry 4.0 will drive operational efficiencies through smart factories and smart supply chains just as develop open doors through advancement and bespoke answers for increment client esteem. They will at last lead to totally new plans of action and administration contri-butions through digitalization.

2.5 CONCLUSION

Present-day fabricating offices of the future are spoken to by exceedingly mechanized creation lines including refined administration and control processing frameworks. Nonetheless, the assembling procedure is as yet reliant on the administrators. The future lies in a mix of expanding gadget self-governance, applying new advances and improving administrators' abilities, in addition to expanding the interoperability all things considered.

It is important to improve cooperation in the man-machine setting, not just by upgrading and presenting savvy innovation on the machine side yet in addition by the human capacities, conceivably utilizing other electronic specific circuits executed legitimately at the administrator. Together with virtualization innovations (prescient reproduction), the previously mentioned conceivable outcomes offer intriguing points of view on the administration of generation forms in the up and coming period. People, associations, and machines will probably impart uninhibitedly and self-ruling, direct exchanges freely, and interface easily and securely. This is the guarantee of blockchain innovation.

Makers influence IoT and prescient investigation in their administration parts production network to proactively fix hardware before it ever separates. "Blockchain can give an expanded degree of perceivability into this procedure, as it would permit a whole worldwide administration store network to see when and where parts are moving to guarantee the fix is set aside a few minutes."

Blockchain is one of many developing advancements that is reclassifying the manner in which organizations work together. The innovation itself, as far as advancement, guideline, and administration, and selection is as yet developing. Be that as it may, as blockchains development and the innovation itself turns out to be increasingly characterized, the production network (and administration inventory network, specifically) could start seeing more genuine types of straightforwardness, responsibility, and productivity than it's at any point seen previously. Producers that go out on a limb and receive new strategic policies and advancements to move past business as usual will be the ones that success.

KEYWORDS

- **artificial intelligence**
- **cyber-physical frameworks**
- **industry-4.0**
- **internet of things**
- **machine-to-machine**
- **operational technology**

REFERENCES

1. Manavalan, E., & Jayakrishna, K., (2019). A review of internet of things (IoT) embedded sustainable supply chain for industry 4.0 requirements. *Computers and Industrial Engineering, 127*, 925–953.
2. Wang, Q., et al., (2019). Blockchain for the IoT and industrial IoT: A review. *Internet of Things*, 100081.

3. Lin, C., et al., (2018). BSeIn: A blockchain-based secure mutual authentication with fine-grained access control system for industry 4.0. *Journal of Network and Computer Applications, 116*, 42–52.

4. Zhao, G., et al., (2019). Blockchain technology in agri-food value chain management: A synthesis of applications, challenges, and future research directions. *Computers in Industry, 109*, 83–89.

5. Sikorski, J. J., Joy, H., & Markus, K., (2017). Blockchain technology in the chemical industry: Machine-to-machine electricity market. *Applied Energy, 195*, 234–246.

6. Andoni, M., et al., (2019). Blockchain technology in the energy sector: A systematic review of challenges and opportunities. *Renewable and Sustainable Energy Reviews, 100*, 143–174.

7. Khaqqi, K. N., et al., (2018). Incorporating seller/buyer reputation-based system in blockchain-enabled emission trading application. *Applied Energy, 209*, 8–19.

8. Atin, A., Benjamin, C., Mahmud, H., & Binil, S., (2018). A case study for blockchain in manufacturing: "FabRec": A prototype for peer-to-peer network of manufacturing nodes. *Procedia Manufacturing, 46th SME North American Manufacturing Research Conference, NAMRC, 46*. Texas, USA.

CHAPTER 3

The Future of Smart Communication: IoT and Augmented Reality: A Review

DEBABRATA DANSANA, S. GOPAL KRISHNA PATRO, and
BROJO KISHORE MISHRA

Department of Computer Science and Engineering GIET University, Gunupur, Odisha, India, E-mails: debabratadansana07@gmail.com (D. Dansana), sgkpatro2008@gmail.com (S. G. K. Patro), brojomishra@gmail.com (B. K. Mishra)

ABSTRACT

Researchers around the world have made it possible for us to conclude that the communication style between us and the real world can be changed with the help of the internet of things (IoT). The IoT concept has made the world smarter and independent. Communication is maintained with the help of sensors installed everywhere which are supervised in a common network via some interfaces. The other side of the coin is augmented reality that has played a great deal in overcoming the fine line of difference between the real world and the internet. Security systems, health monitoring systems (HMS), and home appliances that are smart have already entered the commuter life and generate a large amount of data that are to be effectively handled and monitored. IoT and AR technologies do exist hand in hand in the everyday world. Combining the two technologies can bring about a revolution in future communications. This chapter mainly focuses on the various types of IoT architectures, network configurations and the use of augmented reality systems. The main aim of this chapter is to have insight knowledge of the existing technologies so that analysis can be done of the future endeavor that will help to build us a smarter future.

3.1   INTRODUCTION

Over the past decades, there has been a considerable development in the field of internet. The early realization of the internet was a network of some documents mainly of HTML type linked with each other that gave rise to the concept of the World Wide Web. This collection of HTML pages gradually evolved giving rise to concept of Web 2.0. This was basically a two-way communication that made it possible for the inclusion of human participation. Networking services like blogs, wiki documents made it to the common house and became famous and some of the widely used services of Web 2.0 [1]. Keeping the existing web services aside, researchers all over the world have been working in bringing out web services 3.0 also referred as the semantic web. The main aim behind the design of this web services is to make it possible for the web to communicate with machines. It aims at allowing the machines to behave in a more intelligent than before. This also aimed at allowing the machines to process and share data on own without any human supervision. These two technologies namely internet and the machine network were made to converge giving rise to new possibilities. This development of a framework that allowed direct machine to machine (M2M) interaction made it possible for researchers all over the world to bring about more machines into integration with one another. This concept somewhat gave rise to this paradigm called internet of things (IoT). The name IoT does not have any conventional definition but can be explained as the technique with which devices and things communicate effectively with each other and exchange information over internet. IoT has been serving the fields of industry, security, and maintenance for the past couple of years effectively.

3.2   RECENT MARKET STATUS

IoT can be defined as sensors and actuators in real world that are embedded in natural world and are connected to each other using a common interface called the internet. Internet makes it possible for things to communicate with one other, exchange data and remote control of devices for maintenance system. Along with these devices, personal mobile devices have broadly served everyday activities [2]. These interfaces need expert handlers for their control and coordination. But for non-expert users, the solution is augmented reality. It is possible to realize the real time data and to view the real world environment as well as trace the position and monitor the behavior of various objects via this concept of augmented reality. It basically is an interactive and 3D environment that integrates real and virtual world objects. There is a growing curiosity as to what will happen when augmented reality is made to combine with IoT. Researchers all over the world are trying to assess the way in which the combined technologies will help to change the modern world. The current scenario, therefore, deals with the task of hybridizing AR with IoT giving rise to a new trend called ArIoT.

3.3   A LOOK INTO THE IOT ARCHITECTURE

IoT is on its way to become the driving force that will lead to innovations all around the world, changing market scenarios in different fields of applications. With each passing day, thousands of new applications will pop up in the market, to maintain which we all need a strong interconnection. This interconnection is not only a technological challenge but also includes various aspects like privacy, standardization, and legal issues, etc. Some of the core technical challenges include connection to devices what are heterogeneous in nature and very low computational energy [3]. These challenges needs to be addressed effectively to make way for IoT to be applied in daily life and industries all over the world are striving to reach a viable solution. Hence, the main focus should be on an open and scalable architecture which will act as the backbone of the future IoT development. This new architecture should be able to handle all the previous challenges and ensure reliable functioning for the future IoT connected devices [4].

The first type of IoT architecture referred to is:

1. **Architecture Reference Model (ARM) [6]:** This architecture is more of a set of guidelines that is to be followed while starting with any basic IoT. It provides sufficient information for the operation of various IoT systems connected together. This architecture model actually guides us to penetrate the service layer (Figure 3.1).
2. **Building the Environment for the Things as a Service (BETaaS):** This type of architecture basically defines the overall operation of the system. The actual implementation of this model is not yet available for final version but is still under research stage and efforts are constantly being made to fix the missing events correctly and quickly. BETaaS basically comprises a network of gateways that effortlessly succeeds in integrating heterogeneous systems. This architecture is basically built upon TaaS model which stands for things as a service (Figure 3.2).

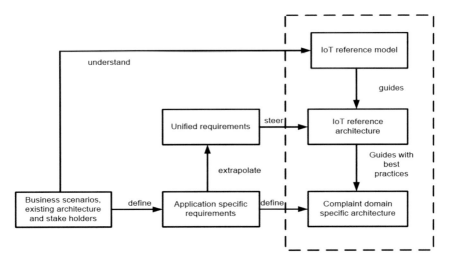

FIGURE 3.1 ARM architecture model of IoT.

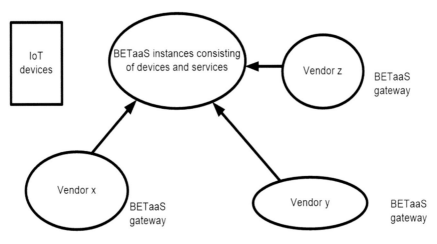

FIGURE 3.2 BETaaS model of IoT.

3. **Open IoT:** This is the third architecture model for IoT services. This basically uses a semantic directory and a sensor middleware model [6]. The entire development of this module was done on the basis of ARM architecture. The sensor middleware handles the data collection part from various sensors attached to physical and virtual objects. These collected data are then stored with the help of CCI (cloud computing infrastructure) which stores information in an

accessible manner. This architecture supports heterogeneous devices to be connected and this is made possible by a distributed deployment model support for semantic models (Figure 3.3).

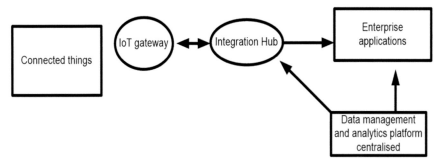

FIGURE 3.3   Open IoT model.

3.4   AUGMENTED REALITY

The concept of augmented reality might date back to older times but the recent times have seen its full implementation and reliability aspect. The recent light on augmented reality has brought to light many factors like precise image identification on IoT enabled devices, fast processing units and effective feature extraction. The extracted features from various objects are stored and accessed instantly. Image displays on devices are done in an elegant way with the help of high definitions devices. The presence of broadband services and wired communication makes it easy for the devices to extract data from servers easily. In a nutshell, augmented reality aims at displaying the real world to the user with the help of computer-generated 3D images on a screen or monitor in real-time [7]. The main framework of augmented reality was constructed on context-aware computing which enables the user to view the same object on 3D as well as on virtual ground. Sensor, visual effects, and image tracking are some of the major classifications of augmented reality.

3.5   BRIEF OVERVIEW ON THE AR WORKFLOW

Augmented reality can be described as an amalgamation of framework software as it comprises of browser interface along with being an application programming. Tools like AR toolkit [8] can be used for easy integration of various AR devices and for creating a suitable interface for the user. AR basically has three major features: detection, tracking, and vision. Detection and tracking is mainly brought about by image recognition, vision rendering and interfacing of camera. The workflow mainly deals with detection of inputs and sending the data to rendering device. This device takes all the inputs from detection and comparing units and then combines them with data from rendering unit to send back a visual feedback to the user (Figure 3.4).

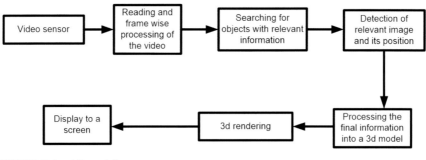

FIGURE 3.4   AR workflow.

3.6   PROCESSING OF AR

The process of augmented reality mainly comprises of five processes namely acquisition to retrieve the images from the cameras. Acquisition is followed by feature extraction which takes care of redundant data and deletes it and prepares a set of useful data. This data set is made keeping in mind some basic data set as reference. The next step that follows is feature extraction that extracts the necessary information from the data retrieved from the images. It basically decides whether the image retrieved has some valid feature to be extracted or is to be made redundant. The next step that comes into play is geometric verification which matches all the images that are related geometrically and creates a database (DB) that acts as a subset of the original AR data set. The associated information retrieval is another feature which follows next. It mainly deals with searching and accessing the data and retrieves vital information. The output of the AIR process is to be displayed on the screen. The AR feature extraction process is by itself comprises of six stages. The first step is the grayscale image generation (GIG) which makes sure that all the images that are captured are converted to greyscale making it easier to impose color modifications on the images.

This is followed by integral image generation (IIG) which means that the greyscale images are to be processed and the sub-regions of the images are to be analyzed. Once this is done, response generation matrix (RGM) analyses the image and points out the Hessian points that give rise to point of interests. This interest points (IP) constructs a map that generates the space scaling of the image. The next step is very important and is mainly used to detect the important points in the images retrieved and to get their response curve based on maxima and minima calculation. This process is known as interest point detection (IPD) method. The identifications of IP are necessary for successful extraction of necessary features from the images. Each and every IP are then reproduced to an orientation suitable enough for the final display of the images (Figure 3.5).

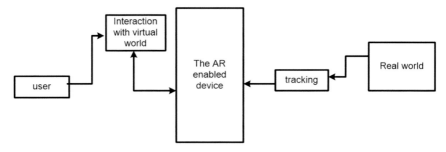

FIGURE 3.5   AR architecture.

3.7   BRIEF OVERVIEW OF AR TRACKING

There are basically two common tracing methods that are employed in AR method to monitor the images as well as the gestures and directions. The general tracking concept is used for indoor applications and the use of GPS (global positioning system) is done for outdoor tracking devices. Paper [20] elaborates another tracking technique which uses image processing to track objects in real life. The indoor tracking logic is basically simple because the movements of indoor objects is within bounded surfaces and are predictable to some extent. Outdoor tracking has limitless possibilities and dimensions of movement. This is the main reason for the use of GPS as tracking technique. The various objects are tracked and their positions in terms of longitude and latitude are noted in DB.

3.8   AR APPLICATIONS CURRENTLY IN USE

Researchers all over the world has been striving to make out the most of augmented reality in various fields of applications. A widespread literature survey has been done and it has been found that augmented reality has made its way effectively into military, medical, and industrial applications.

1. **Use in Educational Purpose:** The use of AR in educational system has been very effective to students as it helped them visualize the virtual objects in real 3D world. This makes education effective and interactive. These also come with audio support and are also used to edit some data and information based on the viewer demands. There are also provisions of an application developed for smart devices that help students to view and educate themselves in a better way [13]. The pros and cons of integrating AR in educational purposes have been discussed and a meta heuristic method for its betterment has been proposed [14].

2. **Use in Surgical Operations Field:** Munich University contributed to the medical sector by coining three systems in 2012 for surgical purposes [9]. The devices were correspondingly named as Freehand SPECT, CamC, and Mirracle all three of them being AR-driven tools that has been a breakthrough in the medical field. An animal study was conducted with and without AR and the accuracy and precision of the surgery conducted was monitored and the pros and cons were pointed out. Suggestions were given to overcome the gap to make AR governed surgeries more viable in future [10].

3. **Use in Military Fields:** AR technologies have been very effective in defense fields to pass valuable information to the soldiers and provide them Intel regarding enemy movement update and position. A wearable computer system has been proposed [15] that has a see through display and a GPS connection that provides visual data while performing any standard tasks. The self-wearable computer is equipped with a built-in navigation package called the map in the hat that has been effective for providing data to soldiers in combat conditions. The ARGOS augmented reality toolkit overcomes any degree of machine autonomy and terrain geography and provides information by creating a partial model for a 3D worksite [16, 17]

3.9   FINAL MEETING OF THE TWO

When we try to integrate the AR with IoT, the first thing that AR benefits us with is that it helps in magnification of the world around us. The IoT e-commerce shopping allows customers to virtually try outfits on them and purchase them according to their choice [19]. This explains what exactly is the job of AR in the world of IoT. The use of AR and IoT in medical field might make it possible for the doctor to get access to a human body on live image. This has a great chance in reducing the risk of the medical hazards and guarantee reliability. The use of smart helmets in construction plants makes it easy for the company to assess and monitor the detailed mental health of the workers all the time. The use of AR in this field might help the engineers to give access to the plant in 3D itself and exercise control actions on the site itself.

3.10   CONCLUSION

The road to future is to be paved by AR and IoT coming hand in hand. The issue of size restrictions is now a myth and can guarantee a dynamic user interface. The integration of these two blooming trends can prove to be beneficial for adoption in large scale. The mixing of the two has a lot of benefits as discussed earlier but also has some limitations which are to be addressed which might make it possible to ensure better future services and manufacturing scenarios.

KEYWORDS

- architecture reference model
- artificial intelligence
- cloud computing infrastructure
- global positioning system
- grayscale image generation
- integral image generation

REFERENCES

1. Whitmore, A., Anurag, A., & Li, D. X., (2015). The internet of things: A survey of topics and trends. *Information Systems Frontiers, 17*(2), 261–274.
2. Tan, L., & Neng, W., (2010). Future internet: The internet of things. In: *2010 3rd International Conference on Advanced Computer Theory and Engineering (ICACTE)* (Vol. 5). IEEE.

3. Datta, S. K., Christian, B., & Navid, N., (2014). An IoT gateway centric architecture to provide novel M2M services. In: *2014 IEEE World Forum on Internet of Things (WF-IoT)*. IEEE.

4. Krčo, S., Boris, P., & Francois, C., (2014). Designing IoT architecture(s): A European perspective. In: *2014 IEEE World Forum on Internet of Things (WF-IoT)*. IEEE.

5. Lanka, S., Sidra, E., & Aisha, E., (2017). A review of research on emerging technologies of the Internet of Things and augmented reality. In: *2017 International Conference on I-SMAC (IoT in Social, Mobile, Analytics, and Cloud) (I-SMAC)*. IEEE.

6. Gazis, V., et al., (2015). Short paper: IoT: Challenges, projects, architectures. In: *2015 18th International Conference on Intelligence in Next Generation Networks*. IEEE.

7. Azuma, R., et al., (2001). Recent advances in augmented reality. *IEEE Computer Graphics and Applications, 21*(6), 34–47.

8. Van, K. D., & Poelman, R., (2007). *Augmented Reality: Technologies, Applications, and Limitations*. Vrije Univ. Amsterdam, Dep. Comput. Sci.

9. Samset, E., et al., (2008). Augmented reality in surgical procedures. *Human Vision and Electronic Imaging XIII* (Vol. 6806). International Society for Optics and Photonics.

10. Navab, N., Blum, T., Wang, L., Okur, A., & Wendler, T., (2012). First deployments of augmented reality in operating rooms. *Computer, 45*(7), 48–55.

11. Huang, Z., et al., (2014). CloudRidAR: A cloud-based architecture for mobile augmented reality. *Proceedings of the 2014 Workshop on Mobile Augmented Reality and Robotic Technology-Based Systems*. ACM.

12. Navab, N., et al., (2007). Action-and workflow-driven augmented reality for computer-aided medical procedures. *IEEE Computer Graphics and Applications, 27*(5), 10–14.

13. Akçayır, M., & Gökçe, A., (2017). Advantages and challenges associated with augmented reality for education: A systematic review of the literature. *Educational Research Review, 20*, 1–11.

14. Radu, I., (2014). Augmented reality in education: A meta-review and cross-media analysis. *Personal and Ubiquitous Computing, 18*(6), 1533–1543.

15. Thomas, B., et al., (1998). A wearable computer system with augmented reality to support terrestrial navigation. *Digest of Papers: Second International Symposium on Wearable Computers (Cat. No. 98EX215)*. IEEE.

16. Drascic, D., et al., (1993). ARGOS: A display system for augmenting reality. *Proceedings of the INTERACT'93 and CHI'93 Conference on Human Factors in Computing Systems*. ACM.

17. Day, P. N., et al., (2005). Wearable augmented virtual reality for enhancing information delivery in high precision defense assembly: An engineering case study. *Virtual Reality, 8*(3), 177–184.

18. Milgram, P., Anu, R., & Julius, J. G., (1995). Telerobotic control using augmented reality. *Proceedings 4th IEEE International Workshop on Robot and Human Communication*. IEEE.

19. http://aircconline.com/ijma/V4N4/4412ijma04.pdf (accessed 20 July 2020).

20. Zhang, X., Stephan, F., & Nassir, N., (2002). Visual marker detection and decoding in AR systems: A comparative study. *Proceedings of the 1st International Symposium on Mixed and Augmented Reality*. IEEE Computer Society.

CHAPTER 4

Heart Rate Monitoring Using IoT and AI for Aged Person: A Survey

SIBO PRASAD PATRO,[1] NEELAMADHAB PADHY,[2] and RAHUL DEO SAH[3]

[1]Research Scholar, School of Computer Engineering (CSE), GIET University Gunupur, Odisha, India, E-mail: sibofromgiet@giet.edu

[2]School of Computer Engineering (CSE), GIET University Gunupur, Odisha, India, E-mail: dr.neelamadhab@giet.edu

3Dr. Shyama Prasad Mukherjee University Ranchi, India, E-mail: rahuldeosah@gmail.com

ABSTRACT

Today the internet of things (IoT) becomes limestone in the world, where lots of daily used objects are interconnected, and they are interacting with the environment for collecting information and automatically perform a specific task. The design and development of wearable systems for health monitoring has garnered lots of attention in the scientific community and in industry during the last years. Mainly motivated by increasing healthcare costs and propelled by recent technological advances in miniature connected devices, smart textiles, microelectronics, and wireless communications, the continuous advance of wearable sensor-based systems will potentially transform the future of healthcare by enabling proactive personal health management and ubiquitous monitoring of a patient's health condition. These systems can comprise various types of small physiological sensors, transmission modules and processing capabilities, and can thus facilitate low-cost wearable unobtrusive solutions for continuous all-day and any-place health, mental, and activity status monitoring. Heart disease becomes critical fatality in the world today. Prediction of cardiovascular heart disease (CHD)—the most significant and critical challenge in the area of medical health care analysis. In this chapter, authors discussed how machine learning (ML) is used for effective in assisting in the making of a decision and predict. The patient health data are collected from a wearable device and how the data storing in blockchain for feature operations. ML now a day are used in various areas of the IoT. Multiple studies give a glimpse into a prediction of heart-related disease with the help of ML techniques. Authors also proposed few methods in this chapter for the automated health monitoring system (HMS) including quantifying patient's heartbeat rate values, different ways for collecting the heartbeat rate of the patient, future prediction of heartbeat rates counting and a proposed ML method for heart disease classification.

4.1 INTRODUCTION

In healthcare clinical domain using connected devices, a vast amount of data are generated and accessed. These data are disseminated regularly. Hence, to store and to distribute these collected data is an important and challenging job. These data are susceptible to limiting factors such as privacy and security [1, 31]. In the area of healthcare domain to make the data safe, scalable, and secure (SSS) data-sharing is extremely important for diagnosis and decision making for these clinical data. The data-sharing concept is highly essential for the doctors to transfer the patient data to the concerned for quick response and look into. The clinical probationers should transfer the patient data in privately, securely, and timely to make ensure all the stockholders have the full and latest information regarding their patient health conditions. E-health and tale-medicine are the two disciplines where the medical health data is remotely delivered to a medicine specialist for an expert opinion. Here the patient data is shared through a "store-and-forward technology" or by online virtual monitoring. Sometimes these online clinical monitoring called telemonitoring or telemetry [2, 3]. With these online health care environments, patients are diagnosed remotely by the doctors by means of exchanged clinical data. In all such cases, the clinical data are required integrity, privacy, preservation. These are all significant challenges to occur due to the case-sensitive nature of the patient data. Thus to exchange the data safely, securely, and scalable is the most critical factor for meaningful clinical communication for the remote patients. This safe and integrated exchange of data provides a network where clinical specialists can share their views and they can gather recommendations for the improvement of diagnosis and effective treatment [4–6]. Such exchanging of clinical data always required substantial, a good collaboration between the entities that are involved. There is specific constraint found in this process, including clinical data, sensitivity, procedures, and agreements for data sharing, various complex algorithms for patient matching, governing rules, and ethical policies. These are a few essential aspects where a mutual understanding required for health care data exchanging [7].

Since a few years, researchers have implemented various applications like artificial intelligence (AI), computer vision, blockchain technology, internet of things (IoT) for facilitating the doctors and clinical practitioners for better treatment of multiple diseases. Blockchain is one of the best applications that can be involved in this scenario for the delivery of safe and secure healthcare data [8, 9], e-health data sharing [10], and biomedical [11].

According to the 2016 Global Burden of Disease Report, heart disease is one of the leading causes of death in India, killing more than 1.7 million Indians in 2016, In India deaths from the heart-related disease among the rural people have surpassed among the urban Indians. Heart-attack is the most dangerous among the heart-related condition. It has three factors that contribute, first one is heart diseases have high recurring rates and a patient suffer heart attack all of sudden without any prior notice. Secondly, the patients who are suffering from this disease the heart attack rescue time is short, only after several minutes, it happens. Thirdly, most of the patients those are suffering from these diseases they are residing at home, instead of residing at hospital. For an emergency, the remote physician is unable to find the physical status of the patient. Besides, patients from a rural area, older people, living alone they face more problem. They are unable to get in-time treatment after a heart attack.

Today, mobile and IoT connected devices technology have brought profound changes in the world [12]. These overwhelming technologies are used to solve heart-related diseases problems. The patients can avail medical resources remotely for diagnosis. With the technology of the connected device, it is an appropriate technology to monitor regularly the vital func-tioning of the patient's body no matter where ever they are available and whatever they are doing. The data is collected from the patient remotely and sent to the physician at a low cost, which ensures to the physicians to keep track the patients physical status continuously in real-time. We proposed in this research paper, an IoT based sensor monitoring system for heart-related diseases. The connected device collected the data from the patient body, including blood pressure, ECG, heartbeat rates, etc. After fetching the data which goes to the cloud and blockchain for data privacy and security. Using a few algorithms, the collected data can be analyzed for finding the patient's health condition.

4.2   AI AND IOT FIGHTS TOWARDS CARDIOVASCULAR DISEASES/ RELATED WORK/LITERATURE REVIEW

Today cardiovascular or heart-related diseases are the common health problems in the world, and this is the most causes for death. American Heart Association said that more than 2300 persons die each day due to cardiovascular disease in the US, and this is a very shocking number on a global level. As per the World Health Organization (WHO) survey, 85% of people death occurs due to heart attack and strokes, which causes CVD. Similarly, 1,500,000 deaths occur each year in China due to cardiovascular diseases. Now, this becomes leading cause of death in China.

Due to the invention of telemedicine, research about virtual and e-shopping for healthcare service has been of great concern. Later the growing of ubiquity and smartphones helped us for initiation of e-health and mobile health (Sonisky and Machael, 2008). Few researchers identified to provide various uses of medical and healthcare education to facilitate the clinical practice with the help of phone applications(Lindquist, 2008), The researcher (Fairhurst and Sheikh, 2008; Armstrong, Watson et al., 2009) said the push features are used for sending notifications, lecture notes, reminders, and medical advice to a patient. But these messages will help only to facilitate patient's self-management only; a doctor can't get clear information of a patient's status remotely. According to (Pop-Eleches et al., 2011) defined the improvement of patient's adherence to antiretroviral treatment. Later, the development of technology SMS, has been taken as one of the best ways to deliver health innervations of a kind. Today the smartphone technology allows connecting with external devices over the internet such as wireless sensors, wirelessly connected devices. These are the new technology implemented in mobile phones to contact a doctor remotely to verify the patient's health records and the conditions of the patient regularly (Klasnja and Pratt, 2012). This health monitoring system (HMS) becomes a hot research topic over the world. A large number of research projects have developed for looking after various diseases from different geographical scopes over the globe (Rofouei et al., 2011). Expect this, and some researcher's implemented technology to control hospital area and to trace the location of hospitals (Pandian et al.,

2007). Research has given attention to getting hearth rate information, ECG, blood pressure (Jin et al., 2009). But to make more analysis by a doctor remotely to a patient, it needs vast parameters.

For these systems with the help of multiple parameters, different physical signs need to be sampled with different frequencies to satisfy the medical requirements. But at the same time due to huge amount data, these sample frequencies may lead to a great burden to the remote server.

4.3   WEARABLE DEVICE (IOT)

To avoid the above said problems, wearable devices (IoT) technology has elected by many researchers due to its ability and accessibility. BAN is the subset of IoT in healthcare domain. The patients are armed with various sensors to sense the heart rate, measure signs in real-time, blood pressure, and temperature are shown in Figure 4.1. Once the data are collected from these connected devices, the data sent to the gateway, then they forwarded to hospitals. Hence, without moving the older people to the hospital, they can cure, and they enjoy their life [13].

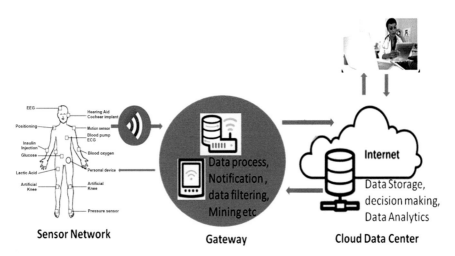

FIGURE 4.1   IoT sensors data processing in cloud data.

In few situations, the IoT server architecture can act as a barrier of failure. That may disrupt the entire network. IoT devices in remote patient HMS are also accessible to DDoS attacks, data theft, hacking, and remote hijacking without any exception. The criminal can attack the system and they can still the personal health care information of the patient. Hence, the data must be strictly protcctcd for these electronic health records (EHRs). Besides few instructions are given to the actuators which are residing inside the IoT device to record and maintain an accurate time of events. To overcome the above-defined issues, few researchers identified the smart contracts on blockchain can be embedded with WBAN system to provide distributed data processing jobs and produce unchanged logs.

Several papers provide reasonable application for blockchain in health care system [14–16] with no much comparable research. IBM identified hyper ledger project for the uses of blockchain and IoT for healthcare, but so far, the operational model are not available. According to the research, paper [17] found that the smart contracts are used to analyze the sensor data. In this research paper, we described the development of smart contracts for analyzing the patient's data (Figure 4.2).

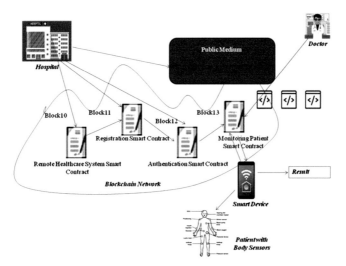

FIGURE 4.2   Blockchain network operation scenario.

This chapter shows how the security and privacy issues of a patient's health records are resolved through a smart contract. It also contains a mechanism, when the data are collected from the wearable devices from a patient and write into a blockchain network shown in Figure 4.2, it alerts to the healthcare providers, and doctors about any abnormal data are collected from the patient's wearable devices. Some few sensors can be attached, which will help to locate the patient for any emergency also.

4.4 BLOCKCHAIN

Blockchain is a P2P (peer-to-peer) network technology. It is an integrated of multi-field network framework. It contains algorithms, cryptography, and few mathematical expressions. These components are used for solving the distributed database (DB) synchronization problems. Blockchain consists of six key elements those are decentralized, immutable, transparent, open-source, autonomy, and anonymity (Table 4.1).

TABLE 4.1   Key Elements of Blockchain

Key Elements	Description
Decentralized	Here database system allows access control for every user over the network for accessing, updating, storing on multiple orders.
Immutable	Here the data once stored in the database are not allowed for modification easily until 51% of control of the node concurrently
Transparent	The data stored on a blockchain is visible to the authorized user, which is allowed for further updating quickly, but sometimes it prevents the data for any modification or stolen.
Open-source	The blockchain network provides an authenticated access to the connected users.
Autonomy	The blockchain data are allowed for access, transfer, modify data by making trustworthy and making it free from any external intervention
Anonymity	Here the data transfers between nodes to the node, hence the identity of the user remains anonymous, and that makes the system more secure and reliable one.

4.5 BLOCKCHAIN NETWORK

The blockchain has to verify for each new transaction made on the network. Due to verification of each transaction in a block of blockchain by the network, it becomes immutable. In coming days, blockchain may help in reliable, secure, and personalized by merging the entire collected health care data and making them secure one. The basic flow of the data in blockchain network defines as in Figure 4.3.

Today in the world, senior citizens health problems are increasing at a higher rate. For these issues, older people are treated and monitored in traditional methods with the help of health care professionals. For health checkup, these older people need to visit the doctor for a diagnosis. At the same time, healthcare professionals should avail in the hospital for monitoring with medical instruments. But it may cause overload to the doctors. Secondly, the older people are unable to go hospital due to unavailability of their family support.

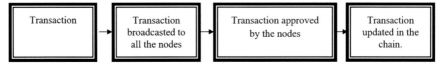

FIGURE 4.3   Basic flow of the data in blockchain network.

4.6   ARTIFICIAL INTELLIGENCE (AI) IN IOT

Today the word 'smart' become quite fascinated, the technology that we have today is still lacking far from being smart as a human. If we take an example of a Smartphone, even it is smart but it can't perform any its operation automatically. For instance, if the Smartphone is silent mode, we can put any notification while driving. It would be better some technology can be enhanced to put some information while the driver is driving the car. This kind of concept can be implemented by some wireless concept which will establish a network between user and device. Similarly, another example can be when the driver of the car gets sick then a push notification can be sent to his family members or hospitals nearby. So that little precaution can be taken earlier before something goes wrong. Similarly, in the physical world, everything will be connected to everything to meet user requirements. To make the entire physical world to connect everything by smart technology, we need AI. AI plays a vital role to handle vast number of connected devices online, and making critical decision on these endless seas of data are streamed from these connected devices. To perform any action by objects automatically, we use two modules. The first is machine learning (ML), and the second one is data analysis (DA) [18]. ML helps the object for learning from the given data and perform and action automatically whereas data analytics is used to analyze from the existing past data and generate some efficient work that can be completed in future. These two techniques are embedded in connected devices or sensors [19] for the smart system.

There are few variations to define the ML algorithms. They are commonly divided into 4 types according their purpose. The main categories are supervised learning, un-supervised learning, semi-supervised learning, and reinforcement learning (RL). In supervised learning, each dataset pair contains an input object and expected output value. The algorithm analyses the given training data and produces an inferred function that can be used for mapping new example. Few standard algorithms are nearest neighbor, decision tree, linear regression, and support vector machine (SVM). For pattern detection and descriptive modeling unsupervised learning is used. Here there are no output categories or labels are identified on which the algorithm can try a model for relationship. The mostly used algorithms are Association rules and K-means clustering.

In supervised and un-supervised learning there are no labels identified in the dataset. The semi-supervised learning comes between these supervised and unsupervised learning. The semi-supervised methods accomplishment the idea that even though the group memberships of the unlabeled data are unknown, this data carries essential information about the group parameters. Finally, the RL method aims at using observations gathered from the interaction with the environment to take actions that would maximize the reward or minimize the risk. It continuously iteratively learns from the environment. Figure 4.4 shows the basic process of patient's health data using AI.

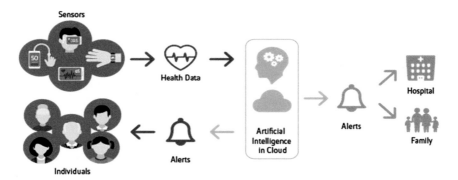

FIGURE 4.4   Healthcare system using artificial intelligence.

Using the above-defined algorithms, a researcher trains the system to analyze the patient's complicated symptoms from the data which are collected from connected devices. Few researchers experienced the machine for identifying lung cancer tissue from a mixture of large samples [20]. But there are many challenges and issues exist using IoT, ML, and blockchain. Personalized diabetic management is one of the most prosperous ones in health care management using IoT and ML [21]. Food usage and insulin response varies from patient to patient. The amount of food to be taken by the patient also varies. A Bluetooth enable blood sugar monitoring device gives information to the patient such as quantity of the food to be taken, dietary instruction on the brunt of unsupervised eating. IoT is a low cost computing device. The IoT device contains microsensors. These sensors communicate with each other to constructs a body area network (BAN), personal area network (PAN), and a few different categories of networks. The combination of IoT and ML gives various benefits in the health care system. They are diagnostic care, assistive care, and regular monitoring and generating alarm. Figure 4.5 describes the ML and IoT work in personal health care.

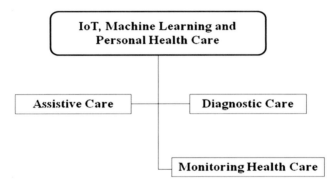

FIGURE 4.5   IoT, ML work in personal health care.

4.7   ISSUES AND CHALLENGES IN PERSONAL HEALTH CARE

The electronic healthcare system is not having many drawbacks. It has a few fundamental problems inherited of ML and IoT. For example, an elderly aged person wears a sensor on his body. The wearing device periodically collects huge amount of health-related data including blood sugar, ECG, heart rate, sugar, blood pressure, etc. Then the collected data send to the cloud. The data which will store in the cloud can be accessible to the users who are present in the DB. In meanwhile in the DB, few ML algorithms are implemented to analyze the aged person risk factor regarding sugar level, high blood pressure, or low blood pressure, etc., after analyzing few sugges-tions will be sent for health improvement. Figure 4.5 shows how the data collected and then it sent to the cloud. But sometimes, each of the steps involved in this scenario contains specific issues and challenges. The data packets which are transmitted into the cloud there may be chances of the drop in the network, access issues, authentication issues, privacy issues, and integrity issues. These all issues by the sensors need to be forward. There are many challenges are need to be taken care:

1. Data security and privacy;
2. Integration: multiple devices and protocols;
3. Data overloaded and accuracy;
4. Cost.

4.8 IOT RELATED ISSUES IN PERSONAL HEALTH CARE

The usefulness would determine the success of technology to the patient. BAN uses some sensors to collect the patient's health information. Hence, the BAN must work successfully for transmitting and transforming the sensed data. At the same time, the collected data from the BAN must ensure it will meet the system requirements. Requirements like integrity, security, and energy efficiency. Later, the data is selectively processed and delivers the information at different levels and appropriate destinations shown in Figure 4.6.

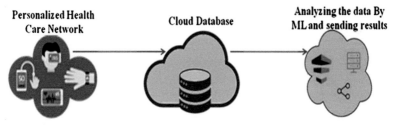

FIGURE 4.6   Data flow in the personal health care system.

4.9 MACHINE LEARNING (ML) RELATED ISSUES IN PERSONAL HEALTH CARE

ML is closely related for statistical analysis; predict future from past data, decision making from the stored data. In health care system, ML helps for analyzing the situation of a patient from the patient's trained dataset. The ML plays an essential role in predicting the future and solves it successfully. But in some condition, the trained dataset may not be diverse to cover a few situations. The noisy data, incomplete data, and dirty data could lead to the least chance to expose and forecast the health-related diagnosis and advisory notice. For example, if we consider sleep tracking, sleep monitoring, sleep patterns, it varies from man to man, age, gender type, and health status. Hence, a complete dataset may not available for sleep tacking, and this may lead to some wrong estimation in personal healthcare. Therefore, ML decisions could be wrong. At a similar time, a few other examples like an autonomous driving car which may lead to accidents due to the wrong decision by the ML-based decision. When an AI machine uses unsupervised learning, few critical questions come that how the decision can be taken. Here a question arises who is responsible for false assertion.

These kinds of problems can limit the uses of ML and IoT in health care applications. It may happen the hospital releases the patients, and he may need to readmit to the hospital for more diagnosis. So it needs few additional efforts, the constant follow-up to avoid such situations. Rothman et al. described the clinical variables and vital signs are more important for predicting readmissions [22].

4.10   SYSTEM LEVEL DESCRIPTION OF SMART HEALTH-LOG

An architectural overview of the proposed IoT-blockchain based HMS shown in Figure 4.7. The system is considered as a patient body which wearing a smartwatch along with smartphone applications. The smartwatch device contains an accelerometer, barometric pressure sensor, ambient temperature sensor, heart rate monitor sensor, oximetry sensor, skin temperature sensor, etc. These sensors sense the body and collect data. Using wireless technology the collected data stores in cloud through the internet or intranet, under the coordination of a integrate microcontroller.

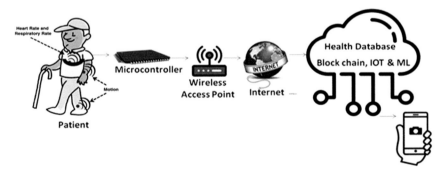

FIGURE 4.7   Proposed IoT-blockchain based health monitoring system.

Figure 4.6 shows how corresponding patient's data are acquired using Smartphone applications. The Smartphone application provides the no of steps walked by the patient, motion tracking, temperature of the body, heart rate, etc. The calculated values and predictions are accessed by Smartphone applications. The data flow of the proposed smart health monitoring shows in Figure 4.8.

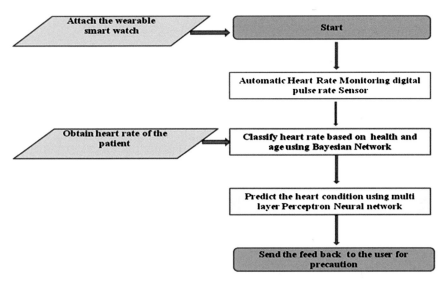

FIGURE 4.8   Data flow for the proposed IoT-blockchain based health monitoring system.

4.11   PROPOSED METHODS FOR AN AUTOMATED HEALTH MONITORING SYSTEM (HMS)

The proposed model automated HMS classified into four significant methods. They are: (a) a method for quantifying a patient's heartbeat rate values; (b) a way for data acquisition in smart health monitoring; (c) a method for future prediction of heartbeat rates counting; and (d) a proposed ML method for heart disease classification.

1. **Quantifying the Patient's Heart Rate Values:** The heartbeat rate varies according to gender. Male heartbeat is less than a female heartbeat. From last few years, the heart-related disease patient's number increased rapidly due to the cause of stress and lack of exercise, inadequate medication, less stress. The heart rate changes with the passage of age shown in Table 4.2 [23].

TABLE 4.2   Body Pulse Rate (BPM) by Age

Age-Wise Body Target and Average Body Pulse Rates			
Years	Average	Good	Excellent
18–25	70–73	62–65	56–61
26–35	71–74	62–65	55–61
36–45	71–75	63–66	57–62
46–55	72–76	64–67	57–62
56–65	72–75	62–67	57–61
65+	70–73	62–65	56–61

The heartbeat rate of a woman is 226 whereas for man heart rate is 220. For healthy human the heartbeat rate is 60–110 bpm but when we check an athlete resting heart rate will be 35–40 which is normal for them. And at same, the bmp varies from gender to gender and person to person sometimes. To measure beat to beat, heart rate detection different methods available like ballistocardiogram (BCG) learn through sensors [24]. An electrocardiogram is used to find the abnormal heartbeats of the children [25].

At first in the smart health care system, we need to quantify the heartbeat rate values of the patient and is initiated immediately after the wearable smart watch, or any intelligent connected devices are placed on the patient body. Figure 4.9 shows the automatic heart rate quantification using smart healthcare device. The load cells helps to calculate the heartbeat rates.

2. **An Approach for Data Recovery in Smart Health Monitoring:** After successfully, the heartbeat count and timestamp of the patient are collected though the connected device sensor board. The next step is to gather the heart rate information (cardiac monitoring). Using optical character recognition (OCR) by the help of Smartphone application, FDA-Cardio tachometer calculates the heartbeat rate, and then the computed data will be updated into the blockchain DB. Once the data are stored in the blockchain DB, the values are used to calculate the heartbeat rate ratio of the patient. To calculate the data for future heartbeat rate, the patient data along with the activeness of the patient body and timestamp are stored under respective patient IDs. The timestamp is recorded to find whether the patient is walking, running, sleeping, or having some heart disease problems.

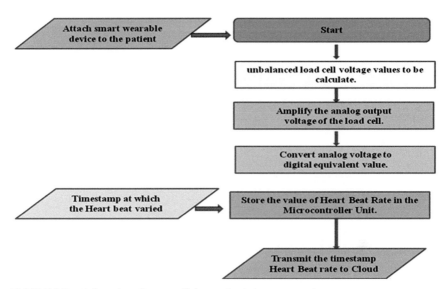

FIGURE 4.9   A flowchart for quantifying patient's heart rate values.

3. **Proposed Method for Future Patient Heartbeat Rate Predictions:**
   After a patient running or walking, the patient's heartbeat rates are
   calculated to quantify the patient's heartbeat rate ratio. If there are
   any anomalies found in the patient cardiac then it needs to identify
   the patient have any cardiac disease, health issues, blood pressure or
   chest problem, etc. To accomplish a fully balanced ratio, the heartbeat ratios for the upcoming heartbeats are calculated. Depending up
   on the inputs, future heart disease are forecasted and suggested as
   feedback to the user (Figure 4.10).

4. **Proposed Machine Learning (ML)-Based Method for Cardiac
   Diseases:** The main goal of this method is to fetch the features of
   patient's information and clustering the data based on the patient
   heartbeat rates, if any problem occurs then send feedback to the
   patient shown in Figure 4.11. To achieve these objectives, we can use
   a Bayesian or ML belief network (BN) algorithm. A BN structure can
   be designed on two approaches. The first approach is score-and-search
   method to search using a scoring function and the second approach
   is a constrain-based method for judgments based on the conditional
   dependencies. For our proposed model, we can take the help of a
   constraint-based process.

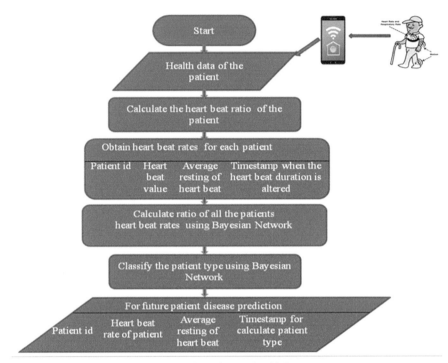

FIGURE 4.10   Data recovery approaches of smart health log.

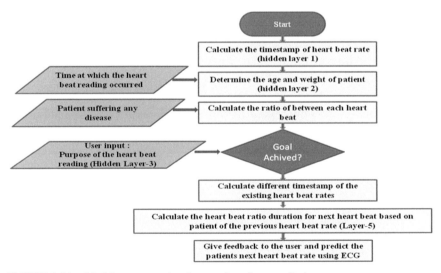

FIGURE 4.11   Model representation for next heartbeat prediction.

When it comes to learning causal graphs from data, two main approaches exist: constraint-based and score-based. Constraint-based approaches query the data for conditional independencies and try to find a DAG/MAG that entails all (and only) the corresponding d-separations. Score-based approaches try to find the graph G that maximizes the likelihood of the data given G (or the posterior), according to the factorization imposed by G. In general, a class of causal graphs, that are called Markov equivalent, fit the data equally well. Constraint-based approaches are more efficient and output a single graph with clear semantics, but gives no indication of the relative confidence in the model. Moreover, they have been shown to be sensitive to error propagation (Spirtes, 2010). Score-based methods on the other hand do not have this problem, and they also provide a metric of confidence in the entire output model. Hybrid methods that exploit the best of both worlds have therefore proved successful in learning causal graphs from data (Tsamardinos et al., 2006).

Cardiovascular disease affects the cardiovascular system [26]. This disease includes coronary heart disease, high blood pressure, peripheral artery disease, congenital heart disease, rheumatic heart disease, and even heart failure also. These types of diseases are treated by cardiologists, thoracic surgeons, neurologists, and radiologists. As the diseases are dangerous, the diagnosis is also an important task for the physicians to diagnose accurately and efficiently. The IoT system is very much important for the physicians to do better diagnosis and treatment. There are many computer-aided diagnosis systems are widely used as classification problems. The main objective is to reduce the number of false decision and increase the true ones. Out of many classification problems Bayesian network are used in cardiovascular disease for predicting the risk. These kinds of networks are selected to produce the probability estimates instead of predictions. The estimated results allow to prediction to be ranked. At the same, time the costs to be minimized. The BN is the ability to represent. It also understands the knowledge. This study evaluates two Bayesian network classifiers; Tree Augmented Naïve Bayes and the Markov Blanket Estimation and their prediction accuracies are benchmarked against the SVM. The experimental results show that Bayesian networks with Markov blanket estimation has a superior performance on the diagnosis of cardiovascular diseases with a classification accuracy of MBE model is 97.92% of test samples, while TAN and SVM models have 88.54 and 70.83%, respectively.

4.12 CONCLUSION

People will be wearing in feature intelligent gadgets, smartwatches, connected IoT devices that will judge and calculate the patient body information. This sounds like a science of fiction, but this is all about the present research. Everything surrounding us will be smart, and those connected devices will be connected to the internet and blockchain DB overcloud. Combination of science, health care, and technology will collaborate to create something new for healthcare. We will have a smart health care revolution in the feature. The past development of ML and its application in medical diagnosis showed that algorithms are used in simple and straightforward. It is possible with the help of connected devices, and a few forms of sophisticated DA can be performed.

An automated smart health log system is presented in this work. This design is cost-efficient with high accuracy of heartbeat rate monitoring. The proposed system monitors patient's physical signs like ECG, SpO2, and blood pressure continuously, and provides different data transmission nodes to balance the health care needs. For heartbeat rate extraction, we used a Bayesian network and for determining the time ratio between each heartbeat after various action of the patient. Based on the ratio, the heart disease can be determined. A smart health log system will be integrated with virtual monitoring mechanism to keep watching user activities and finding accurate automated prediction of heartbeat problems for older aged peoples.

KEYWORDS

- belief network
- body area network
- electronic health records
- heart disease prediction
- internet of things
- machine learning

REFERENCES

1. Yang, C., Huang, Q., Li, Z., Liu, K., & Hu, F., (2017). Big data and cloud computing: Innovation opportunities and challenges. *International Journal of Digital Earth*, *10*(1), 13–53.

2. Houston, M. S., Myers, J. D., Levens, S. P., McEvoy, M. T., Smith, S. A., Khandheria, B. K., & Berry, D. J., (1999). Clinical consultations using store-and-forward telemedicine technology. In: *Mayo Clinic Proceedings* (Vol. 74, No. 8, pp. 764–769). Elsevier.

3. Bhatti, A., Siyal, A. A., Mehdi, A., Shah, H., Kumar, H., & Bohyo, M. A., (2018). Development of cost-effective tele-monitoring system for remote area patients. In: *2018 International Conference on Engineering and Emerging Technologies (ICEET)* (pp. 1–7). IEEE.

4. Castaneda, C., Nalley, K., Mannion, C., Bhattacharyya, P., Blake, P., Pecora, A., & Suh, K. S., (2015). Clinical decision support systems for improving diagnostic accuracy and achieving precision medicine. *Journal of Clinical Bioinformatics*, *5*(1), 4.

5. Zhang, P., White, J., Schmidt, D. C., Lenz, G., & Rosenbloom, S. T., (2018). FHIR chain: Applying blockchain to securely and scalably share clinical data. *Computational and Structural Biotechnology Journal*, *16*, 267–278.

6. Berman, M., & Fenaughty, A., (2005). Technology and managed care: Patient benefits of telemedicine in a rural health care network. *Health Economics*, *14*(6), 559–573.

7. Downing, N. L., Adler-Milstein, J., Palma, J. P., Lane, S., Eisenberg, M., Sharp, C., & Longhurst, C. A., (2016). Health information exchange policies of 11 diverse health systems and the associated impact on volume of exchange. *Journal of the American Medical Informatics Association*, *24*(1), 113–122.

8. Ekblaw, A., & Azaria, A., (2016). Medrec: Medical data management on the blockchain. *Viral Communications*.

9. Souers, A. J., Leverson, J. D., Boghaert, E. R., Ackler, S. L., Catron, N. D., Chen, J., & Huang, D. C., (2013). ABT-199, a potent and selective BCL-2 inhibitor, achieves antitumor activity while sparing platelets. *Nature Medicine*, *19*(2), 202.

10. Kuo, T. T., Kim, H. E., & Ohno-Machado, L., (2017). Blockchain distributed ledger technologies for biomedical and health care applications. *Journal of the American Medical Informatics Association*, *24*(6), 1211–1220.

11. Angraal, S., Krumholz, H. M., & Schulz, W. L., (2017). Blockchain technology: Applications in health care. *Circulation: Cardiovascular Quality and Outcomes*, *10*(9), e003800.

12. Tamrat, T., & Kachnowski, S., (2012). Special delivery: An analysis of health in maternal and newborn health programs and their outcomes around the world. *Maternal and Child Health Journal*, *16*(5), 1092–1101.

13. Sloninsky, D., & Mechael, P. (2008). *Towards the Development of An Mhealth Strategy: a Literary Review.* New York, USA: World Health Organization and Earth Institute.

14. Lindquist, A. M., Johansson, P. E., Petersson, G. I., Saveman, B.-I., & Nilsson, G. C. (2008). The use of Personal Digital Assistant (PDA) among personnel and students in Health care: a review. *Journal of Medical Internet Research*, *10*, e31.

15. Fairhurst, K., & Sheikh, A. (2008). Texting appointment reminders to repeated non-attenders in primary care: randomised controlled study. *Qual Saf Health Care, 17*(5), 373–376.

16. Armstrong, A. W., Watson, A. J., Makredes, M., Frangos, J. E., Kimball, A. B., & Kvedar, J. C. (2009). Text-message reminders to improve sunscreen use: a randomised, controlled trial using electronic monitoring. *Arch Dermatol, 145*(11), 1230–1236.

17. Pop-Eleches, C., Thirumurthy, H., Habyarimana, J. P., Zivin, J. G., Goldstein, M. P., de Walque, D., Mackeen, L., Haberer, J., Kimaiyo, S., Sidle, J., Nagare, D., & Bangsberg, D. R. (2011). Mobile phone technologies improve adherence to antiretroviral treatment

in a resource-limited setting: a randomized controlled trial of text message reminders. AIDS 2011.

18. Klasnja, P., & Pratt, W. (2012). Healthcare in the pocket: Mapping the space of mobile-phone health interventions. *Journal of Biomedical Information, 45*, 184–198.

19. Rofouei, M., Sinclair, M., Bittner, R., Blank, T., & Heffron, J. (2011). A non-invasive wearable neck-cuff system for real-time sleep monitoring. In *Proceedings of International Conference on Body Sensor Networks*, Dallas, TX, USA, 23–25 May 2011.

20. Pandian, P. S., Mohanavelu, K., Safeer, K. P., Kotresh, T. M., Shakunthala, D. T., Gopal, P., & Padaki, V. C. (2007). Smart Vest: Wearable multi-parameter remote physiological monitoring system. *Med. Eng. Phys., 30*, 466–477.

21. Jin, Z., Oresko, J., Huang, S., & Cheng, A. C. (2009). HeartToGo: A personalized medicine technology for cardiovascular disease prevention and detection. In *Proceedings of IEEE/NIH Life Science System and Applications Workshop (LISSA),* Bethesda, MD, USA, 9–10 April 2009.

22. Griggs, K. N., Ossipova, O., Kohlios, C. P., Baccarini, A. N., Howson, E. A., & Hayajneh, T. (2018). Healthcare blockchain system using smart contracts for secure automated remote patient monitoring. *Journal of Medical Systems, 42*(7), 130.

23. Dubovitskaya, A., Xu, Z., Ryu, S., Schumacher, M., & Wang, F. (2017). Secure and trustable electronic medical records sharing using blockchain. In *AMIA Annual Symposium Proceedings* (Vol. 2017, p. 650). American Medical Informatics Association.

24. Ekblaw, A., Azaria, A., Halamka, J. D., & Lippman, A. (2016, August). A Case Study for Blockchain in Healthcare:"MedRec" prototype for electronic health records and medical research data. In *Proceedings of IEEE open & big data conference* (Vol. 13, p. 13).

25. Yue, X., Wang, H., Jin, D., Li, M., & Jiang, W. (2016). Healthcare data gateways: found healthcare intelligence on blockchain with novel privacy risk control. *Journal of Medical Systems, 40*(10), 218.

26. Griggs, K. N., Ossipova, O., Kohlios, C. P., Baccarini, A. N., Howson, E. A., & Hayajneh, T. (2018). Healthcare blockchain system using smart contracts for secure automated remote patient monitoring. *Journal of Medical Systems, 42*(7), 130.

27. Yeh, J. Y. (2008). Applying data mining techniques for cancer classification on gene expression data. *Cybernetics and Systems: An International Journal, 39*(6), 583–602.

28. Monostori, L., Kádár, B., Bauernhansl, T., Kondoh, S., Kumara, S., Reinhart, G., ... & Ueda, K. (2016). Cyber-physical systems in manufacturing. *Cirp Annals, 65*(2), 621-641.

29. Yang, X., Wang, L., & Xie, J. (2017). Energy efficient cross-layer transmission model for mobile wireless sensor networks. *Mobile Information Systems*, 2017.

30. Theodoridis, S. (2009). Pattern Recognition/S. The-odoridis, C. Koutroumbas.

31. Rothman, M. J., Rothman, S. I., & Beals IV, J. (2013). Development and validation of a continuous measure of patient condition using the Electronic Medical Record. *Journal of biomedical informatics, 46*(5), 837–848.

32. https://www.disabled-world.com/calculators-charts/bpm-chart.php (accessed 10 August 2021).

33. Paalasmaa, J., Toivonen, H., & Partinen, M. (2014). Adaptive heartbeat modeling for beat-to-beat heart rate measurement in ballistocardiograms. *IEEE Journal of Biomedical and Health Informatics, 19*(6), 1945–1952.

34. Ho, C. L., Fu, Y. C., Lin, M. C., Chan, S. C., Hwang, B., & Jan, S. L. (2014). Smartphone applications (apps) for heart rate measurement in children: comparison with electrocardiography monitor. *Pediatric Cardiology, 35*(4), 726–731.

35. Silva, R., Scheine, R., Glymour, C., & Spirtes, P. (2006). Learning the structure of linear latent variable models. *Journal of Machine Learning Research, 7*(Feb), 191–246.
36. Aliferis, C. F., Statnikov, A., & Tsamardinos, I. (2006). Challenges in the analysis of mass-throughput data: a technical commentary from the statistical machine learning perspective. *Cancer Informatics, 2*, 117693510600200004.
37. Elsayad, A. M., & Fakhr, M. (2015). Diagnosis of Cardiovascular Diseases with Bayesian Classifiers. *JCS, 11*(2), 274–282.
38. Adhikary, T., Jana, A. D., Chakrabarty, A., & Jana, S. K. (2019, January). The Internet of Things (IoT) Augmentation in Healthcare: An Application Analytics. In *International Conference on Intelligent Computing and Communication Technologies* (pp. 576–583). Springer, Singapore.
39. Jha, S., Kumar, R., Chatterjee, J. M., & Khari, M. (2019). Collaborative handshaking approaches between internet of computing and internet of things towards a smart world: a review from 2009–2017. *Telecommunication Systems, 70*(4), 617–634.
40. Deus, L. A., Sousa, C. V., Rosa, T. S., Souto Filho, J. M., Santos, P. A., Barbosa, L. D., ... & Simões, H. G. (2019). Heart rate variability in middle-aged sprint and endurance athletes. *Physiology & Behavior, 205,* 39–43.
41. Khari, M., Kumar, M., Vij, S., & Pandey, P. (2016, March). Internet of Things: Proposed security aspects for digitizing the world. In *2016 3rd International Conference on Computing for Sustainable Global Development (INDIACom)* (pp. 2165–2170). IEEE.
42. Patro, S. P., Padhy, N., & Panigrahi, R., (2016). Security issues over ecommerce and their solutions. *Int. J. Advanced Research in Computer and Communication Engineering, 5*(12).

CHAPTER 5

BADS-MANET: Black Hole Attack Detection System in Mobile Networks Using Data Routing Information (DRI) and Cross-Check Mechanism

CHETANA KHETMAL,[1] NILESH BHOSALE,[2] UMASHANKAR GHUGAR,[3] and BIKASH RANJAN BAG[3]

[1]*Department of Computer Science, K. C College of Engineering, Mumbai, Maharashtra, India, E-mail: khetmalchetna@gmail.com*

[2]*Principal Software Engineer, Credit Vidya, Hyderabad, Telangana, India, E-mail: nileshxbhosle@gmail.com*

[3]*Department of Computer Science, Berhampur University, Berhampur – 761006, India, E-mails: ughugar@gmail.com (U. Ghugar); bikashbag@gmail.com (B. R. Bag)*

ABSTRACT

These days wireless sensor networks (WSNs) are in demand to every user over communication networks. Every user is using the wireless devices for their daily life such as mobile. There are several threats of attacks in the mobile communication. As per our proposed system, the detection system which detects the black-hole attack in mobile networks using AODV. Here we have considered the relative results of AODV with modification and without modification. We have implemented the simulation in network simulation-2.34 (NS-2). Finally, the result shows the higher accuracy in terms of packet delivery ratio (PDR), normalized routing overhead, and packet dropping ratio (PDPR) by varying number of nodes in the network.

5.1   INTRODUCTION

Basically, MANET is constructed by its own nature as compared to wired network. It is a set-up of autonomous wireless sensor nodes, which is control by itself. These nodes can create the random path for routing over the networks. All the nodes are communicating with each other and work as a router and host [1]. A host node always demands the particular service whereas router node discover the routing path for forwarding the data to neighboring nodes [2, 3]. Traditional routing protocols are used for routing over networks with the mobility and dynamic topology concept [4]. In MANET, since all the nodes are mobile, it does not require central node for network management. Those nodes have capacity to configure themselves [5]. Those nodes are deployed randomly in the networks. So that malicious node may join along with them and affect the whole communication process in the network through packet dropping [6] (Ramaswami, 2007). Nowadays, security is considered the most important part for smooth functionality over network. Mobile nodes are participating in communication process for data transmit and receive over the networks. The attackers always attack to the communication process so that it hampers the routing process. Nowadays, traditional routing protocol does not effectively work in the MANET due to its dynamically changing of nodes position. Hence, tough to build a safe and efficient routing protocol for MANET with network size and traffic density [7]. In MANET, malicious activities are easily done by the attacker due to dynamic topology, lack of central control and management. A number of researchers have published a paper on counter against this malicious attack. Most of the researcher has published papers on intrusion prevention system of routing protocol in MANET as well as WSN with limited power and computational capabilities [8, 9].

To resolve the black hole attacks issue, mechanisms such as test-based routing, intrusion detection system, and hash key mechanism using sequence number comparison and DRI have been proposed [10]. Also, some methods of black hole node detection were based on traditional cryptography, generic algorithm, and fuzzy logic-based techniques introduced by researchers. We are using combination of traditional approach of cryptography along with the proposed authentication techniques and terminologies related with authentication of nodes and path, key sharing, and malicious announcement as discussed above [11]. The main objectives are:

1. BADS-MANET is designed to detect black hole attack using AODV routing protocol. It maintains the authenticity, integrity, and reliability in the communication.

2. Our proposed system is based on AODV routing protocol with DRI and cross checking mechanism; when data packet transmission starts, we introduced proposed algorithm mechanism.
3. The performance of BADS-MANET is analyzed in terms of packet delivery ratio (PDR), normalized routing load (NRL), and packet dropping ratio (PDPR).

5.2 RELATED WORK

In today's world, millions of users are using the mobiles. Everybody wants to communicate with each others. Hence, a MANET has high important role in communication over wireless network. In wireless mobile networks, the data are transmitting and receiving through electromagnetic wave and very popularity due to its characteristics like simplicity, mobility, affordable, and low cost installation [12, 13]. One of the major challenges of WLANs is successful deployment of nodes with low cost infrastructure. So that Ad-hoc networks can resolve these problems. In wireless Ad-hoc network, nodes are communicating with each other without taking the help of centralized node and where each node have capacity to routing itself over the network [14, 15]. In the traditional network, infrastructures can't be deployed but in the case of MANET, it has the ability to deploy his own network setup [16]. At the time of network communication, it disturbs routing part (dataflow) and attacks. The numbers of attacks are available in the network such as black hole attack, wormhole attack, sinkhole attack, selective forward attack, and Sybil attack [17]. As per survey of related work, few numbers of methods for detecting the black hole attacks and a less number of proposed methods has been published on the concept of data routing information (DRI) cross check. Therefore, security in MANET is very important issues.

5.3 NETWORK MODEL

In the network model, it describes about network topology in MANET. The network has been constructed by some mobile nodes and every node sends and receives the data over mobile networks. In this model, every mobile node is checked by two terminologies, i.e., "authentication of node" and "authentication of path." Also, we introduced concept of "sharing public key" with the neighboring nodes; to locate the reliable nodes in the communication by keeping shared key list. Proposed system helps to keep authenticated and reliable transmission of data within nodes in the MANET by modifying legacy working of AODV [18] (Figure 5.1).

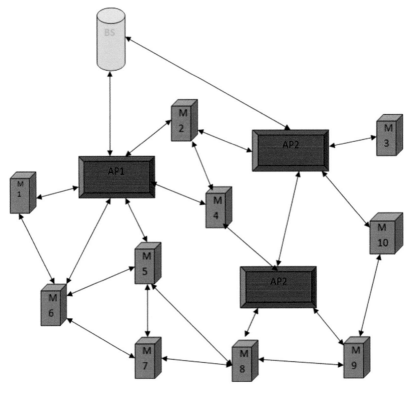

BS: Base Station   AP: Access Point   M: Mobile Node

FIGURE 5.1   Mobile ad-hoc network (MANET).

5.4   ATTACK MODEL

In attack model, black hole attack scenario has been disused, where the malicious node always advertise itself a genuine destination node without the knowledge of actual receiving node. When the source node sending the data to destination node, it always routing the path through route discovery process. In route discovery process, RREQ (route request) and RREP (route reply) messages are used, where low hop count and high sequence number has been considered of a node for checking authentication. The sender node broadcast the route request message to its next neighbor node until the process is continue towards the receiver node. After receiving the RREQ messages from the sender node, the receiver node send back to RREP messages to source node. At that time a malicious node send the false RREP message with low hop count and high sequence number to source node as result, it is treated as a genuine node [19] (Figure 5.2).

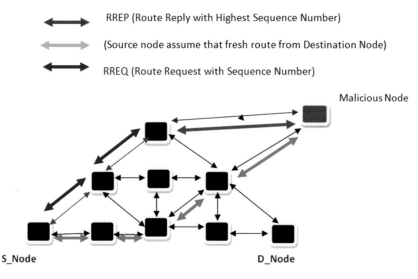

FIGURE 5.2 Black hole in MANET.

When the node is affected by black hole attack, then it violates the routing rules and dropping all the received packet as well as it increases the network overhead, therefore the energy consumption is decreased. Here malicious nodes are always able to achieve their goal through several ways [20].

5.5 PROPOSED WORK

As per the Jaydip [22], he suggested mechanism of DRI table (DT) and cross checking of it while detecting a black hole node in MANET using AODV. We have considered legacy AODV working along with DRI and cross-checking mechanism. Due to this, we are able to find out secure path from sender node to receiver node against black hole nodes.

In our proposed system along with DRI and cross-checking we introduced concepts for authentication and reliability. Through our algorithm, we are checking status of "authentication of node" and "authentication of path" in the network and keeping list of already authenticated nodes. Also, we introduced concept of "sharing public key" with the neighboring nodes; to locate the reliable nodes in the communication by keeping shared key list. The terminologies we introduced are authentication of node (Authn), authentication on path (Authp). Also for secure sharing, we have suggested concepts of key sharing, broadcasting malicious announcements after detecting black hole node. Proposed system helps to keep authenticated and reliable transmission of data within nodes in the MANET by modifying legacy working of AODV.

The proposed algorithm is stated as follows:

_BADS_MANET (S_NODE, D_NODE)_

Input:

S_NODE ← Source Node

D_NODE ← Destination Node

Output:

M ← Malicious Node

1. S_NODE broadcasts RREQ message.
2. Each node 'A' in the network N maintains a DRI table (DT) having the information of NODE_ID and SHARABLE_KEY of the nodes whose RREQs or RREPs are passing through 'A.'
3. Each node 'A' in the network N maintains a Reliable List (RL) by considering the frequency of messages passing through 'A.'
   a. The nodes sending frequently through 'A' are present in RL (A).
   b. Otherwise, the nodes are not present in RL (A).
4. The RL (A) is updated as per the changes in DT that is dynamic in nature.
5. An authentic path between S_NODE and D_NODE is established through the following steps:
   a. After receiving RREQ message from S_NODE, D_NODE sends RREP message to S_NODE by the same path of the first receiving RREQ message but in a reverse direction.
   b. If D_NODE is listed by S_NODE in its RL (S_NODE) then the D_NODE is proved to be authentic and message is received. Set M ← 0.
   c. Else D_NODE is declared as malicious node and broadcasted over the network N. Set M ← D_NODE.
6. Return (M).

5.6 RESULT AND DISCUSSION

We have simulated our proposed model in NS2 (network simulator 2). Here the results from simulation of proposed modified AODV are discussed above with graph (Table 5.1).

TABLE 5.1 Simulation Parameters with Values

Parameters	Values
Total simulation duration	200 sec
Start time and end time of simulation	15 sec to 195 sec
Simulation area	1000 × 1000 m
Number of mobile nodes varying	50 to 100
Movement model	Random waypoint
Traffic type	CBR: Constant bit rate (UDP)
Data payload	2000 bytes/packet
Antenna type	Omni antenna
Number of malicious node	1

The AODV protocol performance can be measured by the following metrics:

1. **Packet Delivery Ratio (PDR):** (Σ No of packet receive/Σ No of packet send) ×100.
2. **Normalized Routing Load (NRL):** (rtr/recv).
3. **Packet Dropping Ratio (PDPR):** [(No of packet send – No of packet received)/No of packet send] × 100.

In Figure 5.3, from the graph we achieved large difference in analysis with_MAODV and without_MAODV. The PDR has drastically increased. As we have seen above, the better performance of protocol is achieved due to mechanism of authentication, i.e., the authentication of node and authentication of path also by using key sharing technique. Using these terminologies, we have implemented secure transmission and eventually we have reached to increased PDR (Figure 5.4).

In Figure 5.4, NRL is the number of routing packets (rtr) transmitted per data packet delivered at the destination. As we have seen due to use of MAODV, routing overhead decreases considerably and remains almost constant (Figure 5.5).

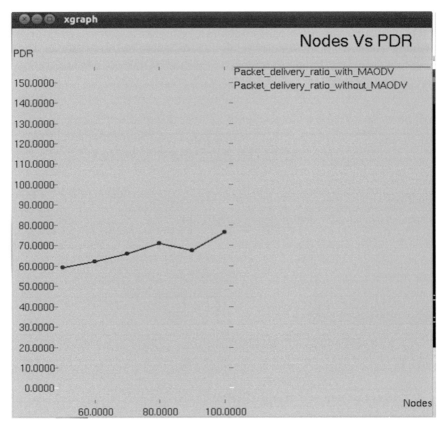

FIGURE 5.3   Number of nodes vs. packet delivery ratio.

In Figure 5.5, here the packet lost values is decreased it gives better performance of the protocol as and the PDPR decreases significantly as we can observe from the graph.

5.7   CONCLUSION AND FUTURE WORK

By keeping legacy AODV working along with it, we added authentication and security mechanism while starting with data packet transmission. Using DRI and cross-checking we able to get the reliable nodes list. With the help of Authn, Authp, and Authkey packet, we are able to maintain authenticated nodes list. Due to key sharing technique, we achieved security in the nodes within a particular period. We tested the AODV with modification and without modification in presence of single black hole node. Varying number of nodes from 50 to 100 and considering other performance metrics such as PDR, normalized routing overhead, and dropping ratio, we successfully implemented authenticated and reliable secure black hole detection system.

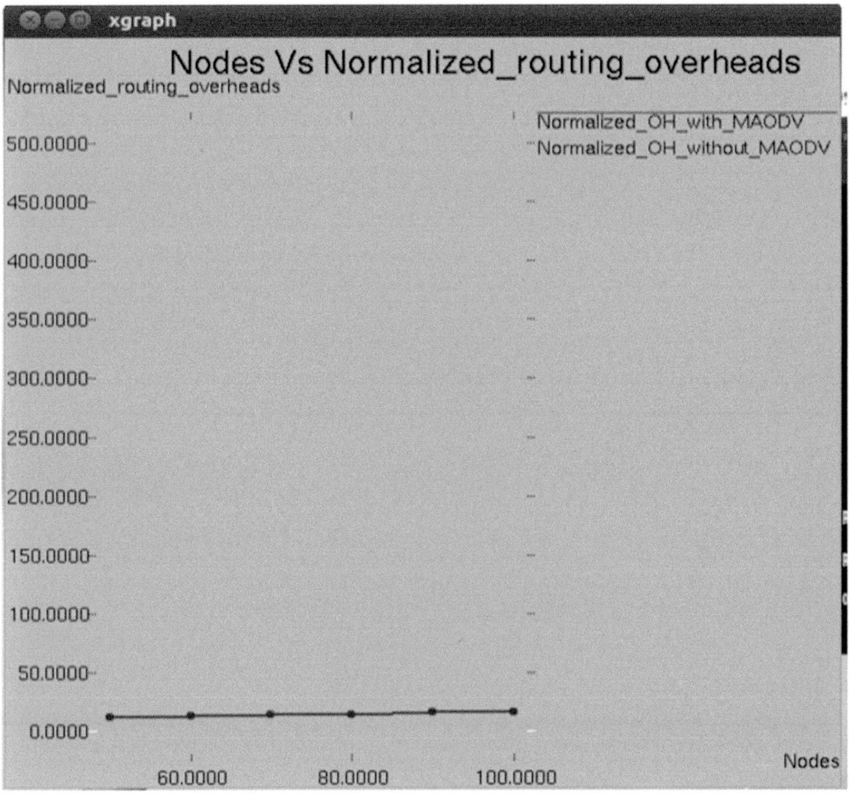

FIGURE 5.4   Number of nodes vs. normalized routing overhead.

The proposed method may be incorporate with trust-based system and internet of things (IoT) in future aspect. So that it can build a safe and secure IDS for tempering, jamming attack, back-off manipulation attack, wormhole attack, sinkhole attack, flooding attack, gray hole attack, and selective forward attack.

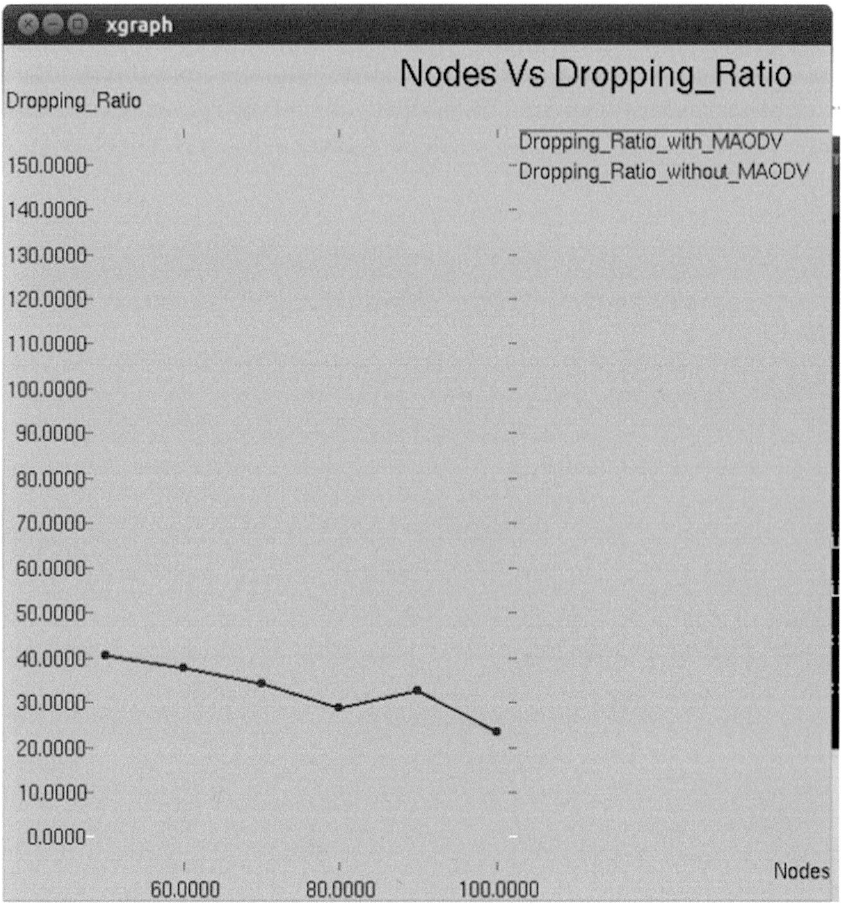

FIGURE 5.5 Number of nodes vs. packet dropping ratio.

KEYWORDS

- **black hole attack**
- **data routing information**
- **mobile ad-hoc network**
- **packet delivery ratio**
- **packet dropping ratio**
- **route reply**

REFERENCES

1. M. A., Yoo, S. M., & Park, S., (2004). Blackhole attack in mobile ad hoc networks. *Association for Computing Machinery Southeast Conference.*
2. Sharma, N., & Sharma, A., (2012). The black-hole node attack in MANET. *Second International Conference on Advanced Computing and Communication Technologies.*
3. Dalal, R., Khari, M., & Singh, Y., (2012). Authenticity checks to provide trusted platform in MANET. In: *Proceedings of the Second International Conference on Computational Science, Engineering, and Information Technology, ACM.*
4. Hassan, Y. K., Abd, M. H., Abd, A. S., & El, R., (2010). Performance evaluation of mobility speed over MANET routing protocols. *International Journal of Network Security.*
5. Dalal, R., Khari, M., & Singh, Y., (2012). Different ways to achieve trust in MANET. *International Journal on Ad Hoc Networking Systems (IJANS).*
6. Supriya, K. M., & Khari, M. M., (2012). MANET security breaches: Threat to a secure communication plat-form. *International Journal on Ad Hoc Networking System (IJANS), 2*(2).
7. Sen, J., Koilakonda, S., & Ukil, A., (2011). A mechanism for detection of cooperative black hole attack in mobile ad hoc networks. *Second International Conference on Intelligent Systems, Modeling, and Simulation.*
8. Dalal, R., Khari, M., & Singh, Y., (2012). *The New Approach to Provide Trusted Platform in MANET.*
9. Garg, N., & Mahapatra, R. P., (2009). MANET security issues. *IJCSNS International Journal of Computer Science and Network Security.*
10. Kumar, D., Srivastava, A., & Gupta, S. C., (2012). Performance comparison of proactive and reactive routing protocols for MANET. *Computing, Communication, and Applications (ICCCA).*
11. Saini, R., & Khari, M., (2011). Defining malicious behavior of a node and its defensive techniques in ad hoc networks. *Int. J. Smart Sens. Ad. Hoc. Netw., 1*(1), 17–20.
12. Dalal, R., Singh, Y., & Khari, M., (2012). A review on key management schemes in MANET. *International Journal of Distributed and Parallel Systems.*
13. Singh, V., Singh, A., & Hassan, M., (2019). Survey: Black hole attack detection in MANET. *Proceedings of 2nd International Conference on Advanced Computing and Software Engineering (ICACSE).*
14. Vijaya, & Rath, A. K., (2011). Simulation and performance evaluation of AODV, DSDV, and DSR in TCP and UDP environment. *IEEE Xplore.*
15. Barakovic, S., Kasapović, S., & Baraković, J., (2010). Comparison of MANET routing protocols in different traffic and mobility models. *Telfor Journal.*
16. Salih, M. A., Sivaram, Y., & Porkodi, (2018). Detection and removal of black hole attack in mobile ad hoc networks using Grp protocol. *International Journal of Advanced Research in Computer Science.*
17. Kannhavong, B., Nakayama, H., Nemoto, Y., & Kato, N., (2007). Security in wireless mobile ad hoc and sensor networks. *IEEE Wireless Communications.*
18. Sheikhl, R., Singh, M., Kumar, C. D., & Mishra, D. K., (2010). Security issues in MANET: A review. *IEEE Xplorer.*
19. Ghugar, U., & Pradhan, J., (2018). NL-IDS: Network layer trust-based intrusion detection system for wireless s sensor networks. *5th IEEE International Conference on PDGC.*

20. Hsun, T. F., Chiang, H. P., & Chao, H. C., (2018). Blackhole along with other attacks in MANETs: A survey. *Journal of Information Process System*.

21. Ramaswamy, S., Fu, H., Sreekantaradhya, M., Dixon, J., & Nygard, K., (2007). Prevention of cooperative black hole attack in wireless ad hoc networks. *Future Generation Communication and Networking (FGCN)*.

22. Jaydip Sen et al., (2012). "Detection of Cooperative Black Hole Attack in Wireless Ad Hoc Networks", *IJSSST*, February 2012.

CHAPTER 6

A Study of Computation Offloading Techniques Used by Mobile Edge Computing in an IoT Environment

DEVPRIYA PANDA,[1] BROJO KISHORE MISHRA,[1] and
CHHABI RANI PANIGRAHI[2]

[1]Department of CSE and IT, GIET University, Gunupur, Odisha, India

[2]Rama Devi Women's University, Bhubaneswar, India

ABSTRACT

Standard cloud computing (CC) with a centralized system faces problems such as high latency, limited power source, and non-adaptive device, because of the fast growth in the use of internet through mobile and applications on the internet of things (IoT). Encouraged to tackle these challenges, a trend is driven by a new technology. This trend incorporates normal CC features into network edge devices (as in mobile edge computation offloading (MECO)). Various authors have suggested different ways to solve the problems found in those scenarios. In this work, different problems with appropriate context have been analyzed and several proposed methods for the respective problems are discussed. Some scopes for future investigation in those areas are also being suggested.

6.1 INTRODUCTION

Nowadays, a large amount of smart devices and items are integrated with sensors, allowing them to retrieve environmental data in real time. This movement resulted in the fascinating concept of internet of things (IoT), in which all intelligent things like intelligent cars, devices being wore by persons, personal computers, sensors, and industrial and utility elements are linked via inter-network. While IoT may possibly benefit society as a whole, there are still many technical challenges to be resolved. One of the several challenges is to make use of the enormous quantity of data being generated by IoT devices at a sky-scraping pace (The European Commission anticipated that 50 to 100 billion internet-connected intelligent devices will be available by 2020 [7]). The data is transferred from the IoT systems to the remote cloud [8]. Big data transfer costs a great deal of bandwidth, power, and time. To address the issues associated with the use of fog computing and software-defined networking (SDN), an efficient and flexible IoT architecture is suggested to gather, classify, and evaluate IoT data streams at the brink of the mobile network [3].

Relative to current IoT, the ultra-dense IoT presents various demands for quality of service (QoS) and quality of experience (QoE) on communication networks [4]. Most of the computations needed by the IoT mobile devices need intensive computing and large amount of energy. But the computation power and battery life of mobile devices are always a constraint because of their size. Mobile-edge computation offloading (MECO) proves to be a viable answer to this problem [5]. The edge servers are deployed at macro base station (MBS) and small cells in the radio access network. There are different approaches being proposed for MECO (mobile edge computation offloading). Some of them focused on single-tier base station scenario and computation offloading between the mobile devices and the edge server connected to the mobile base stations [6]. Some recent approaches of offloading discussed about a multiuser ultra-dense edge server scenario and proposed steps for computation offloading in more than one tier setups [1]. In the second approach, the edge servers can be placed in the small cells. The reason to choose these cells is that they are closer to the mobile devices of the ultra-dense IoT.

By using edge devices' computing, cache, and power resources, computing offloading can decrease computing latency, save battery resources, and even increase computing-intensive IoT applications safety [9]. Energy harvesting (EH) is an encouraging method that extends battery life and offers IoT devices with a satisfactory QoE [11]. The IoT devices are supposed to go through recent EH pattern in order to forecast the future trend [2].

In this work, the authors performed an extensive study of MECO techniques in the IoT scenario. Different approaches used for computation offloading has been studied along with how EH can be useful when edge computing is used in IoT. Several techniques proposed for the above purpose have been analyzed. This analysis can be helpful in proposing a new direction for investigators in the area of MEC.

The contents of the chapter are organized as follows. In the next section, we have surveyed the works related to computation offloading in IoT scenarios. In the subsequent sections, we studied the works related to EH in IoT devices, the use of microwave power to charge IoT devices, and optimal offloading and concluded.

6.2   OFFLOADING TECHNIQUES BY MOBILE EDGE COMPUTING (MEC)

1. **Computation Offloading in Ultra-Dense IoT Network:** Guo et al. suggested about minimization of consumption of user equipments energy in networks with miniature cells using mobile edge computing (MEC) [1]. Another objective of the work is to optimize the computation offloading, distribution of spectrum, power, and also the allocation of computation resource. As a result, the overhead of making the decision whether to offload or where to offload for the use equipments can be removed. A suboptimal algorithm named as "Hierarchical GA and PSO-based computation algorithm" (HGPCA) is used for optimizing the desired parameters. The results of simulations are represented in Figures 6.1 and 6.2.

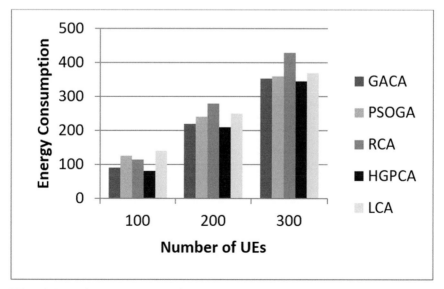

FIGURE 6.1   The energy consumption vs. user equipments [20].

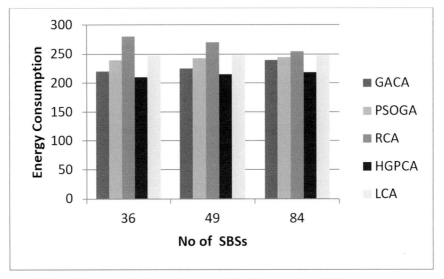

FIGURE 6.2   The energy consumption vs. small cell base stations [20].

Figure 6.1 shows consumption of energy in case of 100, 200, and 300 numbers of UEs. The HGPCA algorithm provides promising results among all other algorithms used.

2. **Computation Offloading for MEC in Ultra-Dense SDN:** Chen et al. proposed an offloading scheme named "software-defined task offloading" (SDTO) as a mixed-integer non-linear program [17]. Being a NP hard problem, it is changed into two subproblems such as task placement sub-problem and resource allocation sub-problem. In contrast to random and uniform task offloading schemes, the proposed scheme proves to decrease the task duration by 20% and increase energy saving by 30%. The simulation results in Figures 6.3 and 6.4 support the correctness of the work.

3. **Offloading of Computation for IoT Devices and Use of Energy Harvesting (EH):** Minghui Min et al., studied offloading of computation task for IoT devices needed in a dynamic MEC network and also considered the aspect of EH [2]. They assumed the following; a device in an IoT network is linked to number of edge devices with separate links. They also considered that the IoT devices can forecast the trend in production of renewable energy with the help of EH model. They provided a computation offloading system based on reinforcement learning (RL) for an

IoT device to identify an edge server. The fraction of computation task to be offloaded may also be decided by the help of the same system (Figure 6.5).

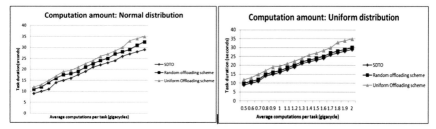

FIGURE 6.3 Data size vs. computation duration [17].

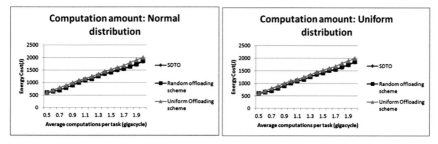

FIGURE 6.4 Average computation vs. energy cost [17].

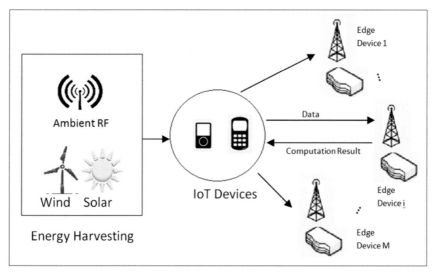

FIGURE 6.5 Energy harvesting enabled IoT devices in MEC [2].

A computation offloading scheme based on reinforcement learning (RLO) has been proposed by the authors [12]. That can be used by an IoT device with EH even if it doesn't have any information of the latency for computation or the energy consumption. This system uses a quality function. This function and the pace of offloading can be used to identify the appropriate offloading policy. Initially to minimize the time to study for the RL a transfer learning technique can be used [12]. It also uses the knowledge gained from previous offloading.

DRLO (D: deep, R: reinforcement, L: learning, O: offloading), which is an offloading scheme [2], has also been suggested by the authors to get better performance. This approach is applicable in a larger scenario. An IoT gadget can anticipate a big quantity of viable battery intensity, a great quantity of sustainable energy produced in a period and a big quantity of possible radio transmission rates for each edge server. This system can be introduced on different IoT systems, which can sustain deep learning [13].

The reinforced learning-based offloading schemes were analyzed in three different MEC circumstances. For analyzing the performance the amount of energy consumed, latency in computation, and task drop rates are being considered. For IoT systems of wireless power flow, experiments are carried out to catch the surrounding radio frequency transmissions from a devoted RF power transmitter to supply energy to the IoT battery for charging it [14]. Simulation findings indicate that the DRLO system saves IoT devices' power usage reduces computational latency and reduces job drop rates, thereby increasing the usefulness of IoT devices. Figures 6.6–6.8 justify the arguments of the authors.

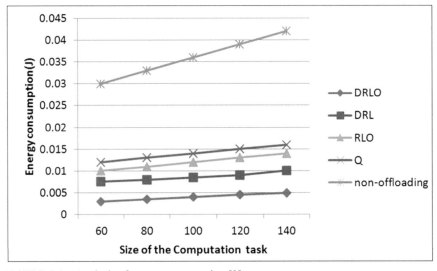

FIGURE 6.6   Analysis of energy consumption [2].

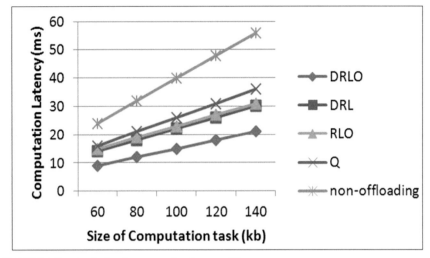

FIGURE 6.7   Analysis of computation latency [2].

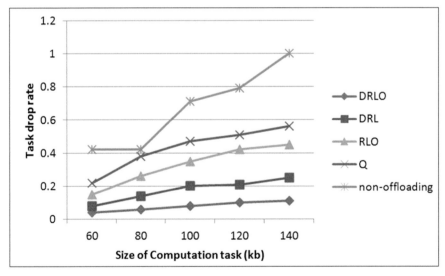

FIGURE 6.8   Analysis of task drop rate [2].

4. **Mobile Cloud Computing (MCC) with Wireless Energy Transfer for Energy Efficiency:** A uni-user system has been considered by You et al. [18]. That system consists of a base station (BS) with multiple antennas used for transmitting microwave power to the mobile. To transfer the data being offloaded, the aforesaid system can

be used. In this system, the mobile device chooses either to compute the data locally or may offload it. For this purpose, it uses harvested energy (Figure 6.9).

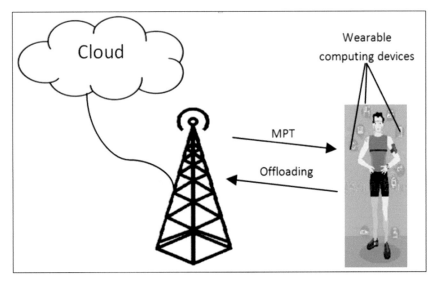

FIGURE 6.9   Mobile cloud computing system powered by the wireless system [18].

The mobile device can work in one of the two different modes. Either it can go for local computation or it can use computation offloading. While using local computing microwave power transmission can be performed concurrently as the system uses CPU cycles which are being controlled. The same can be achieved while using computation offloading scheme also. For that, the computation time is partitioned in an adjustable way. This proposed work makes use of microwave power transmission and mobile cloud computing (MCC) simultaneously in contrast to the work described by Zhang et al. [15], and that lead to fresh theoretical issues. One of these issues is tackled as discussed further. When microwave power transmission is being used, the EH factor forces the optimization problem towards nonconvex. So some technique to relax the convex is introduced.

From Figures 6.10 and 6.11 it is evident that, the suboptimal data allocation scheme suggested by the authors is resulting considerable improvement as compared the equal allocation policy. It is also concluded that the data allocation done using the adaptive scheme is an efficient method to tackle the fading effect in cloud computing (CC).

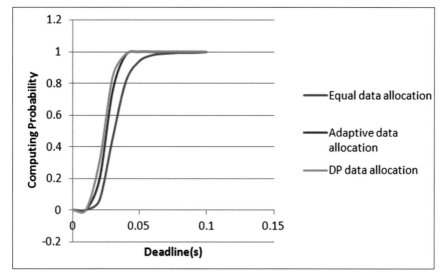

FIGURE 6.10   Effect of deadline [18].

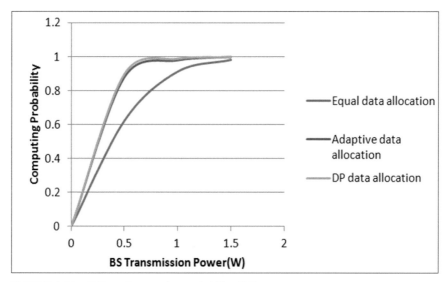

FIGURE 6.11   Effect of computing probability [18].

5.   **Energy-Efficient Resource Allocation Policy for Mobile-Edge Computation Offloading:** Changsheng You et al. have analyzed MECO for a multiuser system [19]. They have proposed resource

allocation in that system which uses TDMA (time-division multiple access) and OFDMA (orthogonal frequency-division multiple access) (Figure 6.12).

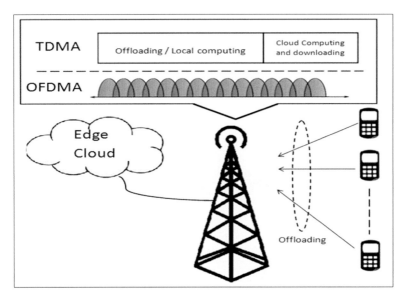

FIGURE 6.12   Multi-user MECO based on TDMA and OFDMA [19].

In this scenario, numbers of mobile devices are needed. It is assumed that these MDs are with different computation loads but the latency is same for all of them. It is assumed that the data used for computation can be split for each different computing. Based on this assumption it can be stated that computation on local system and offloading of data can performed simultaneously by the MDs. There are two cases with CC capacity; either it is with finite or infinite capacity. For both the scenarios, an optimal policy based on threshold is being proposed in the TDMA MECO system. In order to address the resource allocation problem, in particular the mixed integer problem, a sub-optimal algorithm has been suggested by the authors in OFDMA MECO system (Figures 6.13 and 6.14).

The above figures which are being generated by the authors is referring that the suggested approaches (in this case, sub-optimal resource allocation policy) are better than the earlier proposed policies such as equal resource allocation policy. The new approach suggested by the authors [18] is consuming less than half of the energy as compared to the equal allocation policy in two different scenarios.

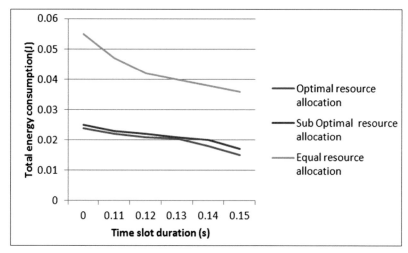

FIGURE 6.13   Effect of cloud computation capacity [19].

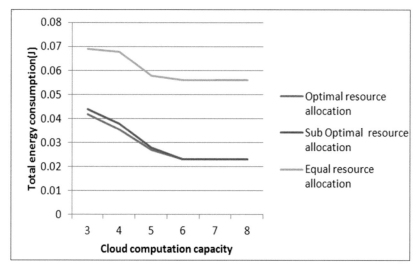

FIGURE 6.14   Effect of time slot duration [19].

6.3   OVERALL DISCUSSION

In the first discussion, the problem of computation optimization in an ultra-dense IoT network with the MECO is considered. The game theory-based greedy algorithm suggested by the authors [1] has been proved to be better than the existing technique. Subsequently, the task offloading in ultra-dense software-defined network is being studied. An optimal allocation policy proposed by the authors [17], which is better as compared to random or uniform allocation of resources. In our next consideration, the authors have finally discussed a deep reinforced learning offloading scheme to use EH for IoT devices while going for computation offloading [2]. And this scheme is proven to be more efficient as compared to the previously proposed schemes. Considering the next work, we got the idea of a sub-optimal policy for data allocation with EH, which is used in a MCC network where the devices are supplied with wireless power [18].

The following issues can be investigated further. Like the authors suggested that when there will be moving mobile users with dynamic status in ultra-dense IoT network the computation offloading is still an issue [1]. Computation offloading with users in mobile mode in complex network is still to be investigated [17]. For self-sustaining IoT devices, different approaches of EH in congestion with offloading techniques are analyzed by some of the authors, but still they have suggested that there can be better approaches for the same purpose [2]. The proposed work in Ref. [18] can be further investigated for full-duplex transmission and for multitask scenarios.

6.4 CONCLUSION

In this chapter, we have tried to study different approaches suggested by several authors, which can be used to enhance the computation offloading schemes for mobile devices in IoT and MEC scenarios. Though the approaches are efficient enough to address the issues picked up by the authors, still there is scope of investigation in the same or the other aspects of MEC and IoT. The discussions in this work might be helpful to the investigators to get better insight of the topics related and should open new directions of investigations in future.

KEYWORDS

- **computation offloading**
- **energy harvesting**
- **internet of things**
- **mobile edge computing**
- **orthogonal frequency-division multiple access**
- **ultra-dense IoT network**

REFERENCES

1. Guo, H., Liu, J., Zhang, J., Sun, W., & Kato, N., (2018). Mobile-edge computation offloading for ultra-dense IoT networks. *IEEE Internet of Things Journal, 5*(6), 4977–4988.
2. Min, M., Xiao, L., Chen, Y., Cheng, P., Wu, D., & Zhuang, W., (2019). Learning-based computation offloading for IoT devices with energy harvesting. In: *IEEE Transactions on Vehicular Technology* (Vol. 68, No. 2, pp. 1930–1941).
3. Sun, X., & Ansari, N., (2016). Edge IoT: Mobile edge computing for the internet of things. In: *IEEE Communications Magazine* (Vol. 54, No. 12, pp. 22–29).
4. Ma, H., Liu, L., Zhou, A., & Zhao, D., (2016). On networking of internet of things: Explorations and challenges. *IEEE Internet Things J., 3*(4), 441–452.
5. Liu, L., Chang, Z., Guo, X., Mao, S., & Ristaniemi, T., (2018). Multi-objective optimization for computation offloading in fog computing. *IEEE Internet Things J., 5*(1), 283–294.
6. Zheng, J., Cai, Y., Wu, Y., & Shen, X. S., (2016). Stochastic computation offloading game for mobile cloud computing. In: *Proc. ICCC* (pp. 1–6).
7. Sundmaeker, H., et al., (2010). Vision and challenges for realizing the internet of things. *EC Info. Soc. and Media, Tech. Rep.*
8. Sun, X., & Ansari, N., (2016). Optimizing resource utilization of a data center. *IEEE Commun. Surveys and Tutorials*. doi: 10.1109/COMST.2016.25582032016, Early Access.
9. Chiang, M., & Zhang, T., (2016). Fog and IoT: An overview of research opportunities. *IEEE Internet Things J.,3*(6), 854–864.
10. Mishra, D., De, S., & Krishnaswamy, D., (2017). Dilemma at RF energy harvesting relay: Downlink energy relaying or uplink information transfer? *IEEE Trans. Wireless Commun., 16*(8), 4939–4955.
11. Wang, F., Xu, J., Wang, X., & Cui, S., (2017). Joint offloading and computing optimization in wireless powered mobile-edge computing systems. *IEEE Trans. Wireless Commun., 17*(3), 1784–1797.
12. Zuo, H., Zhang, G., Pedrycz, W., Behbood, V., & Lu, J., (2017). Fuzzy regression transfer learning in Takagi-Sugeno fuzzy models. *IEEE Trans. Fuzzy Syst., 25*(6), 1795–1807.
13. Lane, N. D., Georgiev, P., & Qendro, L., (2015). Deep ear: Robust smartphone audio sensing in unconstrained acoustic environments using deep learning. In: *Proc. ACM Int. Joint Conf. Pervasive Ubiquitous Comput.* (pp. 283–294). Osaka, Japan.
14. Lu, X., Wang, P., Niyato, D., Kim, D. I., & Han, Z., (2015). Wireless networks with RF energy harvesting: A contemporary survey. *IEEE Commun. Surv. Tut., 17*(2), 757–789.
15. Zhang, W., Wen, Y., Guan, K., Kilper, D., Luo, H., & Wu, D. O., (2013). Energy optimal mobile cloud computing under stochastic wireless channel. *IEEE Trans. Wireless Commun., 12*(9), 4569–4581.
16. Wang, Y., et al., (2015). Energy-optimal partial computation offloading using dynamic voltage scaling. In: *Proc. IEEE Int. Conf. Commun. Workshop (ICCW)* (pp. 2695–2700). London, U.K.
17. Chen, M., & Hao, Y., (2018). Task offloading for mobile edge computing in software defined ultra-dense network. In: *IEEE Journal on Selected Areas in Communications* (Vol. 36, No. 3, pp. 587–597).
18. You, C., Huang, K., & Chae, H., (2016). Energy efficient mobile cloud computing powered by wireless energy transfer. *IEEE Journal on Selected Areas in Communications, 34*(5), 1757–1771.

19. You, C., Huang, K., Chae, H., & Kim, B., (2017). Energy-efficient resource allocation for mobile-edge computation offloading. In: *IEEE Transactions on Wireless Communications* (Vol. 16, No. 3, pp. 1397–1411).
20. Guo, F., Zhang, H., Ji, H., Li, X., & Leung, V. C. M., (2018). An efficient computation offloading management scheme in the densely deployed small cell networks with mobile edge computing. In: *IEEE/ACM Transactions on Networking* (Vol. 26, No. 6, pp. 2651–2664).

CHAPTER 7

AquaTalk: An Intensification of System Influence in Aquaculture

P. SUDHEER BABU and SANJAY KUMAR KUANAR

Department of Computer Science, GIET University, Gunupur, Odisha, India, E-mails: sudheerpunuri@giet.edu (P. S. Babu), sanjay.kuanar@giet.edu (S. K. Kuanar)

ABSTRACT

The internet of things (IoT) is a revolutionary communication paradigm that aims to offer manifold new services in the context of aquaculture. Advancement of embedded chips and computers like Audrion, raspberry pi has provided a great platform to leverage the services in both agriculture and aquaculture. The quality of water parameters plays a critical role in rising aqua organisms like fish and prawns. The data from the sensors like pH, temperature, TDS, etc., are transmitted to the cloud as well as to the farmer. In this chapter, we propose an IoT-based monitoring mechanism for the water quality parameters, a vital parameter for ensuring healthy aquaculture and hence increase productivity. By using PyCharm, we train a model that suggests for a solution on encountering a problem. A new method is proposed for feeding the organisms with the help of a feed sensor. Feed sensor senses the behavior of the hungry fishes and signals the feeder to deliver the foodstuffs to the pond. This reduces majority of the food waste and hence saves money. In this chapter, the quality of water parameters is trained using support vector machine (SVM).

7.1   INTRODUCTION

The clean water prawn cultivation has received augmented interest only in the last twenty years due to its high customer demand. India exports marine products worth about 45,000 crores every year. India had produced over 6,00,000 tonnes of shrimps last year. But due to various reasons like poor water quality and aqua diseases, farmers lose the harvest and do not get the return of their capital money. Today's aqua farmers mostly depend on physical water testing to get acquainted with estimates of the water parameters like pH, salt, hardness, salts, dissolved oxygen (DO) and others. The aqua farmers are facing two primary challenges. Firstly, the farmers manually test the water sample from the pond for the variation of water parameters. But the water parameters change within short span of time. So continuous and effective monitoring of water parameters is required. Secondly, the prawn which dies due to any infection, the dead prawns sink and get settled at the bottom of the pond. This is one of the most challenging tasks presently unidentified by human. As the pond water has turbidity, dead fossils and dead prawns do not appear to human. It is only three to four days after the prawns expire, the fossils float on water. At this point only the farmer is able to identify the damage occurred, that too after a couple of days. In the meanwhile, the infection may also spread to other prawns in the pond, and it becomes too late for the precautionary measures. This is a major issue faced by the aquaculture farmers. To avoid this type of major defies skill should be carrying in which amplifies the output and reduce the cost of production.

Several solutions [3–7] rooted in internet of things (IoT) be projected to computerize the abnormal water conditions for the enlargement of the prawns. A few of them have made available clever mechanism to mechanically make active the actuators to fix the troubles. Item 2 indicates the solutions [3, 7] gives and understanding monitoring with help of sensors but actuators were not controlled. The solutions [1, 4] have dealt with actuator which is managed manually, but they are not forced mechanically. In fact, solution [6] gives unimportant actuator control rooted in trouble-free sensor thresholds. The AquaTalk also provides both manual control and mechanical actuator control by the sensors with non-trivial intellect other than with easy "threshold control" (Tables 7.1 and 7.2).

Table 7.2 shows the range for dissimilar the water quality not in estimate fish parameters for civilizations. If the water quality parameters are specific and go higher than the threshold variety the aqua-farmer will obtain an alert. This alert is also with solution what to do.

TABLE 7.1 The Comparisons by Dissimilar Models

Item	Dolan [1]	Salim [3]	Salim [4]	Chen [6]	Encinas [7]	Tseng [22]	Raju [23]	FishTalk [24]	AquaTalk
1. Sensors	pH, CO_2, O_2, NH_3	pH, DO, Temp.	pH, DO, Temp.	pH, DO, Water level, Temp.	pH, DO, Temp.	pH, DO, Temp., Water Level	DO, Salt, NH_3, Nitrite, Temp., pH, Alkalinity	pH, EC, DO, TDS, Water level, Temp.	pH, EC, DO, TDS, Water Level, Temp.
2. Actuators	Heater	No	Air Pump	Heater, Light, Feeder, Air Pump	No	No	Light	Feeder, Fan, Heater, Light, Air Pump, RO Filter	Air Actuators, RO Filter, Heater, Air Pump.
3. Actuators Controlled by Sensors	No (Manual)	No	No (Manual)	Yes (Simple threshold)	No	No	No (Light is always on at night)	Yes	Yes
4. Smart Feeder	No	No	No	No (manual)	No	No	No	Yes	Yes
5. Video Monitoring	No	No	No	No	No	No	No	Yes	Yes
6. Control Board	No	N/A	Raspberry Pi 3	MSP430	Arduino Uno	Arduino Uno	Raspberry Pi 3	Arduino UNO, ESP8266 ESP-12F, ROHM IoT kit, MediaTek LinkIt Smart 7688 Duo	Arduino Uno

TABLE 7.2 Water Quality Parameter with Threshold Range

Sl. No.	Quality of Water Parameters	Threshold Range
1.	Dissolved oxygen (DO)	(4–10) ppm
2.	Ph	(7.5–8.5) ppm
3.	Temperature	21°C–33°C
4.	Carbonates (CO_3^{2-})	20–40 ppm
5.	Nitrates (NO_2)	(0–0.4) ppm
6.	Ammonia	(0–0.1) ppm

Diagrams of sensors/actuators in AquaTalk's exhibit the aqua proprietor can simply learn how factors of water interrelate in a piece other. Table 7.1 compares the AquaTalk by the earlier solution. Several issues of water management though the trickier to control all through the normal enlargement of the shrimp five to eight months or longer. The shrimp development is greatly depending by the quantity of the foodstuff obtainable. Over and above feeding not only wilds feed, but also surplus organic-matter remains outcomes in deprived water quality. The poor water quality is augmented the aquaculture outlays. As a result, stipulation of give to eat and feeding is vital to watched.

The transparency of water in shrimp pond is mainly pretentious with far above the ground concentration of overhanging specific issue attribute to bio-flocks, phytoplankton, and remains of fee and shrimp waste able to be seen is frequently less than 100 cm. It will be a substantial confront, then to demeanor an underwater ocular examination in such highly turbid water. The underwater video cassette system [10] has been adopted. This UVS helps us in: (1) understand shrimp feeding, (2) to scrutinize the benthic surface for focal pellets and surplus organic substance both of which may outcome in augmented pathogen attentiveness and a fouled bottom; (3) imagine the shrimp in muddy water to approximation their size and performance.

7.2 FACTORS FOR QUALITY OF WATER

7.2.1 TYPICAL PH LEVELS

The representative of pH levels may vary to ecological influences predominantly in alkalinity. The alkalinity of water is different owing to the being there of melted salts and carbonates, in addition to the mineral work of art of the nearby soil. The alkalinity and pH is more high then the lower the alkalinity is the lower then the pH value. The not compulsory pH range for most fish is amid 6.0 and 9.0 with a smallest amount alkalinity of 20 mg/L, with ideal $CaCO_3$ levels amid 75 and 200 mg/L[2]. Figure 7.1 shows the minimum pH level for the different types of aqua organisms.

The oceanic organisms are akin to clownfish and coral needed to high pH levels. The pH levels slighter than 7.6 will cause coral reefs to start to collapse due to the lack of calcium carbonate. As a matter of fact, the sensitive freshwater classes such as salmon like better pH levels amid to 7.0 and 8.0, becoming harshly distressed and suffering physiological injure owing to engrossed metals at levels beneath 6.0.

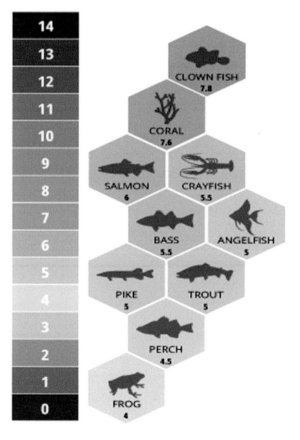

FIGURE 7.1 Minimum pH level required for different species of aqua organisms.

7.2.2   AMMONIA GAS

The vital source for ammonia is just during fish emission. By using the excrement time is articulately allied to provide for speed and protein level in the nourish creature. As the feed's protein is broken down in the fish's corpse, a few of the nitrogen is used to outline protein (muscle), energy, and excreted through the gills as ammonia. The protein in nourish is the ultimate basis of largely ammonia in ponds where fish are nourishing. A further basis of ammonia is the dispersal from the ponds sediment itself. The huge amounts of organic substance are created by algae or added to ponds as nourish. The fecal-solids and deceased algae settle to the pond substructure and start the process of decayed. This process is making ammonia in which disperses as of residue base keen on the water support. The vital procedure is the loss of ammonia throughout the up and about get by algae and erstwhile plants. These plants are used the nitrogen as a nutrient for improvement. Photosynthesis acts similar to a sponge for ammonia is out of support to get the large plant or algae improvement in the ponds can aid the use up the ammonia. Certainly, too much plants are enlargement can have an outcome on the diurnal cycles of melt oxygen levels reason that to go very low in night time hours (Figure 7.2).

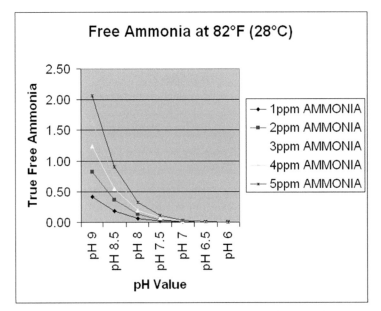

FIGURE 7.2   Ammonia levels are heavily affected by changes in pH and temperature.

7.2.3 CARBON DIOXIDE (CO$_2$)

Free carbon dioxide (CO$_2$) is available in water. CO$_2$ interferes with a key brain receptor in fish's brains affecting their ability to hear, smell, turn, and evade predator. Carbon dioxide levels rising in water change the behavior of aquatic organisms like fish through generations. The fish potential reduction resources by high-CO$_2$ conditions due to the diffusion of atmospheric CO$_2$ into the surface water or direct CO$_2$ injection.

7.2.4 OXYGEN (O$_2$)

The oxygen is necessary to support the livestock. To decreases, the oxygen attentiveness joint with eminent carbon dioxide attentiveness in water leads to suffocation. The fish and shrimp approach to the surface of the water to get sufficient of oxygen. It is very chiefly the shrimps approach to the surface and they rotate. The bodily, if the farmer's timepiece the shaking of the shrimps aerators is underway to make the quantity the oxygen in the water. At this time, we are using routine aerator for making the oxygen in its place of by hand in force the aerator.

7.3 LITERATURE SURVEY

IoT with its appliance is accomplishment to the farmers [1, 2]. Many of documents are in fiction review to deal with the issues in which achieve owing to alter in the quality of water parameters [3]. A lot of papers the intense on sensors like DO, pH, turbidity [4–6], etc., and respond for these problems. The entire the parameters are sensed and transmit to the farmers. But storing the data in the cloud database (DB) helps us in data analyzing using data analytics which can help us to take pro-active measures before the change in water quality parameters. A lot of solutions rigorously sending the data to the farmer but what is the essential action to be taken in the form of an alert message are concentrated. Most of the solutions from different authors did not concentrate on fish feeding. Fish feeding typically accounts for 50 to 80% of a fish farm's overhead costs, but since feeding is a manual task, it's mostly an unmeasured and inexact method with one of two outcomes. Overfeeding means much of the feed goes to waste. On the flip side, under-feeding leads to the die-off of fish. For that reason, we have a solution to address this issue by using a routine smart feeder. Automatic smart feeder dispenses just the right amount of feed by using sensors that measure fish appetite. The system picks up on hungry fish behavior through an in-water vibration-based sensor that can read the movement of a hungry versus a full fish. This rooted in the sensors the nourish release the food. It can be put into auto form like nourish at the usual interval of time. To speak that how we set alarm and wake up.

7.4   FUNCTIONING OF AQUARIUM SENSORS

This chapter suggests an IoT solution namely AQUATALK to manage the melt gases and other issues. AQUATALK has several sensors which including, e.g., the DO, the temperature sensor, the pH, the electrical conductivity (EC), the water level, the gases sensor, feed sensor, and the total dissolved solids (TDS) sensor. It also has a beneath water video system (BVS) as illustrated in Figure 7.1.

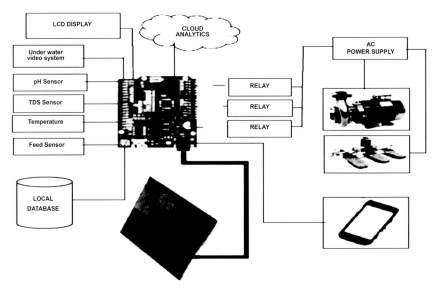

FIGURE 7.3   Block diagram of system.

7.4.1   STAGE I

The data from sensors is composed and stored. The data from the sensors like pH sensor, temperature sensor, DO sensor is engaged and stored in the

DB with timestamp. When pH principles are altering, i.e., diminishing, then the motor begins to pump water into the pond. The new raw water from subversive is added it intensified the pH value of the water. With the data, being stored in the local DB is also a copy of the data is transmitted to the grower into his smart device. By this approach, the farmer requires to call the local testing agency for the test the quality of water parameters. The feed sensor is sensing the vibration of the prawns or fish and the food feeder released the food into the water. This can be complete by setting the food feeder to release food at regular intervals (Figure 7.4).

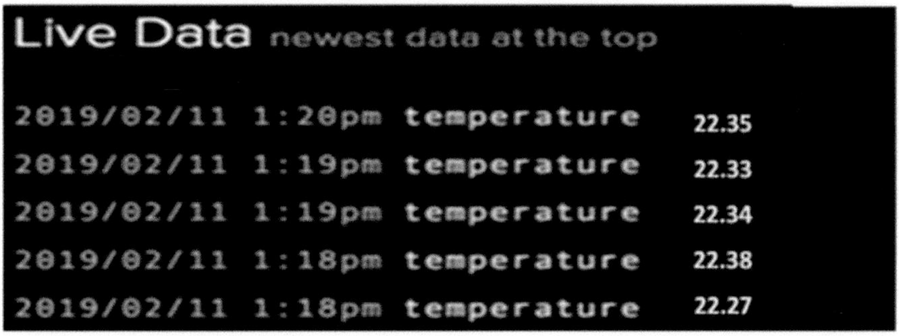

Live Data newest data at the top		
2019/02/11	1:20pm temperature	22.35
2019/02/11	1:19pm temperature	22.33
2019/02/11	1:19pm temperature	22.34
2019/02/11	1:18pm temperature	22.38
2019/02/11	1:18pm temperature	22.27

FIGURE 7.4   Temperature at unlike timestamps.

7.4.2   STAGE II

By using the beneath video system (BVS), the system captures the pictures of the shrimps. The camera pictures are digitized. The following digitized noise elimination is finished and an anti-blur filter is practical with background segmentation. The following accomplishment to this process feature extraction is finished. The features like motion detection, black gill detection and density estimation is done. The purpose of images is shown in Figure 7.5.

The pH, temperature, TDS (totally dissolved solids) data is engaged from the sensors and post to the cloud server. These data for temperature interpretation is shown in Figure 7.4. To give this dataset, we train a logistic regression-based replica using PyTorch that takes the condition of pH, TDS, and temperature as input and forecasts whether these conditions are best for rising healthy fish or prawns. We also train a support vector machine (SVM) which takes the current parameters of water as input and proposes the best conditions in which are essential for rising healthy prawns or fish. Also, proposes the best kind of fishes for the given water circumstances as output.

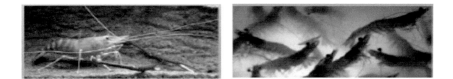

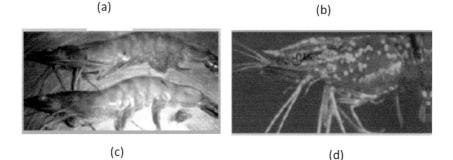

FIGURE 7.5 (a) Shrimp which is active. (b) Shrimps which are in abnormal state. (c) Shrimps which are dead before decay. (d) Shrimps with black-spotted virus.

The data intelligence by the sensors is stored as elements of a Numpy array. The following in every time interval T seconds are the data sensed by the sensors is broadcasting to a cloud server for analysis. The time interval T is a steady value determined based on the bandwidth and the network speed obtainable in device setup vicinity. Based on the bandwidth of the ISP (internet service provider) that is chequered each hour using technology of Python.

The sensor values as input a logistic regression replica are trained and to forecast whether the water conditions to suitable to grow fish and prawns. This replica is also predicting which conditions are not being met. For that reason, the essential steps are taken mechanically. For instance, pH is low, so to increase the pH of the water motor is turned so that new water is approaches into the pond so pH values alter. When oxygen is low than the necessary to oxygen levels then the aerators will be turned so that the melted oxygen augments where it is a vital element for the survival of fish or Prawns. From the UVS we can predict the approximate size of the shrimps where we can come to a conclusion whether to wait for some more time to catch the prawns or not. Size is illustrated in Figure 7.6. Based on the size of the prawns, the weight can be predicted. Once the size looks to be good, enough farmers catch few prawns with net and count the no. of prawns per kilogram. If less no. of prawns comes for 1-kilogram, good profit comes to the farmers.

FIGURE 7.6   This image showing the length of shrimp.

7.5   RESULTS

The proposed model was implemented on aqua ponds with prawns. The results were taken for every consecutive 2 days (Figure 7.7); dissolved oxygen (DO) (Figure 7.8); pH values (Figure 7.9); and CO_3 values (Figures 7.10 and 7.11).

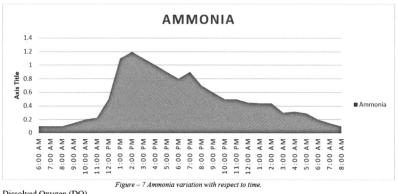

Figure – 7 Ammonia variation with respect to time.

Dissolved Oxygen (DO)

FIGURE 7.7   Ammonia variation with respect to time.

7.6   CONCLUSION

Day-by-day improved the aquaculture especially in Andhra Pradesh, India. It is causal nearly 1.10% of the GDP. This approximated that the fish requirement of the country by 2025 will be 16 million tonnes. The replica in this chapter aids the aqua farmers to monitor the quality of water parameters instead of manual testing in which waste the time and money visiting the labs where the parameters might change in the meantime. IoT has reached the farmers for reducing the risk and in automatic several tasks. Several tasks can be done by monitoring by one person in place of depending on labor to supervise the ponds.

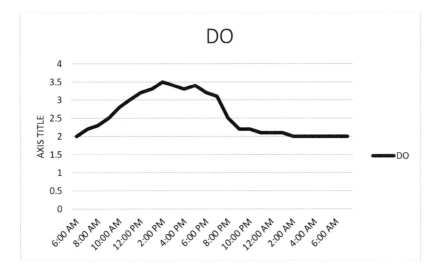

FIGURE 7.8   Dissolved oxygen variation with respect to time.

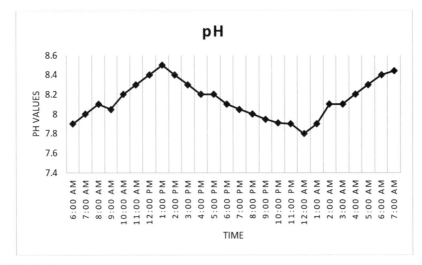

FIGURE 7.9   pH variation with respect to time.

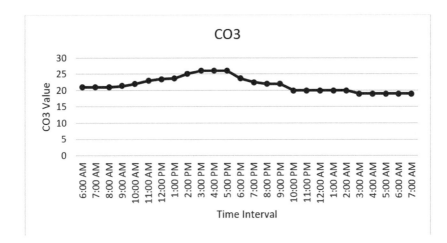

FIGURE 7.10 CO₃ variation with respect to time.

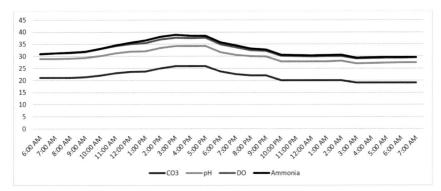

FIGURE 7.11 Different water quality parameters with respect to time.

KEYWORDS

- aquaculture
- dissolved oxygen (DO)
- internet of things (IoT)
- internet service provider
- logistic regression
- support vector machines (SVM)

REFERENCES

1. Sheetal, I., Harshal, M., & Parag, Y., (2015). Application of IoT based system for advanced agriculture in India. In: *International Journal of Innovative Research in Computer and Communication Engineering* (Vol. 3, No. 11).
2. Kawitkar, R. S., & Nikesh, G., (2016). IoT based smart agriculture. In: *International Journal of Advanced Research in Computer and Communication Engineering* (Vol. 5, No. 6).
3. Changhui, D., Yanping, G., Jun, G., & Xinying, M., (2010). Research on the growth model of aquaculture organisms based on neural network expert system. In: *6th International Conference on Natural Computation (ICNC 2010)* (pp. 1812–1815).
4. Kayalvizhi, S., Koushik, R. G., Vivek, K. P., & Venkata, P. N., (2015). Cyber aqua culture monitoring system using ardunio and raspberry Pi. In: *International Journal of Advanced Research in Electrical, Electronics, and Instrumentation Engineering* (Vol. 4, No. 5, pp. 2320–3765).
5. Daudi, S. S., & Shi, F. Y., (2014). Water quality monitoring and control for aquaculture based on wireless sensor networks. In: *Journal of Networks* (Vol. 9, No. 4).
6. Kiran, P., Sachin, P., & Vikas, P., (2015). Monitoring of turbidity, pH, and temperature of water based on GSM. *International Journal for Research in Emerging Science and Technology*, *2*(3).
7. Dolan, A., (2015). *The Effects of Aquarium Size and Temperature on Color Vibrancy, Size, and Physical Activity in Betta Splendens.* In: Technical Report, Maryville College.
8. Yi, B. L., & Hung-Chun, T., (2017). Fish talk: An IoT-based mini aquarium system. *IEEE Access.*
9. Raghu, S. R. R. K., & Harish, K. V. G., (2017). Knowledge based real time monitoring system for aquaculture using IoT. *IEEE 2017, 7th International Advance Computing Conference.*
10. Chin-Chang, H., & Hsu-Kuang, C. (2016). *A Highly Sensitive Underwater Video System for Use in Turbid Aquaculture Ponds.*

CHAPTER 8

A Comprehensive Survey on Mitigation of Challenging Issues in Green IoT

GEETANJALI NAYAK,[1] NEELAMADHAB PADHY,[2] and
TUSAR KANTI MISHRA[3]

[1]*Department of CSE, Gayatri Vidya Parishad College of Engineering,
Visakhapatnam, Andhra Pradesh, India,
E-mail: geetanjalinayak2019vizag*

[2]*Department of CSE, GIET University, Gunupur, Odisha, India,
E-mail: dr.neelamadhab@gmail.com*

[3]*Department of CSE, Gayatri Vidya Parishad College of Engineering for
Women, Visakhapatnam, Andhra Pradesh, India,
E-mail: tusar.k.mishra@gvpcew.ac.in*

ABSTRACT

The green internet of things (IoT) is only the correspondence in machine-to-machine (M2M). "Web of the things" is a system of devoted physical items or gadgets that contain the innovation to detect or cooperate with their inside state or outer condition. In this chapter, an in depth analysis has been made on this emerging topic. Several key issues have been discussed pertaining to the green-IoT domain. A comparative study has also been made between benchmark approaches against certain attributes like energy, focus area, and complexity. Future work has been suggested that would help researchers for active consideration in the same domain.

8.1 INTRODUCTION

The IoT contains a framework that incorporates gadgets, correspondence, applications, and information examination [1] various sorts of items must be interconnected through the web. Therefore, these items get related with one another by means of the instrument of inescapable calculation with heterogeneity in the hinder engineering. Another trouble being the security issue identified with actualizing the total IoT engineering into the real world. Nonetheless, still the essential test for IoT is the enhancement of vitality assets. To determine the equivalent, green-IoT comes into the image. It has been very much received in the field of therapeutic science, where, it is helping with monitoring heartbeat paces of patient at basic circumstances. Other significant application spaces of green-IoT these days are home mechanization, brilliant urban areas, wellness sets, and so forth. By using machine learning (ML) tools and smart grid, the power consumption [2] can be reduced. A novel secure framework can be build for security of data in green IoT.

The green-IoT guarantee of vertical ranches and green IoT using man-made consciousness and AI indoor homesteads kept running by AI and lit by LEDs can be more effective than field farming, however, can they essentially diminish ozone-depleting substance discharges gadgets with computational and organizing abilities are improving our homes, clinics, urban communities, and enterprises. This developing system of associated gadgets or internet of things (IoT) guarantees better wellbeing, upgraded the executives of patients, improved vitality effectiveness, and streamlined assembling forms. Regardless of these numerous advantages, security vulnerabilities in these frameworks can prompt client disappointment (e.g., irregular bugs), protection infringement (e.g., listening in), fiscal misfortune (e.g., refusal of-administration assaults or payment product), or even death toll (e.g., assailants controlling vehicles). Consequently, it is basic to verify this developing innovation upset in an auspicious way.

The IoT [3] is the most recent web development that joins billions of gadgets that are possessed by various associations and individuals who are conveying and utilizing them for their very own motivations. IoT-empowered saddling of the data that is given by organizations of such IoT gadgets (which are frequently alluded to as IoT things) gives phenomenal chances to take care of web-scale issues that have been too enormous and too hard to even consider tackling previously. Much the same as other electronic data frameworks, IoT should likewise manage the plenty of cybersecurity and protection dangers that right now disturb associations and can possibly hold the information of whole ventures and even nations for payment. To understand its maximum capacity, IoT must arrangement viably with such dangers and guarantee the security and protection of the data gathered and refined from IoT gadgets. Notwithstanding, IoT presents a few one of a kind difficulties that make the use of existing security and protection strategies troublesome. This is based on the fact that IoT structures facilitate no compromise of security and security answers for ensuring progress of information passing. Security is also dealt with at the IoT physical layer and the application layer. Thus, security is confirmed at all the stages from origin to the end-stage with utmost guarantee. The proposed security safeguarding procedures presented in [3] use various IoT cloud information stores to shield the protection of information gathered from IoT.

"Web of things" (IoT) [4] is the latest internet progression that incorporates intertwining billions of web-related sensors, cameras, appears, propelled cell phones, wearable, and other splendid devices that pass on by methods for the web (which are aggregately insinuated as IoT things), and (ii) harnessing their data and convenience to give novel keen organizations and things that preferred position our overall population. A progressing figure made by the Gartner adventures IoT and the related organic framework to be a $1.7 trillion market by 2020 and join 28.1 billion related things. IoT is fueling an adjustment in context of a truly related world wherein common things become interconnected, prepared to examine clearly with each another, and arranged to do all things considered giving insightful organizations. Nevertheless, in various such applications the data accumulated by IoT is physiological data assembled by (from time to time wearable) biomedical sensors, imperativeness usage data accumulated by keen meters, and zone data assembled by mobile phones to give a few models. The disclosure of such data should be taken care towards security, insurance, and trust [1], which are seen as among the remaining essential blocks in IoT application improvement. Green IoT fundamentally centers on the vitality, proficiency in the IoT standards. Green IoT is characterized as the vitality productive courses in IoT either to lessen the green-house impact brought about by existing applications or to annihilate the equivalent in IoT itself:

1. IoT framework entrusted with empowering a greener society;
2. Reducing vitality utilization of IoT frameworks while fulfilling mission targets;
3. Green ICT: interchanges, organizing, information handling.

8.2   GREEN COMPUTING

Green preparing, green ICT as per IFG International Federation of Green ICT and IFG Standard, green IT, or ICT viability, is the examination and routine with respect to earth efficient figuring or IT. Green IT "is the assessment and routine concerning arranging, amassing, using, and disposing of PCs, servers, and related substructures, for instance, screens, printers, accumulating devices, and frameworks organization and trades systems capably and sufficiently with irrelevant or no impact on nature." Similarly, bores down the accompanying four corresponding ways along to extensively and viably address the ecological impacts of figuring ought to be tended to for green processing street:

1. **Green Use:** Reducing the vitality utilization of PCs and other data frameworks just as utilizing them in an ecologically stable way.
2. **Green Disposal:** Several old electronic gadgets, computers, phones, etc., are recycled and remodeled for further use.
3. **Green Design:** Designing vitality effective and ecologically stable segments, PCs, and servers and cooling equipments.
4. **Green Manufacturing:** Without effecting adversely to the environment, the electronic products are manufactured.

8.3   GREEN IOT-GAHA ARCHITECTURE

Applying green IoT to green IoT agriculture and healthcare applications (GAHA) architecture before analyzing about green IoT, first we have seen various definitions related to IoT, and it is considered as the accompanying wave in the time of figuring is foreseen to be outside the space of standard work territory. As per this discernment, a novel perspective called IoT rapidly gained ground over the latest couple of years. IoT implies "a general arrangement of interconnected articles phenomenally addressable reliant on standard correspondence conventions." The basic idea behind it is the unpreventable closeness around people of things, prepared to measure, accumulate, grasp, and even change the earth. IoT is fueled by the ongoing advances of an assortment of gadgets and correspondence innovations, however things incorporated into IoT are perplexing gadgets, for example, cell phones, yet they additionally include in this segment, we first observe layout of ICT and empowering green innovations for GAHA are discussed. ICT is an umbrella term that identifies with any office, innovation, application in regards to data and correspondence, empowering clients to get to, store, transmit, and control an assortment of data. We have recorded them beneath, in regards to ID, detecting, correspondence, and calculation which are IoT components. RFID (radio-frequency identification): A little electronic gadget that comprises of a little chip and a reception apparatus, naturally distinguishing and following labels joined to objects.

1. **Wireless Sensor Network (WSN):** A system comprising of spatially conveyed self-sufficient sensors that agreeably screen the physical or ecological conditions (e.g., temperature, sound, vibration, weight, movement, and so on).
2. **Wireless Personal Area Network (WPAN).**
3. **Home Area Network (HAN):** A kind of neighborhood (LANs: local area networks), associating advanced gadgets present inside or inside the nearby region of a home.
4. **Neighborhood Area Network (NAN):** Enabling a user to connect to the internet through local Wi-Fi hotspot connection. This is one among the cheapest configuration and popular also.
5. **Machine-to-Machine (M2M):** Different devices of coherent types are interconnected to each other irrespective of the type of connecting media (may be wired or may be wireless).
6. **Cloud Computing (CC):** A novel computing model for enabling convenient, on demand network access to a shared pool of configurable resources (e.g., networks, servers, storage, applications, services). Integrating CC into a mobile environment, mobile cloud computing (MCC) can further offload much of the data processing and storage tasks from cell phones (e.g., cell phones, tablets, and so forth) to the cloud.
7. **Data Center (DC):** An archive (physical or virtual) for the capacity, the executives, and dispersal of information and data. Green RFID incorporates a few RFID labels and an extremely little subset of label peruses. Encased in a cement sticker, the RFID tag is a little microchip joined to a radio (used for accepting and transmitting the sign), with a one of a kind identifiers. The reason for RFID labels is putting away data in regards to the items to which they are appended. The essential procedure is that the data stream is activated by RFID label peruses through transmitting an inquiry signal, pursued with the reactions of close-by

RFID labels. For the most part, the transmission scope of RFID frameworks is extremely low (e.g., a couple of meters). Moreover, different groups (e.g., from low frequencies at 124–135 kHz up to ultrahigh frequencies at 860–960 MHz) are utilized to perform transmission. Two sorts of RFID labels (e.g., dynamic labels and inactive labels) exist. Dynamic labels have batteries controlling the sign transmissions and expanding the transmission ranges, while the detached labels are without locally available batteries and need to gather vitality from the per user signal with the rule of enlistment.

a. Reducing the measures of RFID labels ought to be considered to diminish the measure of non-degradable material utilized in their assembling (e.g., biodegradable RFID labels, printable RFID labels, paper-based RFID labels), on the grounds that the labels themselves are hard to reuse by and large;

b. Energy-effective calculations and conventions ought to be utilized to enhance label estimation, alter transmission power level progressively, keep away from label crash, and abstain from catching, and so on.

8. **Green Wireless Sensor Networks (WSN):** A WSN, for the most part, comprises of a specific number of sensor hubs and a base station (BS) (e.g., sink hub). The sensor hubs are with low handling, restricted power, and capacity limit, while the BS is ground-breaking. Sensor hubs furnished with numerous on-board sensors, take readings (e.g., temperature, stickiness, speeding up, and so forth) from the surroundings. A normally utilized business WSN arrangement depends on the IEEE 802.15.4 standard for low-power and low-bit-rate interchanges [5]. With respect to WSN, the accompanying systems ought to be embraced [6].

a. Make sensor hubs possibly work when important, while spending the remainder of their lifetime in a rest mode to spare vitality utilization;

b. Vitality consumption [6] [e.g., remote charging, using vitality gathering instruments which create control from the earth (e.g., sun, active vitality, vibration, temperature differentials, etc.)];

    c.   Radio streamlining strategies (e.g., transmission power control, balance streamlining, agreeable correspondence, directional reception apparatuses, vitality proficient intellectual radio (CR));

    d.   Information decrease systems (e.g., total, versatile testing, pressure, arrange coding);

    e.   Vitality proficient directing methods (e.g., bunch structures, vitality as a steering metric, multipath directing, hand-off hub arrangement, and hub versatility).

9. **Green Cloud Computing (CC) [7]:** In CC, assets are treated as administrations, i.e., IaaS (infrastructure as a service), PaaS (platform as a service), and SaaS (software as a service). In light of clients' requests, CC flexibly offers different assets (e.g., high performance processing assets and high-limit stockpiling) [5] to clients. As opposed to owning and dealing with their very own assets, clients share an enormous and oversaw pool of assets, with helpful access. With developing applications moved to cloud, more assets should be sent and more power are devoured, bringing about progressively natural issues and CO_2 discharges[6].

    a.   Adoption of equipment and programming that lessening vitality utilization. In such manner, equipment arrangements should focus at structuring and assembling gadgets which expend less vitality. Programming arrangements should attempt to offer productive programming plans devouring less vitality with least asset usage;

    b.   Power-sparing virtual machine (VM) [8] procedures (e.g., VM solidification, VM relocation, VM situation, VM allotment);

    c.   Various vitality proficient asset allotment systems (e.g., sell-off based asset distribution, tattle based asset allotment) and related undertaking booking systems;

    d.   Effective and exact models and assessment methodologies with respect to vitality sparing arrangements;

    e.   Green CC plans dependent on cloud supporting advances (e.g., systems, correspondences, and so on.).

10. **Green Machine to Machine (M2M):** As far as M2M interchanges, huge M2M hubs which cleverly accumulate the checked information are conveyed in M2M area. In system area, the wired/remote system

transfers the assembled information to the BS. The BS further backings different M2M applications over system in the application area. Concerning green M2M, with the enormous machines associated with M2M interchanges, it will devour a ton of vitality, especially in M2M space.

a. Intelligently modify the transmission control (e.g., to the negligible important level);
b. Design proficient correspondence conventions (e.g., directing conventions) with the use of algorithmic and circulated figuring procedures;
c. Activity booking, in which the goal is to change a few hubs to low-control task/dozing mode with the goal that lone a subset of associated hubs stay dynamic while keeping the usefulness (e.g., information gathering) of the first system;
d. Joint vitality sparing systems (e.g., with over-burden security and assets portion);
e. Employ vitality collecting and the preferences (e.g., range detecting, range the board, impedance moderation, and control enhancement) of CR.

11. **Green Data Center (DC):** The primary occupation of DCs is to store, oversee, process, and disperse different information and applications, made by clients, things, frameworks, and so on. By and large, managing different information and applications, DCs expend immense measures of vitality with high operational expenses and huge CO_2 impressions expanding age of colossal measures of information by different unavoidable and universal things or items (e.g., cell phones, sensors, and so on). About green DC, potential strategies to improve vitality productivity can be accomplished from the accompanying viewpoints:

a. Use inexhaustible or green wellsprings of vitality (e.g., wind, water, sun oriented vitality, heat siphons, and so forth.);
b. Utilize productive unique power-the executives innovations (e.g., turbo help, circle, recurrence scaling) systems and VOVO (fluctuate on/differ off) methods;

TABLE 8.1 Comparative Analysis of Different Green IoT Approaches

Method, Year	Focus Area	Category	Application group	Limitation	Novelty	Energy aspect
[5], 2015	Energy efficient Green IoT	Software	Heterogeneous devices	Minimum number of system	G-IoT and Energy efficient scheme for the IoT devices	Decrease the Energy consumption
[6], 2018	Green Sensor Technology Wireless Network	Hardware, Software	Smart cities	Less quality of service	Embraces toxic pollution	Moderate energy consumption
[2], 2017	Green tag	Hardware	Smart industrialization	Small size	Efficient utilization of energy	Medium Energy
[4], 2018	Sensor-Cloud-Green-IoT	Hardware, Software	Smart Healthcare	No standard communication protocol	Common architecture for IoT model	Less energy
[3], 2017	Reducing RFID size	Hardware, Software	Smart city	Over hearing problem	Green tags, green sensing networks	Low Power Consumption
[9], 2018, 2016	Resource and cooperative layering	Software	Any green IoT component	Not efficient in real time analytic scenarios	Sustainable environment for IoT systems	High Latency
[22], 2017	Sensor cloud, wireless sensor network	Software	Smart world	Sharing the sensor-cloud resources and services	System environment	Energy
[23], 2017	Green Femto cells	Hardware	Smart cities	Achieving enhancement in energy saving and bandwidth utilization	Enhance coverage and capacity	Less energy consumption

    c.   Design epic vitality proficient server farm models (e.g., nano-server farms) to accomplish control preservation;

    d.   Design vitality mindful directing calculations to solidify traffic streams to a subset of the system and power off the inert gadgets [8].

    e.   Construct successful and precise server farm power models;

    f.   Draw support from correspondence and figuring procedures. A comparative analysis of several approaches in the field of green-IoT has been presented in Table 8.1.

8.4   CONCLUSION AND FUTURE WORK

The predominance of green-IoT brings down the obstruction from genuine world to the internet, which leads toward another advanced setting for novel applications and administrations. Creating green organization plans for IoT plays a vital role in its massive implementation. As per the observation, a novel framework can be projected as one among the future work that would minimize the energy consumption. A new secure protocol can be design for security of the green IoT.

KEYWORDS

- **cloud computing**
- **green IoT agriculture and healthcare applications**
- **home area network**
- **infrastructure as a service**
- **local area networks**
- **machine to machine**

REFERENCES

1. Rushan, A., Saman, Z., Munam, A. S., Abdul, W., & Hongnian, Y., (2017). *Green IoT: An Investigation on Energy Saving Practices for 2020 and Beyond (Senior Member, IEEE).*
2. Faisal, K. Z., Sherali, Z., & Ernesto, E., (2017). Enabling Technologies for green internet of things. *IEEE Journal, 11*(2).
3. Mahmoud, A. M. A., Ayman, A. E. S., Muzabir, I., Wael, S., Jusoh, M., Azizan, M. M., & Ali, A., (2017). Green internet of things (IoT): An overview. *IEEE, ICSIMA.*
4. Manas, K. Y., & Ganga, D. B. K., (2018). Green IoT: Principles, current trends. *Future Directions International Journal of Advance Research and Innovation, 6*(3).

5. Sarder, F. A., Md. Golam, R. A., Rim, H., & Choong, S. H., (2015). *A System Model for Energy Efficient Green-IoT Network*. Department of Computer Engineering Kyung Hee University, South Korea. IEEE.

6. Alsamhi, S. H., Ou, M., Samar A. M., & Qingliang, M., (2018). *Greening Internet of Things for Smart Every Things with a Green-Environment Life: A Survey and Future Prospects.*

7. Vinita, T. M. D., (2018). *Green IoT Systems: An Energy-Efficient Perspective Department of Computer Science and Information System*. Noida, India.

8. Zhu, C., Leung, V. C. M., Shu, L., & Ngai, E. C. H., (2015). Green internet of things for smart world. *IEEE Access,3*(11), 2151–2162.

9. Vinita, T., & Mayuri, D. (2018). *Green IoT Systems: An Energy Efficient Perspective*. Department of Computer Science and Information.

10. Abedin, S. F., Alam, M. G. R., Haw, R., & Hong, C. S., (2015). A system model for energy efficient green-IoT network. In: *Proc. Int. Conf. Inf. Netw.* (pp. 177182).

11. Srdjan, K., Boris, P., & Francois, C., (2014). Designing IoT architecture(s): A European perspective. *IEEE World Forum on Internet of Things (WF-IoT)* (pp. 79–84).

12. Atzori, L., Iera, A., & Morabito, G., (2010). The internet of things: A survey. *Comput. Netw., 54*(15), 2787–2805.

13. Shaikh, F. K., Zeadally, S., & Exposito, E. (2017). Enabling technologies for green internet of things. *IEEE Syst. Journal.*

14. Elkhodr, M., Shahrestani, S., & Cheung, H., (2013). The internet of things: Vision and challenges. In: *2013 IEEE Tencon.* (pp. 218–222).

15. Ramaswamy, P., (2016). IoT smart parking systems for reducing greenhouse gas emission. *2016 International Conference on Recent Trends in Information Technology.*

16. Zhou, J., Leppnen, T., Harjula, E., Yu, C., Jin, H., & Yang, L. T., (2013). Cloud things: A common architecture for integrating the internet of things with cloud computing. *Proceedings of the 2013 IEEE 17th International Conference on Computer Supported Cooperative Work in Design* (pp. 651–657).

17. Mahmoud, A. M. A., El-Saleh, A. A., Muzamir, I., Wael, S., Jusoh, M., Azizan, M. M., & Ali, A. (2017). *Green Internet of Things (IoT): An Overview*. Department of Electronics and Communication Engineering, Asharqiyah University, Ibra, Oman.

18. Ashton, K., (2009). That "internet of things" thing in the real world, things matter more than ideas. *RFID Journal.*

19. Whitmore, A., Anurag, A., & Li, D. X., (2015). The internet of things: A survey of topics and trends. *Information Systems Frontiers, 17*(2), 261–274.

20. Sarder, F. A., Golam, R. A. M., Rim, H., & Choong, S. H. (2015). *A System Model for Energy Efficient Green-IoT Network*. Department of Computer Engineering Kyung Hee University, South Korea.

21. Wang, K., et al., (2016). Green industrial internet of things architecture: An energy-efficient perspective. *IEEE Communications Magazine, 54*(12), 48–54.

22. Akshay, G., Ankit, W., Ashish, W., & Shashank, J., (2017). Emerging trends of green IoT for smart world. *International Journal of Innovative Research in Computer and Communication Engineering, 5*(2).

23. Al-Turjman, F., Enver, E., & Hadi, Z., (2017). *Green Femtocells in the IoT Era: Traffic Modeling and Challenges: An Overview IEEE.*

24. Jun, H., Yu, M., & Xue, H. G., (2014). *A Novel Deployment Scheme for Green Internet of Things.*

CHAPTER 9

Shortest Route Analysis for High-Level Slotting Using Peer-to-Peer

DILEEP KUMAR KADALI[1] and R. N. V. JAGAN MOHAN[2]

[1]Research Scholar, Department of CSE, GIET University, GIET University, Gunupur – 765022, Odisha, India, E-mail: dileepkumarkadali@gmail.com

[2]Assistant Professor, Department of CSE, Swarnandhra College of Engineering and Technology, Narasapur – 534280, Andhra Pradesh, India

ABSTRACT

The shorthand distance for the technique of assigning creation to a location within the data warehouse aligned with activity rules and goods choices called slotting. This is typically limited to the options face or online sites. In this chapter, the high-level slotting is the tiniest need of enactment for new-fangled ability. The all-inclusive of detailed slotting is moderately infrequent, which ever novel or present operations. In the uncommon periods that are slotting is stated aligned of effort privileges of vital output gain the made. A picking routing strategy is a strategy by which the picking route through the warehouse is determined. A route could be a path within which you pass all things of associate degree order. The most basic solution is to minimize the cost is to try to use the shortest route possible route. However, in most cases the problem is more complicated. The proposed work expected that automatons could do the same activities as the operators in a traditional warehouse. Thus, one could have a warehouse where both automatons and operators are responsible for replenishment and picking tasks. The task is to make a system where we can learn from the operators in the warehouse (how is the slotting done, which picking shortest routes are operators taken and when).

The idea is to use artificial intelligence (AI) with blockchain technology that learns from operators is to develop a plan in which is used to give directions to the automaton workforce. The experimental result is on shortest route for high-level slotting in which picking the route uses in specific time.

9.1 INTRODUCTION

Blockchain and machine learning (ML) flawlessly complement to each other and greatly are the two masts on which the upcoming novelties are to be built. These two composed are certain to make innovative modernizations in the adjacent future while also making our present more secure. Blockchain and ML are contemporary skills that have appeared in the over the period. ML gives the computer the ability to study ended time without being repetitively programmed and without any human interference. Blockchain and artificial intelligence (AI) have the possible to be much more radical when combined together. Both can supplement to each other's competences in addition to improve stages of transparency, trust, and communication. Blockchain can be able to just about any kind of operation in being. This is the main aim in arrears its speedily rising fame and power. The blockchain is intended exactly to accelerate and make simpler the process of how dealings are noted. This income that somewhat kind of asset can be transparently managed with this fully regionalized system. The vital change at this time is the detail that there's no participation from mediators. As an alternative of it's a colossal association with some prodigious code which meaningfully diminishes defrayal and clearing times to a matter of seconds. This technology is consequently troublesome that it can really transform it. Then, still there are a lot of track race to overwhelm such as safety, interoperability, and rule. In this chapter, ML is a technology that depends on wide slotting of data for perfect construction and accurate prediction. A portion of time is invested in the slotting arrangement and data storage for accuracy. This is where blockchain originates into play as the time taken can be significantly reduced with the help of blockchain skill. This data can be moved directly and securely.

9.2 TRAJECTORY ANALYSIS

In the expansion analysis the trajectory path tracking for particular object, incredible large number of object trajectory data were collected and processed, which can optimize to recognize the best information on that processed to peer-to-peer up to establish a minimized path with time complexity O(n) where n number of peers to be set of trajectories. Trajectory is a proficient technique to scrutinize trajectory data it has been very useful in pattern recognition, data analysis (DA) and ML, etc. Using trajectory to possible to store or collect various kinds of data, which can depend upon type of object or device. We focused to forecasting trajectory clustering technique with considering distance of object moving peer-to-peer from source to target. For example, the path established peer-to-peer like P1 → P2 → Pn. Each existed node path will be examined from source to target based on the distance.

Here, we taken P are a set of peer's trajectories input and produces path distance D is an output.

In approximately precise state of affairs, the object movement's paths are added, such as distance, mean, variance, etc., a sequence of the minimized path will have established from consecutive peers from trajectory, which are similar to a cluster. In the technique of trajectory, data are required to beset as a unified peers distance so that they can be measured. Therefore, expressive trajectories in a unified distance with tiny harm of information are a key initial work of these techniques. This route is called trajectory clustering T_j preparation of path P from source node-S to target node-P (Figure 9.1).

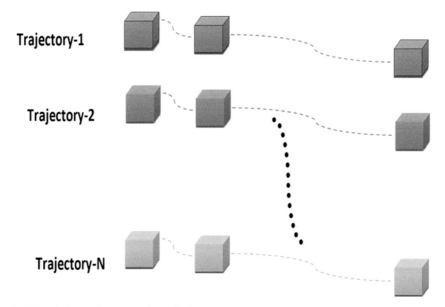

FIGURE 9.1   Trajectory node analysis.

9.3   BRUTE FORCE APPROACH

This algorithm passes on to in doctrinarian method that does not comprise whichever shortcuts to develop recital, however, in its place relies on pure computing authority to try all possibilities until the solution to a difficulty is found. For example, trajectories cluster nodes of shortest route path identification. Assume that a candidate wants to visit node clusters across the trajectory cluster nodes. How does one make a decision the arrange in which node should be visited such that the total distance voyage is minimized? The brute force resolution is just to calculate the overall distance for each attainable route then choose the shortest one. The time-complexity of brute force is O $(n\times m)$. As a result, if search for a node of 'n' nodes in nodes of 'm' trajectory cluster using brute force, it would take usn$\times$ m tries. A brute-force attack engages methodically examination of all probable inputs until the accurate input is established inside cryptanalysis [1]. This plan can in hypothesis be alive used next to some encrypted data [5] by an attacker who is not capable to get benefit of some fault in an encryption method that would or else make his or her job is simple. The input key distance end to end make use of inside the encoding determines the realistic viability of acting a Brute force assault, amid the lengthy input keys exponentially sturdy to fracture than lesser ones. Brute force attacks can be completing less capable by obfuscating the data to be encoded, incredible that makes it more not easy for an attacker to be familiar with fractured the code. One of the gauges of the potency of an encryption system is how long it would hypothetically take an attacker to raise a successful Brute force attack against it.

9.4   SHORTEST ROUTE PATH PROCEDURE

This process is uses to discover the shortest paths from the basis nodes to the entire further nodes in a weighted graph. This depends upon the subsequent thought is a shortest path contains at most n−1 path for shortest route path could not have a cycle. There is no requiring going by a node over again, for the reason that the shortest path to the entire further nodes could be found without the required for a visit for any nodes.

The steps in procedure:

- The outer loop traverses from 0 to n−1 path.
- Loop over all paths, check if the NetNodeDistance>CurrentNodeDistance + PathWeight, in thiscase update the NextNodeDistance to CurrentNodeDistance + PathWeight.

This process is resolute by on the reduction code where the shortest distance for the entire nodes is increasingly replaced by a value that is more accurate until eventually reaching the optimum solution. In the commencement of the entire nodes have a distance of infinity, but only the distance of the source node=0 then update the entire connected nodes with the new distances is pertain to the equivalent concept for the new nodes with new distance and in next to no time.

9.5   IMPROVE SLOTTING AND ALTERNATIVE PATH/ROUTES USING ARTIFICIAL INTELLIGENCE (AI)

The shorthand phrase for the process of allocating manufactured goods is to position in the warehouse along with work-related regulations and manufactured goods features namely slotting. Slotting is typically limited to the preference appearance or online locations. One of the slotting approaches is high-level slotting is a minimum necessity for the execution of a new-fangled capability. In the details of slotting is moderately uncommon in moreover new-fangled or obtainable operations. Slotting is mentioned in line of work privileges of important output to gains the made in rare times. A picking routing strategy is a strategy by which the picking route through the warehouse is determined. A route may be a path within which you pass all things of an order. The most basic solution is to minimize the cost is to try to use the shortest route possible route. However, in most cases, the problem is more complicated. For instance, one has to avoid that a picking location that is used often becomes a bottleneck by balancing the workload. Therefore, in literature, there are many studies about routing strategies. Not all of them are suited for any situation and often-best practices are used within a warehouse. If one looks into the future, it will not take that long robots and AGVs will support those operators. One can see this already in the warehouses where AGVs are used to bring the goods to the operators. Also, the so-called COBOTS has been introduced. These robots can work together with operators. The main goal is predictable that robots can do the same activities as the operators in a traditional warehouse. So, one could have a warehouse where both robots and operators are responsible for replenishment and picking tasks. The assignment is to make a system where we can learn from the operators in the warehouse (how is the slotting is done, which picking routes are operators taken and when). The idea is to use AI that learns from operators, to develop a strategy, which is used to give instructions to the robot workforce (Figure 9.2).

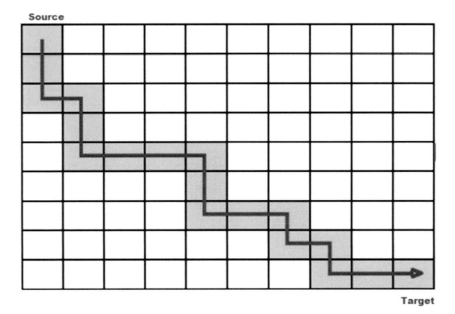

FIGURE 9.2   The starting point of the research could be a basic warehouse design; this has several aisles with picking locations.

However, the solution should be all-purpose. One should be able to use the solution in more complicated layouts as well, like for instance a layout with cross aisles.

9.6   BLOCKCHAIN WITH ARTIFICIAL INTELLIGENCE (AI) FOR SLOTTING

Many arrangements of slotting decisions are taken by AI; this system can be tough for being in the direction of sympathetic the blockchain can lean-to new illumination with the help of path the thinking process and appreciate the decisions. This structure is the exactness of the process flanked by basis and reply on the plank. In that way, even if, the creator unintentionally loses out the different key steps in this aided him or her arrive at a finale. The human being is proficient to evidence in AI is a decision-making process on a blockchain could be a better step towards better clearness. The illustration of blockchain is to provide similar fundamental code with the exemption that information written on later blockchain it can be immutable and permanent. To end with, as awfully helpful in our daily life is usage with the computers is unable to carry out a task without receiving explicit instructions. But the mechanism with a blockchain is entire encrypted data that on a computer require large amounts of processing ability. This approach uses hashing procedure is used to excavate the bitcoin blocks is uses for description to receive a Brute force move to in which enclose in methodically enumerate to the entire probable candidates for the solution and examination whether each candidate satisfies the difficulty report previous to validate a transaction. AI affords us the prospect to go not here from this and deal with errands in a more intelligent and efficient way. To envisage a ML process in which could almost hone its skills in real-time that it can be fed appropriate training data. The blockchain and AI have great potential possess correct to one cannot help but wonder what achieve albeit their joint force be situate to fine use. These expertise's are equally wide-ranging, and might impend to envelop the well-organized in a manner for the huge pact in additional see-through.

9.7   EXPERIMENTAL RESULTS

The experimental result is on queuing model is used the shortest route path for high-level slotting and alternative routes storage allocation of goods. To work out the waiting time for each slot and also find out the shortest route resource needed to allocation of goods. Furthermore, to assess the slotting system recital before actually running the performance tests by applying the queuing hypothesis. To setting up a hypothetical, benchmark next to, which we can run the recital tests to evaluate the slotting system recital relating message infrastructure using blockchain technology. Table 9.1 contains slotting arrangements of each route path for specific time (Figure 9.3).

To plot this, the list of route paths are assigned to one vector namely "Routes S. No." and the corresponding expecting waiting time are assigned to a second vector namely "expected waiting time." In Table 9.1, the vectors are created and the plot used is shown in Figure 9.3.

TABLE 9.1   Route Path for High-Level Slotting and Alternative Routes Storage Allocation of Products

Route No.	Expected Waiting Time
1	−18.21327
2	−17.12134
3	−18.36123
4	−23.18176

TABLE 9.1   *(Continued)*

Route No.	Expected Waiting Time
5	NaN
6	−8.12783
7	−7.12746
8	−6.78143
9	−5.47189
10	−4.86176
11	−3.47815
12	−2.34875
13	−1.45287
14	−0.12576

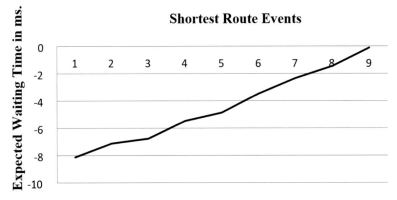

FIGURE 9.3   Route path for high-level slotting and substitute routes: Storage allocation of products.

9.8   CONCLUSION

In conclusion of this chapter, slotting is a tremendously multifaceted replica to work out that drives warehouse efficiencies that counting labor, space, and cycle times. The main point of a slotting set up is to totally notice possible value and house saving. The slotting is to relate in nursing essential a part of the built style and completion. Blockchain can transform supply chains, industries, and ecosystems. Smart contracts will facilitate eliminate pricey delays and waste generated by manual handling of work. From there, new doors may open to faster, more intelligent, and more secure processes and optimized shortest routing decisions using AI.

KEYWORDS

- **artificial intelligence**
- **blockchain**
- **brute force**
- **machine learning**
- **shortest path route**
- **warehouse**

REFERENCES

1. Dilipkumar Khandelwal (2019). https://www.financialexpress.com/industry/technology/artificial-intelligence-blockchain-andmachine-learning-how-companies-can-get-smarter/1558144/ (accessed 20 July 2020).
2. Biryukov, A., & Khovratovich, (2017). Equihash: Asymmetric proof-of-work based on the generalized birthday problem. *Ledger, 2*, 1–30.
3. Borge, M., Kokoris-Kogias, E., & Jovanovic, P., (2017). Proof-of-personhood: Redemocratizing permission less crypto currencies. *Security and Privacy Workshops (Euros & PW), 2017 IEEE* (pp. 23–26). European Symposium on IEEE.
4. Christof, P., Jan, P., & Bart, P., (2010). *Understanding Cryptography: A Textbook for Students and Practitioners* (p. 7). Springer. ISBN: 3-642-04100-0.
5. Crosby, M., Pattanayak, P., & Verma, S., (2016). Blockchain technology: Beyond bitcoin. *Applied Innovation, 2*, 6–10.
6. Dileep, K. K., & Jagan, M. R. N. V., (2014). Optimizing the duplication of cluster data for similarity process. In: *ANU Journal of Physical Science* (Vol. 2). ISSN: 09760954.
7. Dileep, K. K., Jagan, M. R. N. V., & Srinivasa, R. M., (2016). Cluster optimization for similarity process using de-duplication. In: *IJSRD-International Journal for Scientific Research and Development* (Vol. 4, No. 06). ISSN: (online) 2321-0613.
8. Dileep, K. K., Jagan, M. R. N. V., & Vamsidhar, Y., (2012). Similarity based query optimization on map reduce using Euler angle oriented approach. In: *International Journal of Scientific and Engineering Research* (Vol. 3, No. 8). ISSN: 2229-5518.
9. Dziembowski, S., Faust, S., & Kolmogorov, V., (2015). Proofs of space. *Annual Cryptology Conference* (pp. 85–605).
10. Saleh, Fahad, Blockchain Without Waste: Proof-of-Stake (July 7, 2020). *Review of Financial Studies, 34*, March 2021, pp. 1156–1190, Available at SSRN: https://ssrn.com/abstract=3183935 or http://dx.doi.org/10.2139/ssrn.3183935.

11. Gao, Y., & Nobuhara, H., (2017). A proof of stake sharding protocol for scalable blockchains. *Proceedings of the Asia-Pacific Advanced Network* (Vol. 44, pp. 13–16).

12. Gervais, A., Karame, G. O., & Wüst, K., (2016). On the security and performance of proof of work, blockchains. *Proceedings of the 2016 ACM SIGSAC Conference on Computer and Communications Security* (pp. 3–16).

13. Kiayias, A., Russell, A., & David, B., (2017). Ouroboros: A provably secure proof-of-stake blockchain protocol. *Annual International Cryptology Conference* (pp. 357–388).

14. Kim, S., & Kim, J., (2018). POSTER: Mining with proof-of-probability in blockchain. *Proceedings of the 2018 on Asia Conference on Computer and Communications Security* (pp. 841–843).

15. Luong, N. C., Xiong, Z., & Wang, P., (2017). *Optimal Auction for Edge Computing Resource Management in Mobile Blockchain Networks: A Deep Learning Approach.* arXiv preprint arXiv:1711.02844.

16. Mark Burnett, (2007), *Blocking Brute Force Attacks*, UVA Computer Science, University of Virginia (UVA).

17. Zhang, Y., & Wen, J., (2015). An IoT electric business model based on the protocol of bitcoin. *Intelligence in Next Generation Networks (ICIN), 2015 18ᵗʰ International Conference on. IEEE* (pp. 184–191).

CHAPTER 10

Internet of Things (IoT) Based on User Command Analysis and Regulator System

HARSH PRATAP SINGH,[1] R. P. SINGH,[1] RASHMI SINGH,[2] and BHASKAR SINGH[2]

[1]*Sri Satya Sai University of Technology and Medical Sciences, Sehore, Madhya Pradesh, India, E-mail: singhharshpratap@gmail.com (H. P. Singh)*

[2]*Radharaman Institute of Science and Technology, Bhopal, Madhya Pradesh, India*

ABSTRACT

This chapter presents the Internet of things (IoT)-based home automation system which can be regulated and accessed to by the cell phones. This system can perform various activities of home from anyplace of the world. IoT has been marketed as of late; studies are in progress for user-customized services. Appropriately, the services ought to be changed by the attributes of the user instead of the brought together services. Nonetheless, when existing system work naturally, there is an issue of giving a uniform service to all clients without giving customized services. To handle this issue, we propose an IoT-based regulator system for breaking down user directions. For performing such functionality, a functional programming is done through the DTMF Signal is a standard, which permits organizing the information edge originating from the sensors and in this manner regulator of the data. The principle capacity of this undertaking is finished using capacity programming with DTMF signal which permits structuring the data frame originating from the sensors and in this manner the regulate of the data. Through this, it is possible to provide services with improved user convenience and system precision device as a UI. They can speak with home automation system through an Internet gateway, by methods for low power communication protocol like Zigbee, Wi-Fi, and so on. In this propose system, we mainly work on temperature sensor and it is analyzed that our system improves switching time and the operating accuracy of work up to 30%.

10.1 INTRODUCTION

The idea of internet of things (IoT) was presented by the development of the broadly utilized worldwide system known as the Internet alongside the sending of ubiquitous computing and mobiles in smart items which brings new open opportunities for the making of innovative ideas for different parts of life. The idea of IoT makes a system of items that can communicate, associate, and participate together to arrive at a common objective [15]. IoT devices can upgrade our everyday lives, as every device stops going about as a single device and become some portion of a whole full associated system. This furnishes us with the subsequent information to be analyzed for better decision making, following our business, and checking our properties while we are far away from them. The automation is a strategy, method, or system of working or regulating a procedure by electronic devices (EDs) with diminishing social contribution to the base. Basic structure for robotization or automation system for an organization or household is expanding step by step with various advantages. Manufacturers and specialists are attempting to construct productive and reasonableness programmed frameworks to screen and regulate several machines similar to glows, fans, AC dependent as per the need. Automation makes a beneficial similarly as a profitable consumption of the power and water and diminishes a basic bit of the surplus [1]. IoT consents to social orders and kits to be related at whatever point, wherever, with anybody, in a perfect world utilizing any structure and whichever organizations [2]. Automation is alternative basic utilization of IoT headways. This is checking of the essentialness use and the regulating the earth in structures, schools, working conditions, and unquestionable concentrations by utilizing several types of sensors and actuators that regulate lights, temperature, and mugginess. The canny home is known as home automation with the utilization of novel headway, to build the private exercises sensibly useful, charming, secure, and conventional. The home automation system joins fundamental component which are:

1. **UI:** As a screen, PC, or mobile, for example, that can offer solicitations to the regulator system.

2. **Technique for Transmission:** wired communication (model Ethernet) or wireless (radio waves, infrared (IR), Bluetooth, GSM, etc.
3. **Central Regulator:** It is hardware interface that talks with UI by regulating neighborhood organizations.
4. **Electronic Appliances:** AC, light or a radiator, which is great with the transmission manner, and related with the central regulator system.

10.2 HOME AUTOMATION SYSTEM CHARACTERISTICS

As of late, wireless system like remote regulator has turned out to be progressively mainstream in home systems administration. Moreover, in automation frameworks, the businesses of remote advancements give a couple of focal points that couldn't be practiced with the usage of a wired framework figuratively speaking.

1. **Abridged Installation Charges:** Installation charges are fundamentally diminished since no cabling is vital.
2. **Internet Assembly:** Regulator devices from anyplace on the planet with utilize mobiles to regulate brilliant home.
3. **Ascendable and Inflatable:** With the association of wireless, system is particularly valuable as soon as, because of Fresh or transformed prerequisites, an expansion of the system are fundamental.
4. **Security:** Successfully add gadget to make a consolidated clever home security framework and understood security ensures decency of sharp home. Dissected a piece of the primary difficulties looked by home automation system. These combine high movement costs, high congregation costs, high establishment costs, and an extra association and bolster costs and non-appearance of home automation measures, customer curiosity to development, and complex UIs. With the development of time, speedy improvement in advancement and dealing with power which prompts a wide lessening in cost and size of apparatus. These parts have added to the notoriety of electronic device today, so individuals are infrequently again astounded or questionable almost the utilization of the computer, mobiles or tablets. Also the immense measure of home automation shows, communications, and interface rules. The home automation system that utilizations Wi-Fi technology [13]. The application has been made subject to the android computer [3]. An interface card has been made to guarantee communication among the remote client, server, raspberry pi card and the home Appliances.

The application has been introduced on an android Smartphone, a web server, and a raspberry pi card to regulate the screen of windows. Android applications on a smart telephone issue course to raspberry pi card. An interface card has been perceived to resuscitate standard between the actuator sensors and the raspberry pi card. The cloud-based home gadget checking and regulating System, plan, and execute a home passage to accumulate metadata from home machines and transfer to the cloud-set up together information server to accumulate concerning HDFS (Hadoop distributed file system) process them utilizing Map Reduce and usage to offer a watching capacity to remote client [4]. This has finished with Raspberry Pi through investigating the issue of E-mail and figuring. Raspberry Pi shows to be a shocking, cash related and able stage for finishing the astute home automation [5]. Raspberry pi-based home automation is superior to anything additional home automation techniques are a few different means. For instance, in home automation through DTMF (dual-tone multi-frequency) [15], the call commitment is a tremendous trouble, which isn't the situation in their projected technique. Besides, in web server-based home automation, the course of action of web server and the memory space requisite is shot out by this framework, since it basically utilizes the definitively prevailing web server association given by G-mail. The LEDs were utilized to show the exchanging activity. The system is common, effective, and stretchy. Projected Canny Home Monitor and Manager (SHMM), in context on the Zig-Bee, wholly sensors, and actuators are associated by a Zig-Bee remote structure [6]. They composed a basic savvy association, which would remote have the alternative to regulate through Zig-Bee. PC host is utilized as an info gatherer and the expansion distinguishing, all perceiving information are stimulated to the virtual machine (VM) in the cloud. The client can utilize the PC or Android telephone to screen or regulate through the Internet to regulate sparing of the house. Arduino micro-controller to acquire client heading to accomplish through the Ethernet shield. The home system utilized together both remote Zig-Bee and wired X10 impels [12]. This system looked for after unbelievable undertaking orchestrating through experimental for the resource-constrained project scheduling problem (RCPSP). The PDA can be moreover wired to the focal regulator via USB interface or chats with it remotely, inside the level of the home. Arduino comprises the web-server application that goes on via the HTTP appear with web-based Android applications. The structure or system is extraordinarily stretchy and versatile and inflatable. The home system which screens the mechanical congregations and sensors and transfers information to the cloud-based information server which deals with the data and offers associations to clients by transmitting information and getting client direction from versatile application [7]. The proposed system has minded blowing assessed quality and configurability characteristics with incredibly low regulate use in down to earth way. Application made utilizing the Android stage regulated and checked from a remote locale utilizing the sharp home application and an Arduino Ethernet-based downsized scale Web-server [8]. The sensors and actuators or moves are immediate interfaced to the basic regulator. Projected arrangement offers are the regulator of importance directs the structures for example, lightings, warming, cooling, security, fire exposure, and interruption recognizing evidence with alert and email observes. Acquainted structure raspberry Pi with fill in as a communication door between Konnex-Bus (KNX) and cell phone home automation systems [9]. Store the data everything considered and sensors inside a Smart Home, rather than utilizing separate profiles. Guarantees criticalness use could be reduced, showed up contrastingly in connection to a standard work station. Dual tone multi-frequency (DTMF) utilized in phone lines [10]. There are three areas in the structure DTMF recipient and ring locator, I/O interface component, and computer [3]. The interface card has been made to guarantee communication among the remote client, server, raspberry pi card, and the home Appliances. The application has shown on an Android Smartphone, a web server, and a raspberry pi card regulate the screen of Windows. Android application on a Smartphone issue course to the raspberry pi card. An interface card has been seen to enable flag among the actuator sensors and the raspberry pi card.

Cloud-based home machine checking and regulating system. Plan and execute a home method to hoard meta-data from home machines and direct to the cloud-set up together data server to accumulate concerning HDFS, develop them using Map Reduce and utilize to give a watching ability to Remote customer [4]. It has been done with raspberry pi via taking a glimpse at the matter of Electronic mail and figuring. Raspberry pi illustrates to be an astounding, budgetary, and capable stage for understanding the vigilant home computerization [5]. Raspberry pi based home robotization is improved than anything other home automation methods of reasoning is a few unique ways. For example, in home automation through DTMF (twofold tone multi-go over)

[2], the call responsibility is a huge issue, which isn't the circumstance in their projected methodology. Additionally, in Web-server based home automation, the course of action of web-server and the memory space requisite is shot out by this methodology, since it in a general sense usage the convincingly prevailing web server affiliation given by G-mail. LEDs were used to show the trading action. Framework is instinctive, sensible, and adaptable [6]. Projected canny house monitor and manager (CHMM), in context on the Zig-Bee, all sensors, and actuators are connected by a Zig-Bee remote structure. They sorted out a fundamental savvy association, which would remote have the alternative to regulate through Zig-Bee. Computer host is utilized as an information gatherer and the advancement recognizing, all distinguishing information are relocated to the VM in the cloud. The client can utilize the PC or Android telephone to screen or regulate through the Internet to regulate sparing of the house. Arduino micro-controller to get client headings to execute through an Ethernet shield. The home system utilized together both remote Zig-Bee and wired X10 advances [12]. This structure looked for after amazing undertaking, master-minding through exploratory for the Resource-obliged booking issue (RCPSP). The cell phone can be also wired to the focal regulator through USB interface or chats with it remotely, inside the level of the home. Arduino contains the web-server application that goes on through the HTTP appears with web-based android application. The system is phenomenally adaptable and adaptable and inflatable. Home system which screens the mechanical gatherings and sensors and transfers information to the cloud-based information server which agrees with the data and offers associations to clients by transmitting information and getting client orientation from adaptable application [7]. The projected structure has unimaginable assessed quality and configurability attributes with incredibly low regulate utilization in monetarily keen manner. Application made utilizing the Android stage regulated and checked from a remote area utilizing the sharp home application and an Arduino Ethernet-based downsized scale web-server [8]. The actuators or moves and sensors are plainly interfaced to the fundamental regulator. Projected arrangement offers are the regulate of vitality the authorities structures, for example, lightings, warming, cooling, security, fire disclosure and impedance recognizing confirmation with alert and email observes. Acquainted system Raspberry Pi with fill in as a communication passage among PDAs and Konnex-Bus (KNX) home automation structures [9]. Store the data everything considered and sensors inside a Smart Home, rather than utilizing separate profiles. Guarantees centrality use could be reduced, separated from a standard work station. Dual-tone multi-frequency (DTMF) utilized in phone lines [10]. There are three parts in the structure DTMF recipient and ring locator, IO interface unit, and PC.

10.3   INTERNET OF THINGS (IOT) APPLICATIONS

IoT is developing; it is venturing into each part of our lives. This prompts a simpler life through more extensive scope of uses, for example, electronic health care solutions [14] and Smart city idea. The idea of Smart city intends to utilizing resources, expanding service quality overfed to the residents, and reducing the cost of public administrations [15]. Another application is home automation which is the focal point of this project (Figure 10.1).

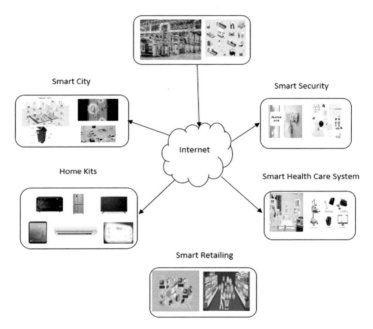

FIGURE 10.1   Different applications of internet of things in all aspects of life.

The IoT is relied upon to be the next revolution following the World Wide Web. It will give new bridge between reality and the virtual world. The internet will never again be simply a system of "human cerebrum," however; will coordinate real object, sensors, and physical exercises. The IoT is characterized by ITU and IERC as a unique overall framework system with self-orchestrating capacities subject to ordinary and inter-operable communication proto-cols where the physical and virtual "things" have characters, attributes, and virtual characters, use wise interfaces and are reliably consolidated into the information arrange. In the course of the most recent year, IoT has moved from being a cutting edge vision-with at times a specific level of promotion to an expanding market reality. The EU has as of now for quite a while put resources into supporting examination and advancement in the field of IoT, eminently in the regions of installed frameworks and digital physical frameworks.

10.4 PROPOSED SYSTEM

It is situated to the realization of a wireless sensor network (WSN), as a major aspect of the execution ought to be viewed as the right platform, perhaps the best service coupled to the IoT stage is the usage of this service will permit giving data the board in a strong way. To do this, work writing function program is done through the DTMF signal is a standard, which permits structuring the data frame originating from the sensors and along these lines the regulator of the data. The advantages of DTMS is that it can be regulate from anywhere, decreases the wastage of electricity and its cost is very less compared to others like GSM. This in a context enabled for the consequent execution of visual interfaces in which the data can be introduced in an all the more methodical way. This system does not use any microcontroller and it is based on dial tone. This system is implemented on temperature sensor, fire sensor, water sensor, moisture sensor, LDR sensor, device switching and for home security system. Motion sensor is an IR human location sensor, which, rather than factory automation sensors that are utilized with manufacturing factory hardware, is intended to be fused into different devices that exist around us in everyday life. Temperature sensor change from straightforward ON/OFF thermostatic devices which regulate a household hot water warming system to exceptionally delicate semiconductor types that can regulate complex procedure regulate heater plants. We recall from our school science classes that the development of projects and molecules produces heat (dynamic energy) and the more prominent the development, the more heat that is created. Temperature sensors measure the measure of heat energy or even chilliness that is created by an object or system, enabling us to "sense" or identify any physical change to that temperature delivering either a simple or advanced yield. Water level sensor, the drinking water emergency in Asia, is arriving at disturbing extents. It may very before long accomplish the idea of worldwide emergency. Thus, it is of most extreme significance to save water for individuals. In the numerous households nearby un-necessary wastage of water because of flood in over-head tanks. Modified water level indicator and regulator can offer a reaction for this issue. The development of water level regulator works upon the way in which that water practices regulate because of the vicinity of minerals inside it.

Therefore, water can be utilized to open or close a circuit. For instance, the water level addition or decreases, various circuits in the regulator send diverse sign. These signs are utilized to execute ON or turn the motor direct according to our necessities. The total aggregate of water open on Earth has been reviewed at 1.4 billion cubic kilometers; adequate to cover the planet with a layer of around 3 km. Approximately 95% of the Earth's water is in the seas, which is unfit for people use. Approximately, 4% is confirmed in the polar ice tops, and the rest 1% incorporates all new water found in conductors, streams, and lakes which is reasonable for our utilization. The assessment assessed that an individual in India eats up on a conventional of 140 liters for reliably. This utilization would move by 40% constantly 2025. This clues the need to guarantee our new water assets. Soil dampness sensors have been used for quite a while to evaluate how much water is held between the dirt particles. This is no straightforward undertaking in light of the fact that not all dirts are made identical, way off the mark. (For more information about soil characteristics, read not all soils are created equally underneath). Many sensors accessible use an assortment of conductive or capacitive innovation. Benchmark's licensed soil dampness sensors use TDT (time space transmission) innovation to give sensors that are significantly increasingly sensitive (which means they are exceedingly open to changes in dampness levels) than officially available sensors—they truly change the standards of the game. TDT sensors are the most excellent sensors open similarly as the most exact and touchy soil sensors accessible. Gauge's protected structure isn't only touchy to little changes in dampness content, they are in like manner intense and trustworthy, less complex to present (over new or existing wire), and they are sans upkeep and viable. A light dependent resistor (LDR) is also called a photograph resistor or a cadmium sulfide (CdS) cell. It is additionally called a photoconductor. It is on a very basic level a photocell that manages the standard of photoconductivity. The uninvolved segment is in a general sense a resistor whose limitation worth reductions when the power of light rots. This optoelectronic gadget is commonly utilized in light fluctuating sensor circuit and light and reduces initiated exchanging circuits. A piece of its applications unite camera light meters, road lights, clock radios, light column alert, smart smoke cautions, and outside timekeepers.

10.5   EXPERIMENTAL RESULTS

To implement our home automation system we have structure a test arrangement as appeared in Figure 10.2, where we utilized Arduino Uno as a primary regulating unit and four channels transfer board to regulate electrical home apparatus. Furthermore, we have incorporated a Wi-Fi module in our framework to interface android and neighborhood Wi-Fi present in the home of client. We have tried it on the different sensor such as temperature sensor, LDR sensor, fire sensor, water sensor, etc. The experimental setup for several loads which are presented in Table 10.1. The analysis of proposed work is done using motion sensor, moisture sensor, temperature sensor, LDR sensor, and water level sensor. The threshold value is set for different sensor which is mentioned in Table 10.1 and analysis of these is shown in Figure 10.3 and it is found that it enhances the switching time and accuracy with different operating mode.

TABLE 10.1   IoT-Based Input Sensor

Sensor	Operating Voltage	Operating Current	Threshold Voltage
Motion Sensor	5 V	0.1 MA	3 V
Temperature Sensor	5 V	0.4 MA	2.3 V
Water Level Sensor	12 V	3 MA	5.6 V
Moisture Sensor	12 V	1.2 MA	7.5 V
LDR Sensor	12 V	6 MA	1.2 V

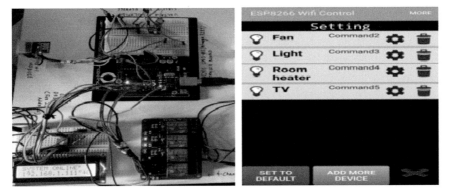

FIGURE 10.2   Experimental setup.

10.6   CONCLUSIONS

As of late, IoT has been utilized in different fields. The smart device can be associated with the system and communication can be performed to work the device or to give the sensor information to the user. Likewise, techniques for giving services in different ways are being considered. Existing system give just a similar service based on the set threshold and activity of the device operation judgment is performed by the earth inside the home as opposed to the user. Additionally, the user cannot regulate the device. To solve this issue, this chapter collects and investigates user's remote regulate command. In view of this, propose an automatic regulator system and analysis is done by utilizing Arduino Uno. The experimental analysis is done among various sensor utilizing operating voltage, operating current and threshold voltage. We mainly work on temperature sensor and for this the threshold voltage is set at the 2.3 V and operating voltage is at the 5 V, if the temperature increased from its threshold the sensor becomes ON and vice versa. After analysis, it is found that the proposed system enhances the switching time and accuracy of work up to 30% than the existing system.

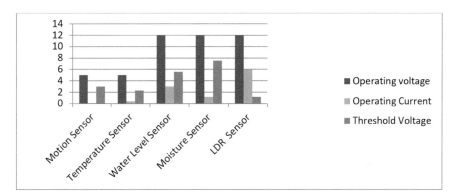

FIGURE 10.3   Graphical representation of IoT-based input sensor.

KEYWORDS

- **data analysis**
- **dual tone multi frequency**
- **Hadoop distributed file system**
- **internet of things**
- **light dependent resistor**
- **smart home**
- **time space transmission**

REFERENCES

1. Rima, H. E. A., Azman, S., & Aln, S., (2013). Provably lightweight RFID mutual authentication protocol. *International Journal of Security and its Applications, 7*(4).
2. Aarti, R. J., & Karishma, K., (2017). Environmental monitoring using wireless sensor networks (WSN) based on IoT. *International Research Journal of Engineering and Technology (IRJET), 4*(1). e-ISSN: 2395-0056.
3. Joeongyeup, et al., (2017). *Embedded IoT System: Network, Plateform, and Software*. Hindwai, Mobile Information System.
4. Raghavender, R. Y., (2017). Automatic smart parking system using internet of things (IoT). *International Journal of Engineering Technology Science and Research, 4*(5). IJETSR, ISSN: 2394–3386.
5. Beny, N., (2017). Analysis of power consumption efficiency on various IoT and cloud-based wireless health monitoring systems. *I. J. Information Technology and Computer Science, 5*, 31–39. Published Online.
6. Sangeetha, C., Gomathi, R., & Anne, R. T., (2018). An IoT based appliance regulate and monitoring system for smart homes. *International Journal of Advanced Information in Engineering Technology (IJAIET)*, 5(5).
7. Salekul, I., (2014). Security analysis of LMAP using AVISPA. *International Journal of Security and Networks, 9*(1), 30–39.
8. Sachin, K. K., (2014). Home appliances regulate system based on android smartphone. *IOSR Journal of Electronics and Communication Engineering (IOSR-JECE), 9*(3), Ver. III, pp. 67–72. e-ISSN: 2278-2834, p-ISSN: 2278-8735.
9. Gurav, & Patil, (2016). IoT-based interactive regulating and monitoring system for home automation. *International Journal of Advanced Research in Computer Engineering and Technology (IJARCET), 5*(9).
10. Rolf, H. W., (2010). Internet of things new security and privacy challenges computer law and security review. *Elsevier Journal, 26*(1), 23–30.
11. Huma, J., Abhilasha, L., & Shashank, S., (2017). IoT based auto irrigation system using soil moisture sensor. *International Journal of Scientific Research in Computer Science, Engineering, and Information Technology*, IJSRCSEIT, 2(3). ISSN: 2456–3307, 2017.
12. Inshik, K., Kwanghee, C., & Hoekyung, J., (2018). IoT automatic regulate system based on user command analysis. *International Journal of Applied Engineering Research*, 13(6), 3369–3372. ISSN: 0973-4562.
13. Rodge, P. R., & Jaykant, P., (2017). IoT based smart interactive office automation. *International Research Journal of Engineering and Technology (IRJET)*, 4(4). e-ISSN: 2395-0056.
14. Reshma, S., & Ritika, P., (2017). Need for wireless fire detection systems using IoT. *International Research Journal of Engineering and Technology (IRJET),* 04(01)*. e-ISSN: 2395-0056 , 4(1).

CHAPTER 11

IoT System with Heterogeneous Servers: Steady-State Solution and Performance Evaluation

P. K. SWAIN and P. K. PANDA

Department of Computer Application, North Orissa University, Baripada – 757003, Odisha, India, E-mails: prasantanou@gmail.com (P. K. Swain), panda3k32003@gmail.com (P. K. Panda)

ABSTRACT

In this chapter, we have analyzed the steady-state queuing system with infinite buffer. Here three numbers of heterogeneous parallel servers are used for an IoT system with different service rates. The arrival of incoming requests is assumed to be with mean rate λ and in Poisson distribution whereas service rates are with different mean values and exponentially distributed. There are three independent servers in the queue and incoming requests are assumed to be waiting in the queue if the servers ware found to be busy. If there is no unit in the queue, it is assumed the next unit to arrive always joins the first server. This model can be used when first come first serve (FCFS) delivery is required. One example of this situation is an IoT-based health care system with servers having different speeds are serving the arrival requests.

11.1 INTRODUCTION

With increased demands for mobile device assisted lifestyle, ubiquitous computing evolved to the next step called internet of things (IoT) where heterogeneous devices share a common platform to communicate with each other. Here internet is the common platform based on TCP/IP protocol stack

[1]. The devices used in home network like closed camera television (CCTV), television, washing machine, air conditioner, etc., are the basic nodes of IoT for data communication and this need a effective buffer management and process scheduling to built a successful ubiquitous network.

Data generated in IoT are of different formats and voluminous in nature which need to be served quickly for a faster decision. Hence, a multiple processor system with adequate buffer and proper scheduling mechanism is needed to handle the network efficiently. Queues, or waiting line in which jobs or customers arrive, wait for service, and leave the system after enjoying the service are part of IoT system. Such queuing phenomenon are observed in real life, in many applications such as congestion control in network, devices with different speed attached to the endpoints of telecommunication system, network supporting communications channels with different rate of transmission, various processes in production environment and manufacturing processes, etc., [2, 3]. Study on multi-server queuing systems usually assumes that the service rates of the servers are same but from practical aspects in modeling real system, it has been noticed that the individual service rate are different for each server in the systems [4].

A lot of work has been done on two heterogeneous servers [5–10] but regarding implementation of a three heterogeneous server with a finite buffer system is not yet addressed with respect to an IoT system. Two server systems with infinite queue have been represented by [5] where requests are arriving to the system with Poisson process and service is done in bulk with exponential distribution. Here the channel busy period is calculated for one channel busy or both cannels are busy along with the queue length. Aloisio, Junior, and Anzaloni [11] proposed a system with three heterogeneous servers but without any buffer capacity. Packets arriving to the system with mean rate λ and based on Poisson distribution. The service times of servers are different and exponentially distributed.

This chapter deals with IoT system having three heterogeneous server points and uses an M/M/3/∞ queuing system, where processing speed is heterogeneous. This model can be used when first come first serve (FCFS) delivery is required. The service rates of three servers are exponentially distributed and mean rates are μ_1, μ_2 and μ_3. We assume that $\mu_1 > \mu_2 > \mu_3$ without loss of generality. It is assumed that the request arrival process is exponentially distributed with rate λ. Here, we proposed a recursive method to calculate the steady state probability distribution of requests in the system. Some performance measures and numerical results having presented as graphs. The maximum size of the buffer is infinity. If there are no requests in the queue, it is assumed that the next request arrives in the proposed system, will serve by the first server.

11.2   MODEL DESCRIPTION AND ANALYSIS

In this chapter, we proposed an IoT system with three servers with M/M/3/∞ queuing system, where servers are giving service at FCFS accordingly. The service rate of three independent servers are with means μ_1, μ_2 and μ_3 and $\mu_1 > \mu_2 > \mu_3$ without loss of generality. When the servers were busy, the incoming requests would wait in the queue. If there is no unit in the queue, it is assumed the next unit to arrive always joins the first server. Here the incoming requests obey Poisson with rate λ and system has a single queue with infinite capacity.

The state probabilities of three servers with an infinite queue can be represented with following terminology.

- **State Probabilities with Empty Queue:**

    $P_{0,0,0}(t) \rightarrow$ Probability of all the three servers are idle at time t.
    $P_{1,0,0}(t) \rightarrow$ Probability that the server 1 is busy and other two servers are idle.
    $P_{0,1,0}(t) \rightarrow$ Probability that the server 2 is busy and server 1 and server 3 are idle.
    $P_{0,0,1}(t) \rightarrow$ Probability that server 3 is busy and the server 1, server 2 are idle.
    $P_{1,1,0}(t) \rightarrow$ Probability that server 1 and server 2 are busy and server 3 is idle.
    $P_{1,0,1}(t) \rightarrow$ Probability that server 1 and server 3 is busy and the server 2 is idle
    $P_{0,1,1}(t) \rightarrow$ Probability that server 2 and server 3 is busy and the server 1 are idle.
    $P_{0,3}(t) \rightarrow$ Probability that all servers are busy and no request in the queue.

- **State Probabilities with Requests in the Queue:**

    $P_{n,3}(t) \rightarrow$ Probability that all the server are busy and $n \geq 0$ customer are waiting in the queue at time t

For a transition of two consecutive arrivals from time period t to (t+1) the relation among states in terms of steady state probability equations represented as:

$$\lambda P_{0,0,0} = \mu_1 P_{1,0,0} + \mu_2 P_{0,1,0} + \mu_3 P_{0,0,1}, \tag{1}$$

$$(\lambda + \mu_1)P_{1,0,0} = \lambda P_{0,0,0} + \mu_2 P_{1,1,0} + \mu_3 P_{1,0,1,} \tag{2}$$

$$(\lambda + \mu_2)P_{0,1,0} = \mu_1 P_{1,1,0} + \mu_3 P_{0,1,1,} \tag{3}$$

$$(\lambda + \mu_3)P_{0,0,1} = \mu_2 P_{0,1,1} + \mu_1 P_{1,0,1,} \tag{4}$$

$$(\lambda + \mu_1 + \mu_2)P_{1,1,0} = \lambda(P_{1,0,0} + P_{0,1,0}) + \mu_3 P_{0,3,} \tag{5}$$

$$(\lambda + \mu_2 + \mu_3)P_{0,1,1} = \mu_1 P_{0,3,} \tag{6}$$

$$(\lambda + \mu_1 + \mu_3)P_{1,0,1} = \mu_2 P_{0,3,} \tag{7}$$

$$(\lambda + \mu_1 + \mu_2 + \mu_3)P_{0,3} = \lambda P_{0,1,1} + P_{1,1,0} + P_{1,0,1}) + (\mu_1 + \mu_2 + \mu_3)rP_{1,3,} \tag{8}$$

$$(\lambda + \mu_1 + \mu_2 + \mu_3)P_{n,3} = \lambda(P_{n-1,3}) + (\mu_1 + \mu_2 + \mu_3)P_{n+1,3}, n \geq 1. \tag{9}$$

Simplifying equations [6] and [7], we get:

$$P_{0,1,1} = (\mu_1 P_{0,3})/(\lambda + \mu_2 + \mu_3) = \mu_1 M_1 P_{0,3,} \tag{10}$$

$$P_{1,0,1} = (\mu_2 P_{0,3})/(\lambda + \mu_1 + \mu_3) = \mu_1 M_2 P0_{,3,} \tag{11}$$

where; $M_1 = 1/(\lambda + \mu_2 + \mu_3)_{,3,}$ $M_2 = 1/(\lambda + \mu_1 + \mu_3)$.

Now using equations [10] and [11] in [4] and [8], respectively, we get:

$$P_{0,0,1} = \mu_1 \mu_2/(\lambda + \mu_3)[M_1 + M_2]P_{0,3,} \tag{12}$$

$$P_{1,1,0} = ((1/r) - \mu_1 M_1 - \mu_2 M_2)P_{0,3,} \tag{13}$$

$$P_{0,1,0} = \mu_1/(\lambda + \mu_2)r)[1 - \mu_2 r M_2 - r M_1(\mu_1 - \mu_3) \mu_1 P_{0,3,} \tag{14}$$

$$P_{1,0,0} = M_3 P_{0,3,} \tag{15}$$

$$P_{0,0,0} = M_4 P_{0,3,} \tag{16}$$

where,

$$M_3 = \frac{1}{\lambda}(\lambda + \mu_1 + \mu_2)\left(\frac{1}{r} - \mu_1 M_1 + \mu_2 M_2\right) - \frac{\lambda\mu_1}{r(\lambda + \mu_2)}(1 - r\mu_2 M_2 - rM_1(\mu_1 + \mu_2)) - \mu_3,$$

$$M_4 = \frac{1}{\lambda}M_3(\lambda + \mu_1) - \mu_2\left(\frac{1}{r} - \mu_1 M_1 + \mu_2 M_2\right) - (\mu_3\mu_2)M_2 \text{ and } r = \frac{\lambda}{\mu_1\mu_2\mu_3}$$

In equation [9], let us define a displacement operator E as $E^j P_{n,3} = Pn + j, 3$.

The differential equation [9] can be simplified as:

$$\left[\lambda E^{-1} - \left(\lambda + \sum_{j=1}^{3} \mu_j\right) + \sum_{n=0}^{3} \mu_j\right] P_{n,3} = 0 \qquad (17)$$

The characteristic equation associated with [17] after simplification reduces to:

$$h(r) = (\lambda - (\mu_1 + \mu_2 + \mu_3)r = 0.$$

Hence root $r = \lambda/(\mu_1 + \mu_2 + \mu_3)$

$$So, Pn, 3 = r^n P0, 3.$$

Using normalizing condition,

$$P_{0,0,0} + P_{1,0,0} + P_{0,1,0} + P_{0,0,1} + P_{1,1,0} + P_{0,1,1} + P_{1,0,1} + P_{n,3} = 1, \qquad (18)$$

Solving equation [18] we get,

$$P_{n,3} = r^n \left[\left(\frac{1}{r(1-r)} + M_3 + M_4\right) + \left(\frac{\mu_1\mu_2(M_1 + M_2)}{\lambda + \mu_3}\right) + \frac{\mu_1}{r(\lambda + \mu_2)}(1 - r\mu_1 M_2 - rM_1(\mu_1 - \mu_3))\right]^{-1}$$

11.3 PERFORMANCE MEASURE

Performance matrices are the important parameters which measures the efficiency of the proposed queuing model based IoT system. Once the state probabilities are solved the performance matrices such as average number of request in the queue (Lq), average waiting time in the queue (Wq), and given by:

$$L_q = \sum_{n=1}^{\infty} n P_{n,3}$$

Or we can write:

$$L_q = \frac{r}{(1-r)^2} P_{0,3}$$

By using Little's rule we can calculate the average waiting time in the queue (Wq) as:

$$Wq = Lq/\lambda,$$

$$Wq = r/\lambda (1-r)^2 \, P_{0,3}$$

11.4   NUMERICAL RESULTS

In this section, the performance results are represented in terms of graphs. In Figure 11.1, we represent the effect of the service rate μ_1 on the average queue length L_q for different μ_2. Here the parameters taken are $\lambda = 2$, $\mu_3 = 0.1$. From the figure, it can be depicted that with increase of μ_1 the queue length L_q decreases for all value of μ_2. It may be observed, for a larger μ_2 the average queue length L_q is more efficient.

Figure 11.2 represents the behavior of mean waiting time W_q with changing service rate μ_1 and μ_2. Here the parameters are taken to be $\lambda = 2$, $\mu_3 = 0.1$. W can observe that W_q decrease monotonically when the values service rate μ_1 increases. Also for larger service rate μ_1, values for Wq converge and will be efficient with the increase values of μ_2.

In Figure 11.3, we show the effect of average queue length Lq on utilization factor ρ. The system parameters are $\lambda = 2$, $\mu_2 = 0.45$. It can be observed that with increase of utilization factor ρ, average queue length increases exponentially.

Figure 11.4 shows the effect of the service utilization factor ρ on mean waiting time Wq. It can be observed that as service utilization factor ρ increases the value of Wq increase exponentially.

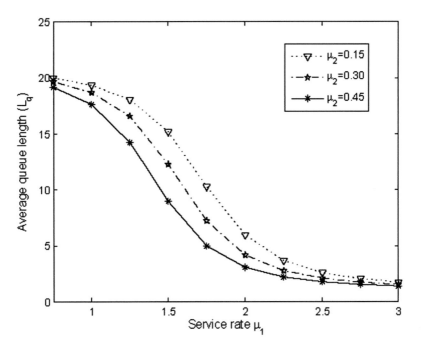

FIGURE 11.1   Impact of μ_1 on Lq for different μ_2.

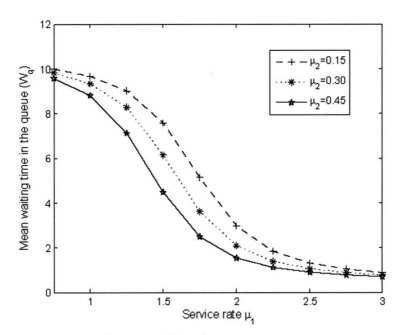

FIGURE 11.2   Impact of μ_1 on Wq with varying μ_2.

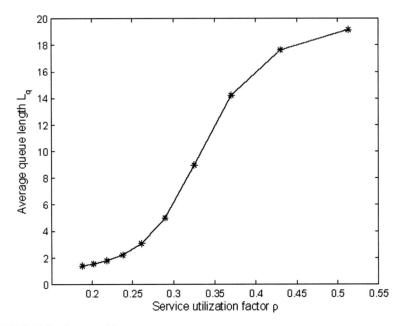

FIGURE 11.3   Impact of Lq on ρ.

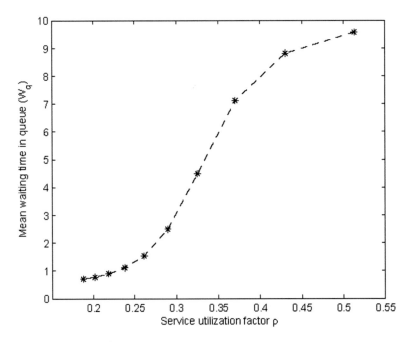

FIGURE 11.4   Impact of Wq on ρ.

11.5 CONCLUSIONS

This chapter represents performance analysis of an IoT system using continuous-time M/M/3/∞queuing system with a heterogeneous server. We have used the displacement operator method; steady-state probabilities are developed and solved by recursive calculation. The performance matrices in terms of graphs are presented to explain the efficiency of the system. This model can be extended to M/M$^{[x]}$/3/N with heterogeneous server in future to handle voluminous IoT data.

KEYWORDS

- **closed camera television**
- **displacement operator**
- **first come first serve**
- **internet of things**
- **Poisson distribution**
- **queuing phenomenon**

REFERENCES

1. Mahamure, S., Railkar, P. N., & Mahalle, P. N., (2008). Communication protocol and queuing theory-based modeling for the internet of things. *Journal of ICT, 3*, 157–176.
2. Gross, D., Shortle, J. F., Thompson, J. M., & Harris, C. M., (2008). *Fundamentals of Queuing Theory*. John Wiley and Sons, New York.
3. Kleinrock, L., (1975). *Queuing Systems: Theory* (Vol. 1). Wiley, New York.
4. Trivedi, K. S., (2000). *Probability and Statistics with Reliability, Queuing, and Computer Science Applications*. PHI, New Delhi.
5. Arora, K. L., (1962). Time dependent solution of two server queue fed by general arrival and exponential server time distribution. *Operation Research, 8*, 327–334.
6. Arora, K. L., (1964). Two server bulk service queuing process. *Operations Research, 12*, 286–294.
7. Krishna, K. B., Pavai, M. S., & Venkatakrishnan, K. S., (2007). Transient solution of an M/M/2 queue with heterogeneous servers subject to catastrophes. *Information and Management Sciences, 18*(1), 63–80.
8. Lin, W., & Kumar, P. R., (1984). Optimal control of a queuing system with two heterogeneous servers. *IEEE Transactions on Automatic Control, 29*(8), 696–703.

9. Satty, T. Y., (1960). Time dependent solution of many servers Poisson queue. *Operations Research, 8*, 755–772.
10. Larsen, R. L., & Agrawala, A. K., (1983). Control of a heterogeneous two server exponential queuing system. *IEEE Transactions on Software Engineering, 9*(4), 522–526.
11. Aloisio, C., Junior, N., & Anzaloni, A., (1988). *Analysis of a Queuing System with Three Heterogeneous Servers* (pp. 1484–1488). IEEE.

CHAPTER 12

IoT-Based Unified Approach to Predict Particulate Matter Pollution in Thailand

FERDIN JOE JOHN JOSEPH

Faculty of Information Technology, Thai-Nichi Institute of Technology, Bangkok, Thailand, E-mail: ferdin@tni.ac.th

ABSTRACT

Particulate Matter pollution (PM2.5) is creating a huge impact in the environment over the past few years. It is responsible for smog and disrupting flights and many business operations. Apart from this business loss, it is also responsible for deadly health hazards in people exposed to PM2.5. This paper addresses the implementation of a system which combines IoT and data analytics. A unified approach in this combination is done to extract data and predict the possible emission of this pollution in the future and to help prevent people getting affected. A support vector regression based method is proposed to predict the air quality index of PM2.5 in the future. The experimental results show that the accuracy and the correlation coefficient of test data taken to be in the positive correlation trend when compared to the existing methodology.

12.1 INTRODUCTION

Particulate Matter (PM) is the most challenging pollution [1] in many developing Asian countries like India, China, Thailand, Vietnam, Malaysia, Indonesia etc. It is a type of pollution arising from industrial effluents, agricultural waste disposal using fire and automobiles. The carbon particles responsible to this pollution less than 2.5 microns in diameter. This micron level carbon particles are responsible for various cardiovascular and respiratory diseases [2]. The rate of propagation of this pollutant is becoming a menace in Thailand over the past few years [11]. There has been so many applications developed to monitor the concentrations of PM2.5 in the past from various countries like Vietnam [3], Chile [2], Thailand [4] and Mexico [1]. Apart from that some studies are done to explore PM2.5 exclusively in Thailand's Chiang Mai [5] and Bangkok [6]. The existing methodologies mentioned above are using data extracted from various sources and it may have different factors affecting the performance. In [7] hidden semi Markov model to predict PM2.5 concentrations in Chicago. Regression and multiple layer neural networks were used in [2] and multiple regression methods were used in [3]. Unsupervised learning methods are used in [1]. Recently an empirical study on the features taken to predict PM2.5 concentration was done by [6] using linear and support vector regression.

From the empirical analysis of features taken, Ref. [6] found that certain features corresponding to humidity and carbon monoxide are having more dominance over other features. These dominant features were obtained by using support vector regression. Still the correlation of data predicted against actual is having a wide gap. In order to rectify this issue, a unified approach is needed to collect data from same location (Figure 12.1).

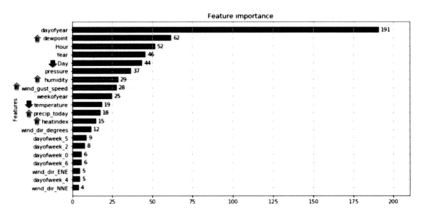

FIGURE 12.1   Feature importance calculated and recommended from [6]. Illustration to show the increase and decrease of feature importance.

An IoT based framework was developed by [8] and a raspberry pi based technique was developed by [9]. There are many Labview based simulations on climate monitoring system like Thilakam et al. [10], Math & Dharwadkar [11], and Shah & Mishra [12]. Microcontroller based approaches done include Alif et al. [13], Strigaro & Cannata [14] and Siva, Ram, & Gupta [15]. As a result of this requirement, an IoT based approach was developed in [16]. The implementation done in the latter uses raspberry Pi. The system developed has SDS011 to calculate PM2.5 and PM10 on an hourly basis, DHT11 to obtain temperature and humidity. Features like relative humidity and preceipitation rate are obtained from the figures obtained from the sensors mentioned above. This paper explores the dominant features identified in [16] and improvise the IoT based methodology in order to obtain a closer predicted values of PM2.5.

12.2   PROPOSED METHODOLOGY

The features obtained using the system of John [16] is in the format of sparse data and are processed using John et al. [17]. There is a need for measuring carbon monoxide to get the features strong enough to get a close correlation. A smoke sensor is used for this purpose. MQ9 sensor is used and it has capacity to record carbon monoxide concentration in parts per million. This sensor is an analog one and it needs to be converted to digital to be feasible enough to be received by Raspberry Pi. There are so many systems to convert analog data signals to digital data from sensors. But in the proposed method Arduino Nano is used. Arduino Nano has analog input pins which are connected to MQ9 sensor and by using the analog to digital conversion module in the microcontroller, output of MQ9 sensor is converted to digital. The digital output obtained is connected to Raspberry Pi through USB interfacing. This modification in [16] is done to make a unified system to collect all the proper features from the designated sensors. Source code used in the proposed methodology is taken from the github repository developed for [16]. The IoT architecture is provided in the Figure 12.2.

The data obtained from the framework in Figure 12.2 is stored in flat files and it is sparse. This sparse database is processed using hover representation [17] and data cleaning is done. Based on the dominance of features, weight is designated and fed to support vector regression. This support vector regression uses radial basis function as kernel and the data set is split into 70% training and 30% testing. The performance obtained using this proposed methodology is given in the results and discussion section.

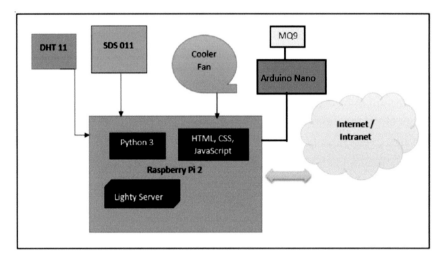

FIGURE 12.2   Framework used in the proposed methodology.

12.3   RESULTS AND DISCUSSION

The proposed framework is implemented and the data is acquired over a period of 6 months. The data is acquired once in an hour using cron tab scheduling in Raspberry Pi. The device is located at 13.737348°N and 100.626023°E at the Data Science and Analytics Lab with no control over the environment. The base system [17] developed uses json format to display it in the front end and database is maintained in the backend as .csv file. This data is currently available in the intranet and it has possibilities to store in an online database to get it accessed from any remote location. The data collected is taken for experimentation to find the relevancy of features and the closeness of correlation with data between the actual and predicted values. The implementation is done using Jupyter notebook with pandas library. The support vector regression with radial basis function is used as a kernel and weights are applied to features like dew point, humidity, relative humidity, precipitation and heat index. Support vector regression is chosen over linear or logistic regression based on the dominance of features obtained by support vector regression as shown in [6]. Data collected is shown in graphical format in the Figure 12.3.

The regression data obtained accuracy got more than [6] and the correlation coefficient is much better. This portrays the justification of the fact reported by the dominance of features and its experimentation. The features extracted are taken in a correlated subspace and normalized using HoVer representation. This reduced underfitting of the curves between actual and predicted values. The correlation coefficient tends towards the positive correlation in the proposed methodology and is clearly evident from the Table 12.1. The existing methods listed in the introduction are not feasible enough to compare with those in the Table 12.1. This is due to various factors like experimental setup, features selected and the external environment setup.

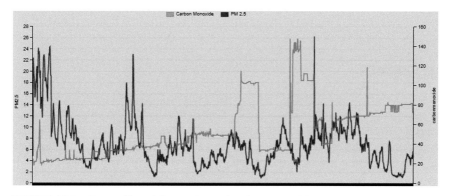

FIGURE 12.3   Data collected for PM 2.5 and carbon monoxide.

TABLE 12.1   Performance Evaluation of the Proposed Methodology

Sl. No	Methodology	Accuracy	Correlation
1.	John [6]	68%	−0.435
2.	Proposed Methodology	75.18%	+0.316

12.4   CONCLUSION

Thus it is clearly evident that the weather monitoring data from a unified setup gives better performance than those collected from various sources. The selection of features also increased the accuracy of the methodology and the correlation coefficient of the magnitude in predicted data against actual data is found to be much better. However, this data has to be checked during the next occurrence of PM2.5 pollutions by the end of 2019 and the beginning of 2020. This approach provided interesting insights in the prediction of PM2.5 values and the Air Quality Index.

KEYWORDS

- **air quality index**
- **carbon monoxide**
- **HoVer representation**
- **internet of things**
- **particulate matter**
- **raspberry pi**

REFERENCES

1. Alif, Y., Utama, K., Widianto, Y., Hari, Y., & Habiburrahman, M. (2019). Design of Weather Monitoring Sensors and Soil Humidity in Agriculture Using Internet of Things (IoT). *Transactions on Machine Intelligence and Artificial Intelligence*, *7*(1), 10–20.
2. Dong, M., Yang, D., Kuang, Y., He, D., Erdal, S., & Kenski, D. (2009). Expert Systems with Applications PM 2.5 concentration prediction using hidden semi-Markov model-based times series data mining. *Expert Systems with Applications*, *36*(2009), 9046–9055.
3. Hien, P. D., Bac, V. T., Tham, H. C., Nhan, D. D., & Vinh, L. D. (2002). Influence of meteorological conditions on PM 2.5 and PM 2.5 x 10 concentrations during the monsoon season in Hanoi , Vietnam. *Atmospheric Environment*, *36*, 3473–3484.
4. JOSEPH, F. J. J. IoT Based Unified Approach To Predict Particulate Matter Pollution In Thailand. In *5th Thai-Nichi Institute of Technology Academic Conference TNIAC 2019* (pp. 385–388).
5. Joseph, F. J. J. (2019). IoT Based Weather Monitoring System for Effective Analytics. *International Journal of Engineering and Advanced Technology, 8*(4), 311–315.
6. Joe, J. F., Ravi, T., Natarajan, A., & Kumar, S. P. (2010, July). Object recognition of Leukemia affected cells using DCC and IFS. In 2010 Second International conference on Computing, Communication and Networking Technologies (pp. 1–6). IEEE.
7. Kanabkaew, T. (2013). Prediction of Hourly Particulate Matter Concentrations in Chiangmai, Thailand Using MODIS Aerosol Optical Depth and Ground-Based Meteorological Data. *Environment Asia*, *6*(2), 65–70.
8. Math, R. K. M., & Dharwadkar, N. V. (2018). IoT Based Low-cost Weather Station and Monitoring System for Precision Agriculture in India. In *2018 2nd International Conference on I-SMAC (IoT in Social, Mobile, Analytics and Cloud) (I-SMAC)I-SMAC (IoT in Social, Mobile, Analytics and Cloud) (I-SMAC), 2018 2nd International Conference on* (pp. 81–86).
9. Ordieres, J. B., Vergara, E. P., Capuz, R. S., & Salazar, R. E. (2005). Neural network prediction model for fine particulate matter (PM 2.5) on the US e Mexico border in rez (Chihuahua) El Paso (Texas) and Ciudad Jua. *Environmental Modelling & Software*, *20*, 547–559.

10. Pe, P., Trier, A., & Reyes, J. (2000). Prediction of PM concentrations several hours in advance using neural networks in Santiago , Chile. *Atmospheric Environment*, *34*, 1189–1196.
11. Pollution Control Board, T. (n.d.). Thailand's air quality and situation reports. Retrieved from http://air4thai.pcd.go.th/webV2/index.php
12. Ray, P. P. (2016). Internet of Things Cloud Based Smart Monitoring of Air Borne PM2.5 Density Level. In *International conference on Signal Processing, Communication, Power and Embedded System (SCOPES)* (pp. 995–999).
13. Shah, J., & Mishra, B. (2016). IoT enabled environmental monitoring system for smart cities. In *2016 International Conference on Internet of Things and Applications (IOTA)* (pp. 383–388).
14. Shete, R., & Agrawal, S. (2016). IoT Based Urban Climate Monitoring using Raspberry Pi. In *International Conference on Communication and Signal Processing* (pp. 2008–2012).
15. Siva, K., Ram, S., & Gupta, A. N. P. S. (2016). IoT based Data Logger System for weather monitoring using Wireless sensor networks. *International Journal of Engineering Trends and Technology (IJETT)*, *32*(2), 71–75.
16. Strigaro, D., & Cannata, M. (2019). Boosting a Weather Monitoring System in Low Income Economies Using Open and Non-Conventional Systems : Data Quality Analysis. *Sensors*, *19*(5), 1–22.
17. Thilagam. J, S. T., Babu, T. S., & Reddy, B. S. (2018). Weather monitoring system application using LabVIEW. In *2018 2nd International Conference on I-SMAC (IoT in Social, Mobile, Analytics and Cloud) (I-SMAC)I-SMAC (IoT in Social, Mobile, Analytics and Cloud) (I-SMAC), 2018 2nd International Conference on* (pp. 52–55).
18. Vinitketkumnuen, U., Kalayanamitra, K., Chewonarin, T., & Kamens, R. (2002). Particulate matter, PM 10 & PM 2.5 levels, and airborne mutagenicity in Chiang Mai, Thailand. *Mutation Research/Genetic Toxicology and Environmental Mutagenesis*, *519*(1–2), 121–131.

CHAPTER 13

An Internet of Things-Based Novel Approach to Rescue Children from Borewells

VIKRAM PURI,[1] BHUVAN PURI,[2] RAGHVENDRA KUMAR,[3] and CHUNG VAN LE[1]

[1]*Duy Tan University, Da Nang, Vietnam*

[2]*D. A. V. Institute of Engineering and Technology, Jalandhar, Punjab, India, E-mail: bhuvanpuri239@gmail.com*

[3]*Department of Computer Science and Engineering, LNCT Group of College, Jabalpur, India*

ABSTRACT

Nowadays, borewell is the main source of freshwater and also eliminates the water scarcity problem. On the other side, many children's deaths are reported from the last few years due to borewell fall. During the time of the borewell incident, the immediate rescue operation is required to save the life of the child but it is also a challenging situation because the environment inside the borewell is totally unpredictable. Many solutions are already developed and tested. However, these systems have some pitfall during the rescue operation. To overcome these challenges, internet of things (IoT)-based robotic suction gripper is proposed. Suction gripper enables with the camera, ultrasonic sensor, and oxygen level detector as well as controlled in the presence and absence of the Internet. Novel suction techniques help to rescue the children without harm.

13.1 INTRODUCTION

Due to the rapid demand of fresh water, the water scarcity has become a threat for the sustainable growth of society. World Economic Forum annual list report categorize water scarcity as the global endangered in terms of potential affect (World Economic Forum: *Global Risk*). Rising demand of freshwater are depending on some factors namely rapidly increasing population, ameliorate living standard, extended techniques for agriculture irrigation [15]. 4 billion people live under severe fresh water scarcity problem and faced 1 month of every year from water scarcity [1]. Borewells use as medium to overcome the problem of water scarcity. From the recent years, India faced most tragic and helpless incidents that attract everyone to look. More than 30 children died due to the stuck in the borewell from 2006 to now [2]. Nowadays, borewell make a death trap and waiting for the children who are playing near the borewell. However, government, and contractor are also known about these facts and figures but the number of deaths are continuously increasing. Inside of Borewell is same as congested black environment, lack of oxygen that's take life slowly before reach of rescue operation. Some incidents happened from last few years are discussed as in Table 13.1.

TABLE 13.1 List of Children Felt on the Borewell

Person Felt in Borewell	How Old?	Year	Rescue Operation Persons and Operation Pass or Fail
Prince [3]	5 year old boy	2006	Indian Army and operation successful
Sarika [4]	1 year old girl	2007	Unsuccessful and Died
Amit (boy Amit felt borewell) [5]	2 year old boy	2007	Unsuccessful and Died
Vandana [6]	3 year old girl	2008	Unsuccessful and Died
Mahi [7]	4 year old girl	2012	Indian Army and L&T but unsuccessful and Died
Fatehveer (2 year boy felt) [14]	2 year old boy	2019	Unsuccessful and Died

Some researchers have been proposed some solution to overcome the challenges faced at time of rescue children. Raj [8] proposed a pneumatics based robotic arm for rescuing the child from the borewell. This robot arm enabled with teleconferencing system for setup a communication between the rescuer and victim child. Sridhar [9] designed a gripper arm system for rescuing purpose and equipped with the atmospheric air quality such as temperature, humidity, pressure, oxygen, voice recognition, as well as carbon dioxide level. In Ref. [10], smart child rescue system named SCRS has been designed. This system enabled with the sensors on the top of the borewell to sense the child activities. It addition, Autonomous sires will be activation if the distance between the system and child is nearly five feet depth. A rescued robot has been proposed by the Kurukuti [11] that can be resized according to the diameter of the borewell. Moreover, it is equipped with two different robotics arms: one is for holding the baby with aim of camera and the other one is for rescue the child from the borewell. This prototype is developed with the aid of 3D printing technology. Rajesh [12] developed a rescue system called Prosthetic borewell system (PBRS) for rescuing the children from the borewell in less time. This system is based on the safe holding as well as detection the condition such as distance between the system and children, temperature, pressure, and gases with the aid of sensors without camera. Also, detect the crack inside the borewell or pipeline. In 2018 [13], adjustable robot is proposed to rescue the children that can adjust its diameter according to the borewell diameter and also enabled with the camera, air filler as well as oxygen supplier. It also equipped with an infrared (IR) transmitter and receiver to measure the distance between the child and system as well as temperature and gas sensor is used to measure any toxic gas.

In the above-mentioned previous studies, researchers are mostly worked on the robotic arm and sensors to measure the environmental condition. However, robotics arm build up from the metal or some kind of plastic which can create an injury while rescuing children from the borewell. In our proposed study, we develop an internet of things (IoT) based rescue system to extricate the child from borewell. Our proposed system is based on the suction pump to fetch the child from borewell as well as also equipped with ultrasonic sensor, camera, and oxygen measure sensor. The data is controlled through the internet and wireless technology that help to monitor real time conditions inside the borewell.

13.2 METHODOLOGY

13.2.1 ARCHITECTURE

The proposed study categorized into three different sections: (1) enabled the suction pump with the camera; (2) monitor the depth distance between the system and children; and (3) remotely monitor through the cloud server or wireless technology (see Figure 13.1).

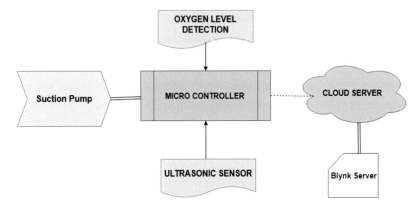

FIGURE 13.1   Architecture of proposed system.

Suction pump is similar to vacuum pump that can be controlled through the use of microcontroller. In the previous studies, researcher proposed robotics arm to rescue the children from the borewell but robotics arm build from the metals or plastic material which creates some injury during the rescue operation. To overcome these limitations, suction pump based technique propose to fetch children from the borewell. IR based camera also interfaced with suction pump for ease to rescue the children. Microcontroller (Arduino nano) is the backbone of the proposed system. It monitors distance between the victim and the system. Arduino Nano is ATmega328 based microcontroller board that's work on the Mini USB-B cable power supply. It has in-built 32kb of flash memory and 2kb of SRAM with clock speed of 16 MHz. Moreover, 22 digital input output pin and 6 PWM pins are integrated which is suitable for our proposed system. To measure the oxygen level inside the borewell, Oxygen sensor (MQ-2) is connected to the microcontroller through the analog pins (see Figure 13.2).

MQ-2 sensor basically a gas sensor that's have the capability to detect oxygen level inside the borewell. MQ-2 works on the principle of change resistance according to the gases concentration and connects via variable resistor in series to make voltage divider circuit. The concentration of this sensor varies from 300–10,000 ppm. If the level of oxygen is low, children will die due to suffocation. Suction pump works according to the instructions of microcontroller means it creates a sufficient pressure for suction to rescue the child without any harm. Ultrasonic sensor (HC-SR04) also integrated with the proposed system to calculate the distance between the child and proposed system. HC-SR04 works on the concept of transmit ultrasonic wave and reflect back on the strike of object. In this process, speed, and time are already calculated and with the aid of distance formula (see Eq. 1), the distance between the child and the system will be calculated.

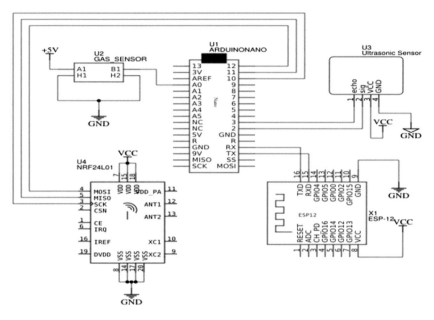

FIGURE 13.2   Circuit diagram of proposed work.

$$\text{Distance} = \text{Speed x Time} \qquad (1)$$

After processing, data is directly sent to the cloud server and remotely monitor data via Blynk app. ESP32 and NRF module works as a bridge to communicate between system and cloud server. Moreover, if there is less connectivity of internet, data will be transferred with the aid of wireless technology (NRF module). ESP32 and NRF module connects online and offline mode, respectively.

13.2.2 SUCTION PUMP GRIPPER

Suction pump gripper or robotic gripper designed for our proposed system. Suction pump consists of rubber material at the outer layer, filled with granular material and vacuum pump. When the rubber is inflated, granular material inside the rubber are lose their contacts with neighbors and easily move around the children head or other body part. When the air is sucked out from the rubber, material are attached to the body of children and tightly gripped from all sides. This process names as "jamming." This process is based on the friction between the granular materials to jam them in a particular place.

Figure 13.3 shows the concept of our proposed work. Suction pump gripper goes inside the borewell via pulley fixed at top of borewell top. Granular material helps to fetch the children out.

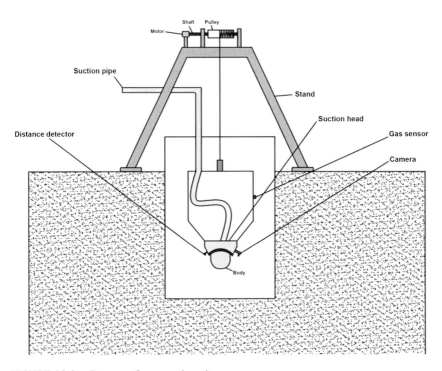

FIGURE 13.3   Concept of proposed work.

13.3   RESULT AND ANALYSIS

In this section, Figures 13.4 and 13.5 represent the outcome of the results from our proposed work. Figure 13.4 shows mobile app for the distance monitoring and oxygen level through the Blynk server. Figure 13.5 shows the real image of proposed system equipped with the camera and ultrasonic sensor. The most noteworthy point of this proposed system is that it works remotely through the use of Internet. Red rubber is the suction rubber filler with granular material as well as black pipe is for suction pipe which connects to motor.

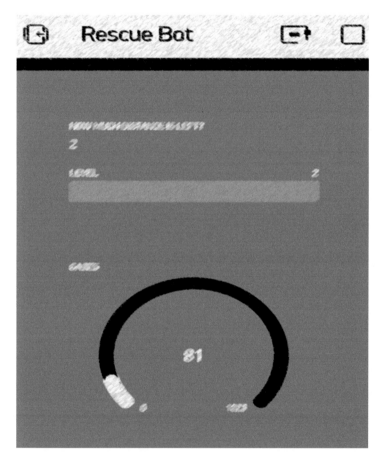

FIGURE 13.4   Blynk server output.

13.4   CONCLUSION

Borewells use as a medium to fulfill the demand of water scarcity but on the other side it makes a death trap for the children. There is a rapid increase in the children death count due to fall on borewell. Many researchers have been proposed numbers of solution to rescue the children from borewell but majority of the studies are based on the robotics arms enabled with sensors. In this chapter, robotic gripper based on suction pump enabled with oxygen sensor and camera is being proposed. This technique is better than other robotics arm to avoid any harm due to the sharpness of metal during the rescue operation. In addition, Robotic gripper is controlled with the aid of IoT and NRF technology. The result shows that the proposed system performance is based on real time and accurate.

FIGURE 13.5   Proposed robotic suction gripper.

KEYWORDS

- **borewell**
- **granular material**
- **internet of things (IoT)**

- **robotics**
- **safety machine**
- **suction gripper**

REFERENCES

1. Mekonnen, M. M., & Hoekstra, A. Y., (2016). Four billion people facing severe water scarcity. *Science Advances*, *2*(2), e1500323.
2. *Children Borewell Death*, (2012). https://www.aljazeera.com/indepth/feat ures/2012/06/201262518453409715.html(accessed 20 July 2020).
3. *Prince Rescued from the Borewell*, (2006). https://www.indiatoday.in/magazine/ indiascope/story/20060807-5-year-old-boy-falls-into-abandoned-borewell-in-haryana-784895-2006-08-07 (accessed 20 July 2020).
4. *Sarika Stuck on Borewell*, (2007). https://www.pressreader.com/ (accessed 20 July 2020).
5. *Amit Felt on Borewell*, (2007). https://www.indiatoday.in/latest-headlines/story/child-falls-into-150-ft-borewell-in-agra-31291-2008-10-09 (accessed 20 July 2020).
6. *Vandana Felt on Borewell*, (2008). http://archive.indianexpress.com/news/twoyearold-sonu-brought-out-dead-from-bore/372771/ (accessed 20 July 2020).
7. *Mahi Died Due to Suffocation*, (2008). https://www.indiatoday.in/india/story/mahi-dies-from-suffocation-after-falling-into-borewell-106861-2012-06-25 (accessed 20 July 2020).
8. Raj, M., Bansal, A., Makhal, A., Chakraborty, P., & Nandi, G. C., (2014). An approach towards rescue robotics in borewell environment. In: *2014 International Conference on Communication and Signal Processing* (pp. 1097–1100). IEEE.
9. Sridhar, K. P., & Hema, C. R., (2015). Design and analysis of a borewell gripper system for rescue. *ARPN Journal of Engineering and Applied Sciences*, *10*(9), 4029–4035.
10. Kavianand, G., Ganesh, K. G., & Karthikeyan, P., (2016). Smart child rescue system from borewell (SCRS). In: *2016 International Conference on Emerging Trends in Engineering, Technology, and Science (ICETETS)* (pp. 1–6). IEEE.
11. Kurukuti, N. M., Jinkala, M., Tanjiri, P., Dantla, S. R., & Korrapati, M., (2016). A novel design of robotic system for rescue in borewell accidents. In: *2016 International Conference on Robotics and Automation for Humanitarian Applications (RAHA)* (pp. 1–5). IEEE.
12. Rajesh, S., Suresh, G., & Mohan, R. C., (2017). Design and development of multipurpose prosthetic borewell system-an invincible arm. *Materials Today: Proceedings*, *4*(8), 8983–8992.
13. Arthika, S., Eswari, S. C., Prathipa, R., & Devasena, D., (2018). Borewell child fall safeguarding robot. In: *2018 International Conference on Communication and Signal Processing (ICCSP)* (pp. 0825–0829). IEEE.
14. *Two Year Old Boy Felt on Borewell*. https://www.indiatoday.in/india/story/borewell-child-punjab-sangrur-dead-fall-1546357-2019-06-11 (accessed 20 July 2020).
15. Ercin, A. E., & Hoekstra, A. Y., (2014). Water footprint scenarios for 2050: A global analysis. *Environment International*, *64*, 71–82.
16. World Economic Forum, (2015). *Global Risks 2015* (10th edn.). World Economic Forum, Geneva, Switzerland.

Part II
Blockchain Techniques and Methodologies

CHAPTER 14

Blockchain: A New Way to Revolutionize CRM Systems

TANAY KULKARNI, PURNIMA MOKADAM, and KAILAS DEVADKAR

Department of Information Technology, Sardar Patel Institute of Technology, Mumbai – 400058, Maharashtra, India,
E-mail: tanaykulkarni06@gmail.com (T. Kulkarni)

ABSTRACT

Customer relationship management (CRM) is a process of tracking, storing, and analyzing all the interactions a business has with its customers and prospects to earn profits. In the current CRM Model, the customer data is hosted on cloud which leads to security issues and problems like data duplication and corruption. The major concern that traditional CRM encompasses is that it being a centralized system has a single point of failure. This chapter proposes the use of Blockchain in CRM to overcome the problems related to the traditional CRM systems.

14.1 INTRODUCTION

For any business, customers are very important and thus customer relationship management (CRM) becomes necessary to retain customers and drive sales growth. The CRM helps the company to understand the customer needs and expectations and helps to find potential future customers. A CRM system is a great tool for syncing the process of sales and marketing, it also helps in jobs like task management, project management, and lead nurturing. However, some drawbacks that the traditional CRM systems are that of security issues ease of use, gaining customer trust, system integration, data migration, high cost, and the requirement of trained professionals. The CRM database (DB) hosted on cloud contains a lot of outdated information due to lack of use of tools for data cleansing and thus analysis of such data becomes a challenge.

One of the first CRM platforms available over the internet is the salesforce. Salesforce is a CRM system hosted on cloud. It is software as a service (SaaS) provider which means it hosts the applications on cloud which is available to everyone over the internet. It is a platform where you can connect to and manage your customers, leads, and partners. It helps to track the customer's activity in real time, solves issues faster, increase productivity, and makes sure that customer expectations are fulfilled. As the DB is hosted on cloud, it can be accessed anywhere and anytime. It is also available on phone that means you can run the entire business on phone. It provides different cloud platform like IoT cloud, community cloud, marketing cloud, analytics cloud, etc.

The major drawback of CRM systems is that, being centralized it has a single point of failure. Blockchain in CRM helps to overcome this by making the system decentralized. Also, there is no requirement of middlemen for verification and authentication of data as the data in itself is secure. Also, other advantages of transparency, security, efficiency, and speed of transaction make blockchain a desirable application in the field of CRM.

Blockchain being a new technology brings immense advantages in its integration with various existing technologies and thus understanding it becomes important. Blockchain basically means chain of transactions which are represented as a series of chained blocks put together and distributed among all the users or participants in the network. An important feature of these blocks is that the storage of any kind of data is permitted in these blocks which make it more usable.

14.2 TERMINOLOGIES

14.2.1 CUSTOMER RELATIONSHIP MANAGEMENT (CRM)

CRM is a term that encapsulates all the functions that a company makes use of, to guide the interactivity and interactions with the clients or customers and enhance relationships with them. It not only includes the guidelines and rules for how interactions should take place directly, but also includes systems that organize and track all the information that is relevant to the company like meetings, sales conversions, etc.

A CRM system takes into consideration a lot of data and arranges it to portray business of a company and shows how well or bad, the relationships with customers are being handled, and what effect does that have on business. So in simple words, a CRM system is a system made for managing client relationships with the company.

CRM is very important for any business. Businesses don't usually have a system that manages their relationships with the customers and this role of management is often left out for the sales people to deal with and in their own different ways. These customers can be individuals (B2C) or even other businesses (B2B) and thus a lack of generalized approach towards customers and a lack of direction in a company lead to failure in managing customer relations.

The CRM system is of three types: operational CRM, analytical CRM, and collaborative CRM [1]. The operational CRM is related to generation of leads and converting them into customers using lead score. The analytical CRM system analyzes the customer's behavior. In collaborative CRM, all teams work together towards one goal. The information from one team is shared with other team so that they can work collaboratively [2].

The current CRM systems have many issues like the knowledge management capability, lack of skilled employees, training plans and issues [3]. Customer segmentation is very important in CRM [4] where the customers are divided into different departments depending upon their interests, which has not been given much importance in the traditional CRM systems. The most common modules of CRM are sales, marketing, and service. The importance of each module depends upon the type of company and how the company operates.

14.2.1.1 SALES

Company usually wants their business to grow. The sales module keeps track of new leads and opportunities associated with them. Thus, the sales module is very important for a company that is looking for new customers or clients. Finally, it all comes down to evaluating the prospects gathered by the company and getting more information from them.

14.2.1.2 MARKETING

The marketing of various products and services of the company comes under this module of CRM. Newsletters, campaigns, social media, and marketing are the things that fall under this module. These contacts that the company got from leads can be added to the mailing list for such marketing campaigns and deals that the company is offering.

14.2.1.3 SERVICE

This module is very important for a service-based company where all the service requests to the company are handled by various customer service representatives. The services that a company provides to the customer decide its growth.

14.2.2 SALESFORCE

Salesforce is a cloud CRM platform and its important components are lead, contact, account, and opportunity. Leads are potential future customers. These leads are then further converted into account, contact, and opportunity, where account is the user, opportunity is the business needs of the user, and contact contains the contact information about the user. For example, if the user is an IT company then opportunity is procurement of computers and therefore the sales team of different companies tries to reach out to clients depending on the business needs.

The salesforce CRM also supports various other features like deal management, task management, email marketing, tracking, and analysis, web engagement, feedback management, etc.

14.2.3 BLOCKCHAIN

Blockchain is a chain of blocks which contain encrypted transactions making it immutable and tamper-proof. Some of the prominent characteristics of Blockchain are as follows:

1. **Distributed Public Ledger:** Each participant in the network has a copy of the ledger, which essentially means all participants have a record of all transactions.
2. **Hash Encryption:** All the data in the Blockchain is encrypted making it very secure.
3. **Decentralized Nature:** It does not have a single point of failure.

Blockchain essentially has a distributed ledger which is accessible to all. This ledger records all the transactions between parties. These records are immutable and verified. The transactions recorded in the blockchains are immutable and they do not require validation from any exterior authority, to check the authenticity of the block and integrity of the data in the block. Once the data is written in the block, it cannot be changed. Each block contains information, its own hash value, and the hash of previous block so that chain can be formed. Whenever a new block is created, all the nodes verify the block. After the consensus, this verified block is appended to the chain. All blocks contains smart contract a digital contract. It is a program code which is automatically triggered when there is a transaction related to that block.

The prominent advantages of blockchain, which enables its application in the field of CRM, are its features of security, transparency, immutability, and most importantly decentralization of system. A decentralized CRM system can prove to be more effective, efficient, and impactful compared to the traditional CRM systems. By using blockchain in CRM, the outdated data no longer remains in the CRM database [5].

Blockchain are of two types, i.e., public and private blockchain. In public blockchain, the distributed ledger is open to every participant that means all the participants in the network can view the ledger. There is no privacy in this network. Whereas in private blockchain, the participant which have permission to view a particular data can only view it.

The following chapter proposes the use of Hyperledger Composer. Hyperledger composer is the open source development toolset to develop blockchain applications. It is used for developing blockchain CRM network for addressing various business problems. It is built over a Hyperledger Fabric environment. Hyperledger Fabric is responsible for making network administrators and assigning unique network card identifiers to the participants for their authenticity and also performs various other functions.

Business network established using the Hyperledger composer is a private blockchain containing:

1. **Participants:** They are members of the business network.
2. **Assets:** They can be real things like car or any untouchable thing like a contract, i.e., tangible, and intangible.
3. **Chaincode:** It is a business logic written in JavaScript. It is also called smart contract or digital contract.
4. **Transactions:** Transactions are used for modifying the asset using chaincode. These transactions are recorded on the distributed ledger.

5. **Event:** Events are triggered when any transaction happens. Emitting an event is done through chaincode.

14.3   IMPLEMENTATION

The implementation shows a small part of Blockchain in CRM using Hyperledger Composer, the use case can include any service driven company which uses CRM systems to drive customers, for example say a telecom company, ABC Ltd. Here every data is represented in JSON format. In this scenario we have considered only three participants Lead queue manager, sales department and marketing department from ABC Ltd. These participants are identified by unique participant Id and their name. Figure 14.1 shows the participants in the network with their names and their unique identity numbers. The unique id 123 identifies the lead queue manager, the unique id 11 identifies with the marketing department, and the unique id 12 identifies with the sales department.

ID	Data
123	``` { "$class": "org.example.basic.lead_queue_manager", "manager_name": "Mr. Pandey", "participantId": "123" } ```

ID	Data
11	``` { "$class": "org.example.basic.marketing", "name": "Rani Deshmukh", "participantId": "11" } ```

ID	Data
12	``` { "$class": "org.example.basic.sales", "name": "Mrs Kale", "participantId": "12" } ```

FIGURE 14.1   Participants in blockchain network.

The lead is an asset. Lead is identified by a unique lead id. Other attributes of lead are lead name, lead status, and lead owner as shown in Figure 14.2. The data type of lead status is lead status which is an enum, similar to the data-structure in the C programming language. The lead status can be new, contacted, nurturing, qualified, and unqualified. At the start, the lead status is new and lead owner is the lead queue manager.

FIGURE 14.2   Asset: Lead.

The blockchain defines three transactions to modify the attributes of asset. In the transaction 'change_state' which is Figure 14.3, the status of the lead is changed in the series new-contacted-nurturing-qualified. In this transaction, we have to put the ID of the asset whose state we have to change. The logic of this transaction is written in the script file.

FIGURE 14.3   Transaction I: change_state.

During the nurturing of the lead, the owner finds that this lead is not worth qualifying then he can execute the second transaction, i.e., 'change_ state2,' where the status is changed from nurturing to unqualified. This transaction executes only if the lead is in the nurturing state (Figure 14.4).

The third transaction is to change the owner of the lead and it is named as 'Transfer Lead.' Initially, when the lead is created, the owner is the lead queue manager. Once the lead is qualified, it needs to be transferred to different departments in a company. The lead queue manager first decides which department is appropriately needed for a particular lead and accordingly executes this transaction and changes the lead owner. Here the input is the lead id and the new owner's id. This is shown in Figures 14.5 and 14.6.

Transaction Type   change_state2   ⌄

JSON Data Preview

```
1  {
2      "$class": "org.example.basic.change_state2",
3      "lead": "resource:org.example.basic.Lead#21"
4  }
```

FIGURE 14.4   Transaction II: change_state2.

Transaction Type   TransferLead   ⌄

JSON Data Preview

```
1  {
2      "$class": "org.example.basic.TransferLead",
3      "lead": "resource:org.example.basic.Lead#21",
4      "newOwner": "resource:org.example.basic.marketing#11"
5  }
```

FIGURE 14.5   Transaction III: Transfer lead.

ID	Data

21	`{`
	`"$class": "org.example.basic.Lead",`
	`"leadId": "21",`
	`"lead_name": "Mr. Raj",`
	`"leadStatus": "CONTACTED",`
	`"leadOwner": "resource:org.example.basic.marketing#11"`
	`}`

FIGURE 14.6   Changed owner.

Here, as seen in Figure 14.6, the ownership of the lead has been granted to the marketing department as the lead queue manager transfers it to the marketing department.

Figure 14.7 is the distributed ledger that is shared with all the participants in the network and whenever a new entry is registered in the ledger, all the participants verify the entry, and only after the consensus of all the participants, the entry is registered in the ledger. In this case, if there is an entry from the transaction done by the lead queue manager, then the consensus of the sales and the marketing department will register the entry in the ledger.

Date, Time	Entry Type	Participant	
2019-06-27, 21:41:22	TransferLead	admin (NetworkAdmin)	view record
2019-06-27, 01:32:56	change_state	admin (NetworkAdmin)	view record

FIGURE 14.7   Distributed ledger.

If you view the records, you can see all the attributes of transaction with transaction id which is a hash value and the timestamp which indicates when the transaction was done as shown in Figure 14.8. If an event is associated with the transaction, one can see that too with its event ID. Events are mainly used for triggering notifications.

FIGURE 14.8   Transaction record.

14.4   ADVANTAGES OF BLOCKCHAIN IN CRM

14.4.1   INCREASED SECURITY

A blockchain basically means a chain of blocks where the blocks contain a specific number of transactions or transactional records. These blocks are connected to the blocks that come before it or after it in a manner such that it is cryptographically secured and only those users will be allowed to access these blocks that have the network access key assigned to them and approved by all members of the network. Thus, the possibility of unauthorized entries and tampering of data almost becomes impossible.

The traditional CRM database where all the information about the customers is stored exists on cloud and thus the security of these cloud DBs is a prominent concern that the traditional CRM systems face. Also, because of this data being stored at one centralized place on cloud there exists a huge risk of single point of failure. The companies thus risk their data by keeping it at a single centralized store.

Blockchain being a decentralized and distributed architecture thus overcomes the problem of single point of failure as the data is distributed with all peers in its network. Blockchain is a synchronized peer to peer (P2P) network thus each peer in the network has a copy of blockchain which has blocks in the same order. This makes the data stored in the blockchain tamper-proof as it would require gigantic amount of computation to change data in blocks of each peer of the network which is practically impossible. This also helps in reducing the chances of repudiation and fraud.

14.4.2   REDUCES DATABASE (DB) TRACKING EFFORT

There are often issues encountered by the CRM systems regarding the duplication and the corruption of data of the users and thus tracking down the places of redundancy and changed or missing data is quite a tedious job in the traditional CRM systems. Blockchain tackles these problems of data corruption and duplication by enabling a customer to have their own single block which can help the participants in a network get a standard and unified image of user's personal details, transactional history, previous activity and other vital information. Thus, the CRM database is now is free from duplication and corruption, which is something that the blockchain inherently provides. Blockchain also helps in getting rid of outdated, inaccurate, and obsolete information and thus it plays a major role in data cleansing of CRM databases. The speed of CRM processes and the analysis of customer behavior and their business engagement becomes more accurate with the use of blockchain in the CRM systems.

14.4.3   BETTER TRANSPARENCY

The traditional CRM systems required companies to take into consideration financial institutions like third party broker or the bank to facilitate the process of dealing with the customers which made the customers data to cross many entities in the network and thus there was a risk of customer data falling into the wrong hands. Blockchain being decentralized in nature enables companies to deal with customers in a more transparent and secured manner without relying on these middlemen. By eliminating all these intermediaries blockchain helps in achieving greater speed and efficiency in all the interactions of the company with customers.

14.4.4   IMPROVED PRIVACY CONTROL

The security of private data has become one of the most important issues in the world of information technology with the cases of viruses, breaching of data and malware attacks, it is very important that the CRM deals more effectively with private data of the customers.

Blockchain being decentralized enables CRM vendors to apply the concept of self-sovereign identity which means that customers now has the sole ownership of their digital identities and are in complete control over how their data is shared and used. This also adds an additional security layer allowing The Identity holder to show the other entity only the necessary data for any given transaction or interaction.

14.5   CONCLUSION

Thus, the traditional CRM systems have many drawbacks and shortcomings like its centralized nature, lack of security, etc., which can be mitigated using the blockchain technology which is decentralized, highly secure and transparent. Blockchain will not only change the approach of CRM systems but also revolutionize the architecturing of business models in general.

KEYWORDS

- **blockchain**
- **customer relationship management**
- **database**
- **information technology**
- **peer to peer**

REFERENCES

1. Daif, A., Eljamiy, F., Azzouazi, M., & Marzak, A., (2015). Review current CRM architectures and introducing new adapted architecture to big data. *International Conference on Computing, Communication, and Security (ICCCS)*.
2. Chun, N. L., & Xiao, W. Z., (2009). A study on CRM technology implementation and application practices. *International Conference on Computational Intelligence and Natural Computing*.
3. Luís, H. B., Marco-Simó, J. M., & Joan, A. P., (2014). An initial approach for improving CRM systems implementation projects. *9th Iberian Conference on Information Systems and Technologies (CISTI)*.
4. Xiaojing, Z., Zhuo, Z., & Yin, L., (2011). Review of customer segmentation method in CRM. *International Conference on Computer Science and Service System (CSSS)*.
5. Julija, G., & Andrejs, R., (2018). The advantages and disadvantages of the blockchain technology. *IEEE 6th Workshop on Advances in Information, Electronic, and Electrical Engineering (AIEEE)*.
6. Alireza, F., Chen, W., & Elizabeth, C., (2010). Intelligent CRM on the cloud. *13th International Conference on Network-Based Information Systems*.
7. Al-Mudimigh, A. S., Saleem, F., Ullah, Z., & Al-Aboud, F. N., (2009). Implementation of data mining engine on CRM-improve customer satisfaction. *International Conference on Information and Communication Technologies*.
8. Anuradha, M., & Ankit, C., (2017). Sales force CRM: A new way of managing customer relationship in cloud environment. *Second International Conference on Electrical, Computer, and Communication Technologies (ICECCT)*.
9. Fabrice, B., Shai, H., & Tzipora, H., (2018). Supporting private data on hyper ledger fabric with secure multiparty computation. *IEEE International Conference on Cloud Engineering (IC2E)*.
10. Nor, H. M. A., Abd, R. H., Khairuddin, O., & Norjansalika, J., (2012). Customer relationship management (CRM) implementation: A soft issue in knowledge management scenario. *IEEE Colloquium on Humanities, Science, and Engineering (CHUSER)*.
11. Pinyaphat, T., & Chian, T., (2018). Blockchain: Challenges and applications. *International Conference on Information Networking (ICOIN)*.
12. Sophie, W., Marcel, H., & Simon, T., (2013). Evaluation of the di.me trust metric in CRM settings. *Second International Conference on Future Generation Communication Technologies (FGCT 2013)*.

CHAPTER 15

Blockchain-Enabled Decentralized Traceability in the Automotive Supply Chain

SONALI SHIRISH PATWE

PhD Research Scholar, Symbiosis International University, Lavale, Pune, India

ABSTRACT

With the evolution of Industry 4.0, the automotive supply chains are getting not only technology-enabled but also technology-centered. However, the current implementations of these supply chains are centralized in nature and are designed specific to the needs of the automotive OEMs. Such implementations have their own limitations like lack of transparency, provenance tracking, end to end visibility, audibility, etc. Hence, they may not be capable of handling the exponential growth of IoT enabled automotive supply chains expected in near future. Also due to such limitations, trust-building among the supply chain stakeholders becomes a further challenge. Blockchain-based supply chain traceability is a promising transformation for just-in-time, safe delivery with transparency in the automotive industry. This chapter presents a decentralized traceability solution with blockchain for the automotive supply chain use case in the context of transparency, visibility, provenance, and audibility. Blockchain can positively impact and improve the current ways of working in automotive supply chains. The decentralized traceability architecture is proposed which will lead to increased efficiency, reliability, speed for manufacturers, suppliers, and logistics service providers.

15.1 INTRODUCTION

With increased connectivity and globalization, the world has come closer. Various goods and services are exchanged across different states, countries, continents, etc. The supplier, procurers, manufacturers, distributors, and consumers are spreading across various industries, geographies, and product categories. With the uprising of Industry 4.0, automotive supply chains are becoming not only technology-enabled but technology-centered by connecting devices, machines across the supply chains and striving towards increasing the efficiency [1]. Adaption of the technologies like internet of things (IoT), augmented reality, drones, etc., is giving automotive supply chains competitive advantage. The IoT enabled supply chains to provide various advantages like improved resource utilization, higher, enhanced customer satisfaction, and proactive response and satisfying customer demand.

However, majority of the digital supply chain implementations are centralized client-server applications or cloud-based systems or enterprise applications [3]. Some of the stakeholders in supply chain are still maintaining the data in excel sheets or conventional databases (DBs). Such centralized systems are designed specific to the needs of the organization and lack transparency and inherently has security threats such as, data lock-in, availability, auditability, and confidentiality [4]. Data sharing across the heterogeneous stakeholders is also a challenge. For example, an automotive OEM having it parts distribution via logistics partner across the globe is implemented specific to the needs of that company. The identification of the devices, authentication of users, data storage, etc., is centralized to the OEM servers and isolated from the logistics partner. So, if these two stakeholders want to share the parts distribution traceability information to the users, there is no common platform to share the information which is trusted by both the parties. Similarly, in the complete automotive supply chain scenario, there are several stakeholders like suppliers, vendors, dealers, manufacturers, etc., who are hesitant to share the information due to lack of trust and a common platform. Such implementations pose various challenges like lack of coordination across the industry, unorganized inventory management, mismatch between demand and supply, higher shelf time, artificial shortage, lack of shipment visibility and traceability, etc. Also, in the case of automotive supply chains, the shipment goes through various countries and uses various commute mediums. The process involves various shipment companies, government, manufacturers, consumers, legal entities, etc. These entities have their own centralized systems to manage the supply chain. This process is time consuming, inefficient, and faces many visibility and traceability challenges. It also involves lot of paperwork. Current supply chain implementations are also vulnerable to the frauds which cost loss of billions of dollars [5]. The automotive spare parts sold in developing counties are fake or substandard, leading to car accidents and product recalls. This results in loss of trust from the consumers as well as other supply chain stakeholders. In short, it is clear that there are multiple challenges that current automotive supply chains are facing. Hence, there is a need of a common platform to handle heterogeneous stakeholders across the supply chain, which will bring trust, transparency, traceability, visibility, and auditability to the current and upcoming digital supply chains. Blockchain technology comes to the rescue to address such problems.

Blockchain is a widely known distributed ledger technology which enables tracking of the goods and services across multiple parties. Blockchain offers a secure and synchronized record of transactions for all the involved participants. The distributed ledger in blockchain records every sequence of transaction from the first to the recent one. Every transaction is recorded into the block. Every block is connected to the previous and next block in the chain. The group of such transaction added one after the other forms an irreversible chain [5].

15.2   AUTOMOTIVE SUPPLY CHAIN TRACEABILITY WITH BLOCKCHAIN

The current supply chains are moving towards digitization and automation. However, still majority of the supply chains operations are tracked and maintained in excel sheets, centralized ERP systems, enterprise-level DBs, centralized cloud, etc. [7, 8]. The following diagram shows the current implementations of the centralized nature of the automotive supply chain (Figure 15.1).

However, such implementations pose various challenges [6, 7, 11] enlisted below:

1. **Lack of Visibility across Supply Chains:** The current implementations of supply chain management do not give the required visibility to the existing stakeholders.
2. **Loss of Products across the Supply Chains:** There are many cases of loss of products or containers across the supply chains. Due to lack of visibility, transparency, and auditability, it becomes difficult to track the loss.

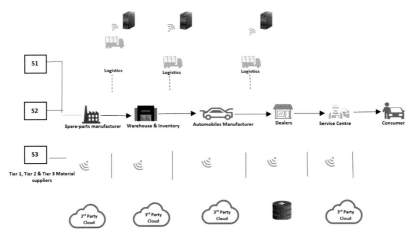

FIGURE 15.1   Automotive supply chain-current scenario.

3. **Counterfeit Products:** Especially in Automotive supply chains, it is very common that one in ten spare parts products are counterfeit and it causes car accidents or product recalls.

4. **Trust:** Due to the heterogeneity of the stakeholders, lack of transparency, and poor visibility, building a trust among the stakeholders become a challenge. They find it difficult to collaborate with the new stakeholders and prefer to transact and trust the older trusted parties.

5. **Security:** Lack of transparency among the supply chain stakeholders makes it vulnerable to the security threats like accessibility, data lock-in, privacy, etc., [4]. The centralized system faces the challenge of tampering the data, fraudulent transactions, monopoly of the certain third parties. This creates a security a challenge in supply chain use cases.

6. **Less Traceability:** The centralized cloud-based systems are not favorable for traceability. Traceability is complex and dynamic in supply chains because they involve multiple parties.

7. **Less Transparency:** The centralized systems [4] are no useful for the data and transaction transparency. Relying on third party to maintain centralized cloud system has the possibility of fraud, inherent bias, and single point of failure.

8. **Less Auditability:** As mentioned in Ref. [4] heavily centralized cloud architecture presents lack of auditability.

9. **Lack of Provenance Tracking:** Tracking the origin of the product and giving that visibility to the customer helps in building the trust.

The current implementations of supply chain lack the mechanism of provenance tracking.

To overcome above-mentioned challenges, blockchain is an ideal solution to bring all the stakeholders involved in the automotive supply chain on the common, trustworthy platform [10–12]. Figure 15.2 schematic of how blockchain can bring all the stakeholders on the common platform.

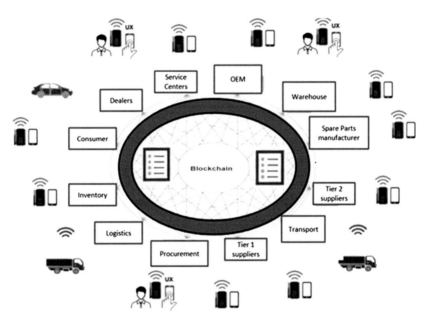

FIGURE 15.2   Blockchain as a common platform for automotive supply chain stakeholders.

Table 15.1 summarizes the employability of blockchain in candidate IoT enabled automotive supply chain use cases.

15.3   PROPOSED ARCHITECTURE

Based on the findings, there is a need for the common platform for the heterogeneous stakeholders involved in the IoT enabled supply chains [15, 18]. The centralized cloud-based architecture will not be capable of handling the future digital supply chains. Hence, this chapter proposes a blockchain platform based architectural framework for the IoT enabled automotive supply chains as shown in Figure 15.3.

TABLE 15.1 Impact Factors for Automotive Supply Chain

Impact Factor	Description
Traceability [5, 6, 13]	Traceability is an important impact factor as it gives exactly where the product is located currently. The traditional ERP/logistics systems trace the product with the accuracy of 'in transit' or 'with courier,' etc. However, with Blockchain-enabled supply chain IoT; it will trace exact location in real time. This will also help in avoiding the delays due to vehicle failure, weather conditions, etc.
Auditability [8, 14]	As Blockchain provides the important feature of having the tamper-proof records, it becomes easier to go back at any point in time and audit the transactions over the Blockchain. This auditability feature is useful in building the trust across the stakeholders. This also helps in auditing the track record of unknown vendors and becomes easy to collaborate with them
Provenance [13, 15]	The origin of the products can be tracked with blockchain. The genesis block shows the very first transaction on the blockchain. These blocks are linked to previous and next blocks in the chain. So, it becomes easier to track from where the product has originated. This is helpful in tracking original spare parts.
Transparency [16]	Blockchain brings transparency to the multi-party applications. With the traditional implementations of supply chains, it is difficult to understand what is happening internally within the organization. There are also many cases of frauds, corruptions, etc. However, with blockchain, all the transactions become transparent and anyone can go back and check the blocks in the history as well. Hence, the application becomes very transparent. The decentralized and tamper-proof nature of the Blockchain makes the application secure. Also, the transactions are added to the chain only with the consensus of the participants. So, it makes it trustworthy. The security is important for the multi-party applications as the data is being shared among various organizations.
Efficiency [5, 13, 17]	Blockchain makes supply chains more efficient. The approvals on the transactions are based on predefined consensus algorithms and approvals are quite quick. In traditional supply chain systems, lot of time is consumed in paper work, approval of central authorities, government regulatory bodies, etc. Blockchain brings such participants on a shared ledger and there is no need of central authority. All the records are permanently recorded in a block which cannot be modified. Hence, blockchain makes information flow across the supply chain more efficient.

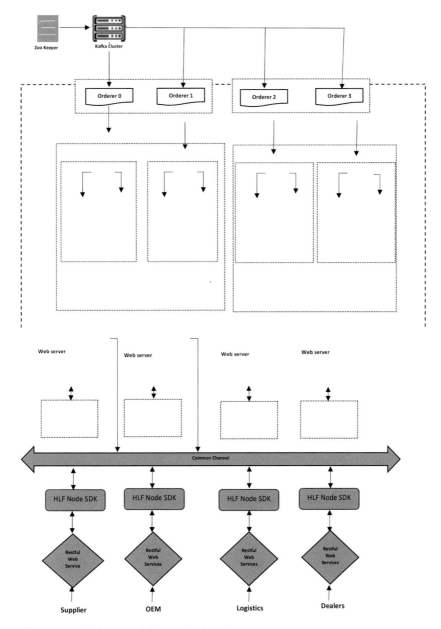

FIGURE 15.3 Architectural framework for the proposed system.

15.4   BENEFITS OF USING BLOCKCHAIN FOR AUTOMOTIVE SUPPLY CHAIN (TABLE 15.2)

TABLE 15.2   Benefits of Blockchain for Automotive Supply Chain IoT

Benefits of IoT Enabled Automotive Supply Chains	Shortcomings of IoT Enabled Automotive Supply Chains	What Blockchain Offers to Automotive Supply Chain IoT?
Connectivity	Centralized models	Decentralization [15]
Speed [21]	No interoperability	Interoperability [20]
Real-time tracking and visibility	Less security	Security [21]
Remote monitoring	No auditability [4]	Auditability in permissioned and public blockchain [23]
Increased efficiency	Lack of Transparency	Transparency [7]
Less or no paperwork	Lack of trust due to lack of trust, transparency, and centralized systems	Tamper-proof [24]
Real-time inventory tracking	Limitations on Scalability	Scalability [5]

15.5   CONCLUSION AND FUTURE WORK

This chapter identifies and discusses the challenges of the centralized nature of the current automotive supply chains. Blockchain is proposed as a solution to the decentralized traceability of the IoT enabled automotive supply chains. After looking at the benefits, blockchain looks like a promising solution. Architecture for the proposed system is given in the chapter. It offers traceability, visibility, auditability, provenance tracking, etc. However, there is a potential need of having a common platform to increase in order to build trust among the supply chain stakeholders. This chapter also elicits the architectural framework for such system.

KEYWORDS

- **auditability**
- **automotive supply chain**
- **blockchain**

- **decentralization**
- **internet of things**
- **traceability**

REFERENCES

1. Gülçin, B., & Fethullah, G., (2018). Digital supply chain: Literature review and a proposed framework for future research. *Computers in Industry, 97*, 157–177.
2. Kitchenham, B., & Charters, S., (2007). *Guidelines for Performing Systematic Literature Reviews in Software Engineering.*
3. Forsström, S., Butun, I., Eldefrawy, M., Jennehag, U., & Gidlund, M., (2018). Challenges of securing the industrial internet of things value chain. In: *2018 Workshop on Metrology for Industry 4.0 and IoT*, (pp. 218–223). Brescia.
4. Lee, J., & Pilkington, M., (2017). How the blockchain revolution will reshape the consumer electronics industry [future directions]. In: *IEEE Consumer Electronics Magazine* (Vol. 6, No. 3, pp. 19–23).
5. Kshetri, N., (2017). Can blockchain strengthen the internet of things? In: *IT Professional* (Vol. 19, No. 4, pp. 68–72).
6. Singh, S., & Singh, N., (2016). Blockchain: Future of financial and cybersecurity. In: *2016 2nd International Conference on Contemporary Computing and Informatics (IC3I)* (pp. 463–467). Noida.
7. Abeyratne, S. A., & Monfared, R. P., (2016). Blockchain ready manufacturing supply chain using distributed ledger. In: *International Journal of Research in Engineering and Technology* (Vol. 5, No. 9, pp. 1–10).
8. Apte, S., & Petrovsky, N., (2017). Will blockchain technology revolutionize excipient supply chain management? In: *Journal of Excipients and Food Chemicals* (Vol. 7, No. 3, pp. 76–78).
9. Ben-Daya, M., Hassini, E., & Bahroun, Z., (2017). *Internet of Things and Supply Chain Management: A Literature Review* (pp. 1–24).
10. Cortés, B., Boza, A., Pérez, D., & Cuenca, L., (2015). Internet of things applications on supply chain management. *Int. J. Computer Electronics. Autom.,* 2204–2209.
11. Chen, R. Y., (2015). Autonomous tracing system for backward design in food supply chain. *Food Control, 51*, 70–84.
12. Christidis, K., & Devetsikiotis, M., (2016). Blockchains and smart contracts for the internet of things. In: *IEEE Access* (Vol. 4, pp. 2292–2303).
13. Ahram, T., Sargolzaei, A., Sargolzaei, S., Daniels, J., & Amaba, B., (2017). Blockchain technology innovations. In: *2017 IEEE Technology and Engineering Management Conference (TEMSCON)* (pp. 137–141). Santa Clara, CA.
14. Lee, J., (2018). Blockchain technologies: Blockchain use cases for consumer electronics. In: *IEEE Consumer Electronics Magazine* (Vol. 7, No. 4, pp. 53–54).
15. Dorri, A., Steger, M., Kanhere, S. S., & Jurdak, R., (2017). Blockchain: A distributed solution to automotive security and privacy. In: *IEEE Communications Magazine* (Vol. 55, No. 12, pp. 119–125).

16. Conoscenti, M., Vetr, A., & De Martin, J. C., (2016). Blockchain for the internet of things: A systematic literature review. *IEEE/ACS 13th International Conference of Computer Systems and Applications (AICCSA)*, 16.

17. Banerjee, M., Lee, J., & Choo, K. K. R., (2018). A blockchain future to internet of things security: A position paper. *Digit. Commun. Networks*, 149–160.

18. Fernández-Caramés, T. M., & Fraga-Lamas, P., (2018). A review on the use of blockchain for the internet of things. *IEEE Access*.

19. Dorri, A., Kanhere, S. S., Jurdak, R., & Gauravaram, P., (2017). Blockchain for IoT security and privacy: The case study of a smart home. *Pervasive Computing and Communications Workshops (PerCom Workshops) IEEE International Conference*, 618–623.

20. Jesus, E. F., Chicarino, V. R., De Albuquerque, C. V., & Rocha, A. A. D. A., (2018). A survey of how to use blockchain to secure internet of things and the stalker attack. *Security and Communication Networks*, *2018*.

21. Huh, S., Cho, S., & Kim, S., (2017). Managing IoT devices using blockchain platform. *Advanced Communication Technology (ICACT) 2017 19th International Conference*, 464–467.

22. Khan, M. A., & Salah, K., (2018). IoT security: Review blockchain solutions and open challenges. *Future Generation Computer Systems, 82*, 395–411.

23. Restuccia, F., Doro, S., & Melodia, T., (2018). Securing the internet of things in the age of machine learning and software-defined networking. *IEEE Internet of Things Journal.*

24. Lee, B., & Lee, J. H., (2017). Blockchain-based secure firmware update for embedded devices in an internet of things environment. *Journal of Supercomputing*, *73*(3), 1152–1167.

CHAPTER 16

Understanding the Concept of Blockchain and Its Security

MRINAL,[1] ANJANA MISHRA,[1] and BROJO KISHORE MISHRA[2]

[1]Department of Information Technology, CVRCE, CVRCE, Bhubaneswar, India, E-mail: anjanamishra2184@gmail.com (A. Mishra)

[2]Department of Computer Science and Engineering GIET, Gunupur, Odisha, India

ABSTRACT

Blockchain has been a great topic of interest in the recent times and it has gained this because of the limitless possibilities it creates, and the main reason behind its popularity is the way it works. Blockchain is very useful to track records and digital transactions and due to this property this technology has been used in bitcoin.

The chapter discusses about blockchain, bitcoin, and the pillars of security of blockchain. What are the techniques used to increase the security in a blockchain. What are the security concerns that it carries, since the technology is used in bitcoin and once the protocols are set, it becomes very difficult to manipulate them, so this compares that their robustness is very valuable and so is its security. So here, we will try to explain and discuss about blockchain, bitcoin, and its security.

16.1 INTRODUCTION

16.1.1 BLOCKCHAIN

Blockchains are distributed ledger that is open to everyone and allows the participants to maintain a set of global states. It can be used as a crypto currency recording all transactions in a distributed public only ledger. Each participants participating in the process agrees to terms like existence, values, and the histories of state.

The security in blockchain is the most important part. Data that gets recorded in a blockchain is very hard to modify [1].

If we take a closer look at a single block, it mainly consists of three parts data, hash of previous and next block. There are many types of blockchains available and the type of data inside a blockchain is solely dependent on what kind of blockchain it is.

16.1.1.1 WORKING

Blockchains are connected with each other using hashes, each block contains: Data, its own hash value, hash of the previous block which is used to link the blocks together. But not all blocks have reference to the previous blocks the first or the initial block also termed as genesis block does not have any reference to the previous block. It is the first block in any blockchain-based protocol [2] (Figure 16.1).

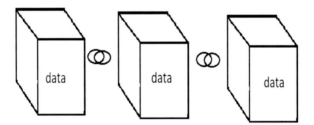

FIGURE 16.1   Structure of blockchain.

16.1.1.2 STRUCTURE

Blockchain node contains three parts:

1. **Data:** This is the main content that is inside a blockchain, the stored data depends upon what type of blockchain we are using. If we take an example of bitcoin, then the data would be details about the transactions such as sender, receiver, amount, etc. (Figure 16.2).
2. **Hash:** Hash can be compared with a very unique key. It is used to differentiate between blocks and its contents. The hash is a unique value and is always different for any two given blocks.

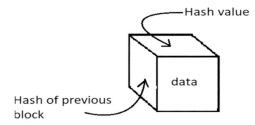

FIGURE 16.2   Structure of a block.

3. **Previous Block Hash:** This is used to link the previous block using the hash value. It is responsible for connectivity of the blockchain.

16.1.1.3   TYPES

1. **Public Blockchain:** Blockchain can be accessed by everyone. Such blockchains are completely decentralized, in which the information present is completely disclosing.
2. **Private Blockchain:** There are two types of permission in a blockchain: read and write. In private blockchain, the access to write will remain in one organization, while the access to read can be public which can even be limited further.
3. **Hybrid Blockchain:** Blockchain whose consensus process is governed by pre-selected nodes. This block has properties between private and public blockchain. The right to read the blockchain can be limited to some selected participants or can even be public. These are even called partially decentralized [3].

16.1.2   BITCOIN

Bitcoin is a cryptocurrency that has a feature in which all transactions gets recorded in a distributed append only public ledger called blockchain.

Bitcoin security depends on proof of work-based distributed protocol. These protocols are run by the network nodes called miners. In this chapter, we cover the security and privacy issue of bitcoin.

The advantage of bitcoin is more when compared to other services. If we take a scenario of money transfer, any international transaction takes multiple business days to complete and even goes through multiple sources and many financial bodies which take the extra money for a simple transaction in which bitcoin and digital transactions comes handy [4].

16.2   SECURITY IN BLOCKCHAIN

16.2.1   PROOF OF WORK

Proof of work is responsible for some amount of security. But can we only rely on proof of work for the security. Let's first see what proof of work is and how it works.

The POW is used to perform two major tasks firstly; it confirms transactions and secondly adds new blocks to the chain.

When you perform a transaction in a blockchain, multiple processes happens simultaneously. First thing that happens is the transactions are bundled together in a block. Then there are some mathematical puzzle solvers called miners, they verify the transactions by solving a mathematical puzzle known as POW. Then the miners earn some rewards who claim to solve the puzzle first among others. And finally, all transactions are verified by the consensus and stored in public blockchain [5] (Figure 16.3).

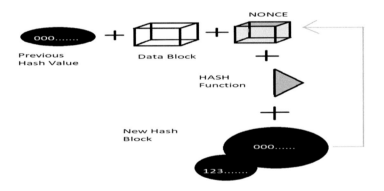

FIGURE 16.3   Proof of work working.

16.2.2   PUBLIC AND PRIVATE KEY

Bitcoin and all other major cryptocurrency are built upon the public-key (PK) cryptography. Bitcoin uses two keys: private and public. Public keys are known publically, can be accessed by anyone, and are used for identification. On the other hand, private keys are used for authentication, encryption, and are not open, they are kept secure. The generation of bitcoin address starts with the generation of private key.

16.2.2.1 PUBLIC-KEY (PK) CRYPTOGRAPHY

Security of blockchain is an issue in what we need to focus, and PK cryptography has an essential part in it, but nowadays elliptic curve cryptography is in main focus [6].

There are several processes in the working of public-key cryptography. In the beginning, the process of generating message signature occurs and it is done using private key, in which the signer has no permissions to deny signing it. Then the address exchange process occurs in which there is a receiving address and public key is treated as the receiving address. Finally, protection of the cryptocurrency is done using private key.

16.2.2.2 ELLIPTIC-CURVE CRYPTOGRAPHY

A technique that is used to create much faster, smaller, and much more efficient keys maintaining the security compared to other such algorithms. Elliptic curve theory is the main base for this encryption technique [3]. Security is governed by intractability of elliptic curve logarithmic problem [6]. Before this encryption technique traditional method of generating key was used in which key was generated through taking product of two large prime numbers but now we use elliptical curve cryptography algorithm, in which generates key by properties of equation of elliptical curve.

It is similar to RSA, having some slight differences as RSA has a better evolving capacity as it has alternative ways for researchers of such algorithms. The commercially acceptance of the EEC algorithm gives it a lot of focus and popularity. The EEC algorithm has also been adopted by major standardizing bodies such as ANSI, IEEE, ISO, and NIST [3].

16.3 CONCLUSIONS

Blockchain is a reliable technology and the security level is getting better day-by-day there are many new algorithms that are being developed. The main components of its security are proof of work, public key, private keys, public-key cryptography, elliptic curve cryptography, etc.

Blockchain has many uses across all industries like real estate, voting, identity solutions, marine insurance, food safety, FX transactions and bilateral payments and many more.

The current scenario of bitcoin security is very high and needs very much computational power and time at the same time to crack the security which is very tough to manage in the given time. The elliptical curve cryptography and PK cryptography is the main pillars of security in bitcoin after proof of work and serves the purpose of security.

KEYWORDS

- **bitcoin**
- **blockchain**
- **private key**
- **proof of work**
- **public key**
- **public key-cryptography**

REFERENCES

1. Das, S., Kolluri, A., Saxena, P., & Yu, H., (2018). (Invited paper) on the security of blockchain consensus protocols. *Information Systems Security*, 465–480.
2. Chatterjee, R., & Chatterjee, R., (2017). An overview of the emerging technology: Blockchain. *2017 3rd International Conference on Computational Intelligence and Networks (CINE).* doi: 10.1109/cine.2017.33.
3. Amara, M., & Siad, A., (2011). *Elliptic Curve Cryptography and its Applications.* International Workshop on Systems, Signal Processing and their Applications, WOSSPA. doi: 10.1109/wosspa.2011.5931464.
4. Hsieh, Y., Vergne, J., Anderson, P., Lakhani, K., & Reitzig, M., (2018). Bitcoin and the rise of decentralized autonomous organizations. *Journal of Organization Design, 7*(1).
5. Kumar, G., Saha, R., Rai, M., Thomas, R., & Kim, T., (2019). Proof-of-work consensus approach in blockchain technology for cloud and fog computing using maximization-factorization statistics. *IEEE Internet of Things Journal, 6,* 6835–6842.
6. Gao, Y., Chen, X., Chen, Y., Sun, Y., Niu, X., & Yang, Y., (2018). A secure crypto currency scheme based on post-quantum blockchain. *IEEE Access, 6,* 27205–27213.

CHAPTER 17

Sentiment Analysis for Prediction of Bitcoin Response

SUDHANSU BALA DAS and DIPAK DAS

Integrated Test Range, Defence Research Development Organization (DRDO), Chandipur, Odisha – 756025, India

ABSTRACT

Bitcoin has appeared as the highest fortunate cryptocurrency from the last few years. The social network platform Twitter has become a beneficial point of user sentiment. This chapter mainly focuses on the integration of sentiment analysis to predict the response about bitcoin in social media, i.e., Twitter. Through different tweets in Twitter, we have computed the sentiment score. In this chapter, we have taken a survey for knowing what people think about bitcoin from their tweets in Twitter from a single month of both the year, i.e., 2018 and 2019 and compared those. Our result seems to confirm that the trend of Bitcoin price may be predicted from the sentiment without incorporation of complex business models.

17.1 INTRODUCTION

Social networking websites have revolved the netting into a broad warehouse of statements of opinions on many subjects, creating a possible origin of data for different research. The opening of major computerized social data from different sites has already been taken the limelight in the field of research. The information which we are getting from social media platform helps us to predict the sentiment of the users. In simple way, sentiment can be defined as "a look at or the viewpoint that is being conveyed." But, according to economics, market sentiment can be classified as being positive, negative, or neutral about a certain investment. However, determining market sentiment using different sites is a little tricky. Today, enormous things have happened worldwide that attract the market and one of them is about cryptocurrencies. In today's world, the domain of cryptocurrency [1] is very recent and emerging. Cryptocurrencies are digital currency that are generated and handled using leading encryption methods and used by the private parties or groups. It creates the rise from being a scholastic approach to in existence with Bitcoin in the year of 2008. This was proposed in 2008 in a Ref. [2] originated by someone behind a pseudonym of Satoshi Nakamoto. His connection with the project completed in 2010, but many other developers from the Bitcoin community have also contributed to it. Even though Bitcoin used to get popularized from consecutive years, it grabs meaningful focus by media in 2013 for a sudden surge of its value. Bitcoin is a type of e-money with no interference of the third party which can be used for online transactions. It authorizes all works such as currency insurance, business undertaking, and authentication to be accomplished jointly by the network. While this decentralization contributes to Bitcoin to be liberated from authority control or conflict, the contradiction is that there is no main control to guarantee that the things work very well or to back the value of a Bitcoin. It is built digitally using a 'mining' method over a cloud of secure and intelligent computation using complicated algorithms and encryption.

FIGURE 17.1   Physical tokens (cash).

From different surveys, it is found that when there is a growth in price, people uploads positive messages over their social media account, and when the prices fall down people shows negative reaction with negative posts. From the beginning, we have been using physical tokens for the means of payment. In this type of scenario, there is a direct communication between dealer's goods and purchaser's tokens which grant them to perform an instantaneous settlement as shown in Figure 17.1. But this choice is inaccessible, when both the users are not available in the identical place, necessitating the usage of digital tokens. The medium of settlement in digital currency system is commonly, a string of bits. It is difficult to prevent the purchaser by continuing the same bit string repeated again and again. This is known as the double-spending problem.

This issue can be resolved in presence of a trustworthy third party who would supervise both parties and helps them to transfer balances into their respective accounts as shown in Figure 17.2. But mostly, it would be difficult to adopt due to lack of trust on third party. Under such scenario, when there is an absence of a trusted third party, Bitcoin is used as a digital means of payment in a distributed network as shown in Figure 17.3.

FIGURE 17.2   Exchange of credit using a third party.

FIGURE 17.3   Digital currency in payment.

17.2   LITERATURE SURVEY

Satoshi Nakamoto [1] projected the idea Bitcoin on 31 October 2008 in an article named "Bitcoin: A Peer to Peer Electronic cash system." Later in 2009, Nakamoto put Bitcoin as open-source code to become with one decentralized peer-to-peer cryptocurrency. In January 2009, the Bitcoin network came into permanence with the first Bitcoin clients and the first issuance of Bitcoin was witnessed with Satoshi Nakamoto mining the first block of Bitcoin known as the genesis block, and receiving 50 Bitcoin reward. Hal Finney, a programmer is one of the first defender, adopters, contributors, and receivers of the first Bitcoin transaction; he received 10 Bitcoin from Nakamoto [1]. Mankiw [5] outline three norms of successful currency: a medium of exchange, a unit of account and a store of value. The evaluation of Bitcoin matching with this outline is explained by Yermack [6], Lo, and Wang [7] or Kancs, Ciaian, and Rajcaniova [8]. Earlier research articles [8, 9] explained about the supply-demand connection, global macroeconomic and financial growth. Similarly, Pang et al. [12] and Kristoufek [9] display the consent proof about the connection between the social activity and Bitcoin price. Abbasi et al. [11] define sentiment as any misperception that can lead to mispricing in the fundamental value of an asset. In this context, Bitcoin is exposed to sentiment. On top of that, Bitcoin works on a new platform Blockchain, which is not common in the field pertaining to economics. Moreover, big institutions are not a part of the Bitcoin market to increase the trust in Bitcoins. On contrary, the Bitcoin economy consists rather of small business and individuals who use Bitcoins as a medium of exchange. Therefore, in parallel with theory of finance, speculative investments in Bitcoin are more likely driven by retail or individual investors called noise traders because big institutional investors are not yet partaking in this market. According to behavioral finance research [5, 6], noise traders are prone to behave according to less rational factors like sentiment. Other researchers have specifically studied the efficacy of sentiment analysis of tweets [3, 4, 13]. R. Piryani et al. found that standard natural language processing (NLP) techniques such as sentence level and document level sentiment scoring was ineffective due to the short nature of tweets and uniqueness of language used [14]. Alexander Pak and Patrick Paroubek showed that separating tweets into positive, negative, or neutral categories could result in effective sentiment analyses [15]. O'Connor et al. showed that the sentiment found in tweets was reflective of public opinion on various topics in national polling [16].

17.3   SENTIMENT ANALYSIS AND BITCOIN

The scrutiny explained later in this chapter needs an analysis of where and why the data has been collected and also about how crypto currencies may vary from standard fait currencies or stocks in companies to traditional stock markets. Sentiment Analysis from texts (uttered/drafted) is done through NLP method which mainly takes a content which is drafted in natural language and then extract the meaning of those contents. Using it, we can analyze and draw out facts from the contents; and can allow companies or organizations to perceive the social sentiments of their variety, goods, or assistance while following online communications.

Currently, the sentiment is a vast active aspect in the cryptocurrency market. It has been on the rise from last few years. Concluding market sentiment using social media analytics outcome is little bit tricky. The strength of currency generally depends upon various aspects such as market assurance, acknowledgment, and public anticipation. Unlike fiat money, Bitcoin is not regulated through any government administration which can influence its cost and supply. This approach removes third-party issues such as involving any banking system. Basically, it is an open-source software program, executable on system computers, i.e., nodes. The single way to boost the supply of Bitcoin is to partake in transaction calculations, which impels for growth of Bitcoin supply and pays for the infrastructure. At the same time, the budgetary expense of Bitcoin is affected by the same variables as that of fiat currency. Such a blockchain method is an increasingly accepted basis for the growth of a new business idea providing access to users of a commodity to be a factor of the value chain and giving priority for peer-to-peer actions.

Nowadays, the blockchain method is appreciated to transfer value, to gather and distribute data as well as distribute data, and to disperse many companies by pushing the normal people away from the system. Hundreds of industries pop up daily introducing benefits based on blockchain technology [2]. These benefits contain tour, real estate, logistics, transmit, matchmaking, leisure, and more. The main objective is to decrease costs, to increase efficiency and to provide transparency. Normally in Twitter, users post their views regarding interesting topics and show their positive or negative response about them. Mostly 80% of the world's digital information is unorganized and information achieve from social media is not omission to that. As the data is not arranged in any sequential order, so it's tough to sort and determine. But, still due to NLP, it's possible to develop different models that acquire information from examples and can be used to refine and organize text data. Twitter sentiment analysis systems grant us to arrange huge quantity of tweets and to identify the action of each word undoubtedly the most important part of this is that it is speed and simple which preserve company's important time and also help them to target on function where they can give a great impact.

17.4 ANALYSIS METHODOLOGY

Sentiment analysis is a utilization of machine learning (ML) in which a part of text is examined using a computer model to decide the 'sentiment' of the message (tweet) such as positive or negative about an individual content. This needs a platform or way for implementation. Python is considered as the platform for examining any type of source script due to its NLP capability to gather with libraries and packages. First, we need to accumulate data taken out from the chosen place, i.e., in our case its Twitter data. The data need to go through different steps which prepare it more realistic for machine then its earlier state. The tweets can be examined and distinguished according to emotion of the users. We try to categorize the tweets in the form of positive and negative aspects. But, if the tweet has the pair of positive and negative aspects, the more assertive sentiment needs to be selected as concluding label. So, to start with python, first, we need to install libraries like 'tweepy,' 'pandas,' 'textblob,' 'matplotlib,' and 'numpy' (Figure 17.4).

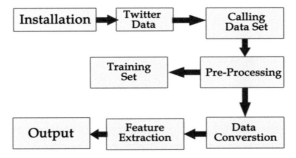

FIGURE 17.4   Procedure for sentiment analysis.

Then for Twitter data analysis (DA), we need to go to the Twitter Developer website and log into our account. Then, for authentication, we need to give the following data such as consumer key, consumer secret, access token, and access secret. As tweets, contain so many unnecessary words, i.e., slangs, and unnecessary punctuations. So, we need to filter those prior to the use for training the ML model as shown in Figure 17.4.

17.4.1   COLLECTIONS OF TWEETS AND PROCESSING

We have done sentiment analysis based on the tweets about Bitcoin from past two years where we have considered a single month for both the years, i.e., for July 2018 and July 2019 (flowchart is shown in Figure 17.5). The proposed work is done in two aspects, i.e., (i) collecting data set, and (ii) training data set. In the first step, first we have collected the Twitter data about the Bitcoin. The collected Twitter data is not in correct format which is required to find out the sentiment analysis. So, we convert this data in CSV format. Then, for training data sets, we remove the unnecessary tweets from it. The first step in selection of the desired tweets was to find the hashtag for the cryptocurrencies. For this, we utilize Tweepy-an open-source Python library for accessing the Twitter API, to collect Twitter data. Tweepy allows for filtering based on hashtags or words. There are multiple ways in which the cryptocurrencies of interest may be referred to in tweets. The most direct way is by using a hashtag ("#") followed by "Bitcoin." For this work, first information is extracted as tweets with the name of Bitcoin.

In our case, we have collected data for past two years of same month, i.e., July 2018 and July 2019. Those are filtered and properly formatted for training the ML model and finally, we processed that data as shown in Figure 17.6. Then, we have to mention the total numbers of tweets which need to be analyzed for sentiment analysis. It shows the average result of user's reaction and sentiments. The tweets are examined based on their sentiment polarity as shown in Figure 17.7.

17.5   RESULTS

For July 2018, we analyzed 3034 Tweets, and labeled 1376 as SPAM or USELESS, and 1658 as decent QUALITY or USEFUL as shown in Figure 17.8. Then by using AI (artificial intelligence) we have found out sentiment analysis for July 2018 and wherein marked 10.31% tweets as Angry, 3.01% as Fear, 2.41% as Bored, 0.18% as Sarcasm, 36.30% as Excited, 4.16% as Sad, and 45.17% as Happy which is shown in Figure 17.9. Finally, the sentiment for Bitcoin is shown in pie chart as shown in Figure 17.10.

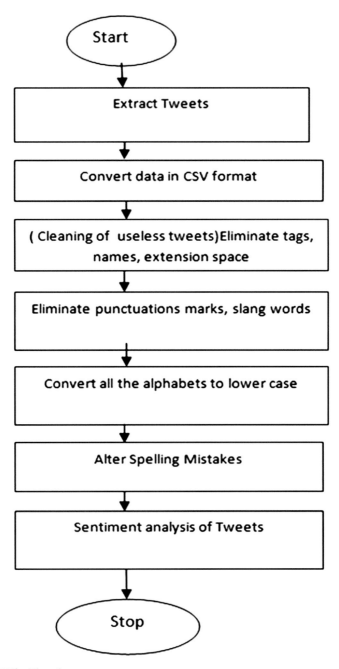

FIGURE 17.5   Flowchart.

```
def clean_tweet(self, tweet):|
    return ' '.join(re.sub("(@[A-Za-z0-9]+)|([^0-9A-Za-z \t])|(\w+:\/\/\S+)", " ", tweet).split())
```

FIGURE 17.6   Cleaning of useless tweets.

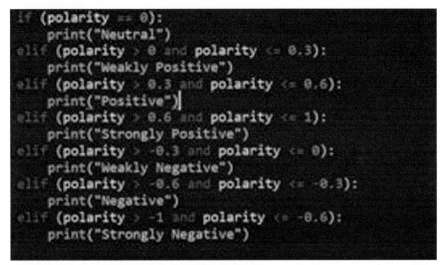

```
if (polarity == 0):
    print("Neutral")
elif (polarity > 0 and polarity <= 0.3):
    print("Weakly Positive")
elif (polarity > 0.3 and polarity <= 0.6):
    print("Positive")|
elif (polarity > 0.6 and polarity <= 1):
    print("Strongly Positive")
elif (polarity > -0.3 and polarity <= 0):
    print("Weakly Negative")
elif (polarity > -0.6 and polarity <= -0.3):
    print("Negative")
elif (polarity > -1 and polarity <= -0.6):
    print("Strongly Negative")
```

FIGURE 17.7   Sentiment polarity of different tweets.

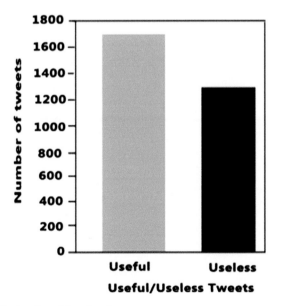

FIGURE 17.8   Ratio of useful and useless tweets.

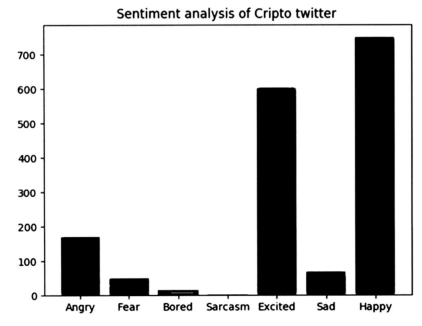

FIGURE 17.9   Bar graph for representing different sentiment of tweets.

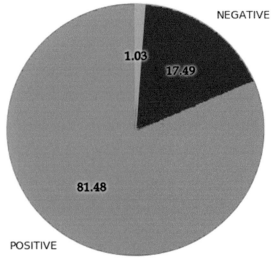

FIGURE 17.10   Pie chart for representing tweets in the form of percentage.

Similarly, for July 2019 have taken the tweets and did sentiment analysis for that tweets where the program analyzed 2030 Tweets, and labeled 1230 as SPAM or USELESS, and 800 as decent QUALITY or USEFUL as shown in Figure 17.11. Our AI for sentiment analysis marked 0.54% tweets as angry, 2.81% as fear, 48.28% as bored, 23.00% as sarcasm, 4.63% as excited, 8.62% as sad, and 12.07% as happy as shown in Figure 17.12. Finally, the sentiment for Bitcoin is shown in pie chart as shown in Figure 17.13.

From the above results it is found the words mostly used in Twitter about bitcoin are good, bad, market, rate, and currency. We also find out a downward-trend in number of tweets pertaining positive emotions.

The trend of Bitcoin price is determined comparing its variations with the number of tweets, with the number of tweets with positive mood, and with Google Trends results. Our result seems to confirm that volumes of exchanged tweets may predict the fluctuations of Bitcoin's price. Furthermore, the comparison between tweets with a positive mood and trend of Bitcoin's price seems to prove this behavior as shown in Figure 17.14.

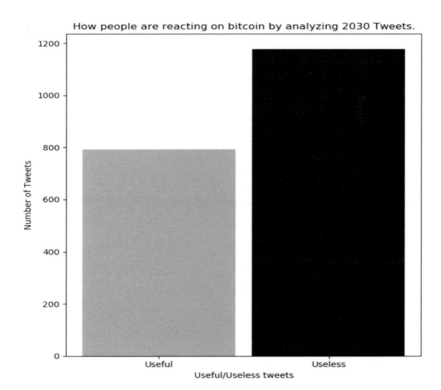

FIGURE 17.11   Ratio of useful and useless tweets.

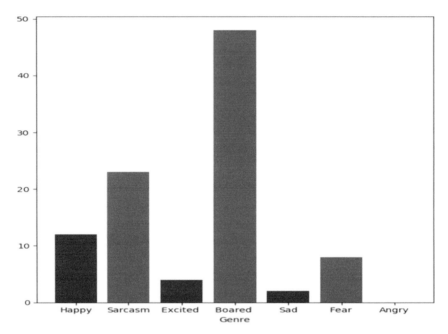

FIGURE 17.12   Bargraph for representing different sentiment of tweets.

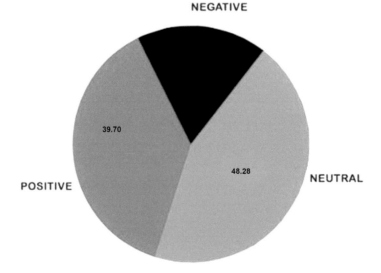

FIGURE 17.13   Piechart for representing tweets in the form of percentage.

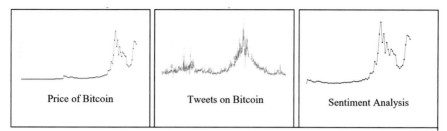

FIGURE 17.14   Price trend of bitcoin matching to sentiment analysis.

17.6   CONCLUSION

In this chapter, we examine a relationship between user's reaction in social media and cryptocurrency. This chapter also describes different emotions of people about bitcoin, i.e., sad, angry, excited, happy. The tweets, being analyzed, prove that the trend of cryptocurrency can more or less be predicted through them. For future work, more data will be analyzed not only from Twitter but also from other social media platform. This will pave the way to predict various trends of cryptocurrency without applying any complicated model of economics and market trend.

KEYWORDS

- **artificial intelligence**
- **bitcoin**
- **blockchain**
- **natural language processing**
- **sentiment analysis**
- **Twitter data**

REFERENCES

1. Santoshi, N., (2008). *Bitcoin: A Peer-to-Peer Electronic Cash System.* https://bitcoin.org/bitcoin.pdf (accessed 10 August 2021).
2. Antonopoulos, A. M., (2017). *Mastering, Bitcoin: Programming the Open Blockchain.* O'Reilly Media, Inc.

3. Georgoula, I., Demitrios, P., Christos, B., Dionisios, S., & George, M. G., (2015). *Using Time-Series and Sentiment Analysis to Detect the Determinants of Bitcoin Prices.* Available at: SSRN: 2607167.

4. Li, T. R., Anup, C., Xander, F., Nicholas, R., & Feng, F., (2019). Sentiment-based prediction of alternative cryptocurrency price fluctuations using gradient boosting tree model. *Frontiers in Physics*, 7.

5. Baker, M., & Wurgler, J., (2007). Investor sentiment in the stock market. *Journal of Economic Perspectives, American Economic Association, 21*(2), 129–152.

6. Yermack, D., (2015). Is bitcoin a real currency? *The Handbook of Digital Currency (Elsevier),* 31–44.

7. Lo, S., & Wang, C. J., (2014). Bitcoin as money? *Federal Reserve Bank of Boston, Current Policy Perspectives*, 14–24.

8. Nai, F. I., Steri, G., Fontana, A., Ciaian, P., Kancs, D., & Nordvik, J. P., (2015). *On Virtual and Crypto Currencies: A General Overview, from the Technological Aspects to the Economic Implications*. JRC Technical Report JRC99978.

9. Kristoufek, L., (2015). What are the main drivers of the bitcoin price? Evidence from wavelet coherence analysis. *PLoS One*, 10.

10. Barber, B., & Odean, T., (2013). The behavior of individual investors. In: Constantinides, G., Harris, M., & Stulz, R., (eds.), *Handbook of the Economics of Finance* (Vol. 2, pp. 1533–1570A). Elsevier, Amsterdam.

11. Abbasi, Chen, H., & Salem, A., (2008). Sentiment analysis in multiple languages: Feature selection for opinion classification in web forums. *ACM Transactions on Information Systems*, *26*(3), 1–34.

12. Pang, B., & Lee, L., (2008). Opinion mining and sentiment analysis. *Foundations and Trends in Information Retrieval, 2*(1/2), 1–135.

13. Ji, F., & Bi, C., (2011). Incorporating lexicon knowledge into SVM learning to improve sentiment classification. In: *Proceedings of the Workshop on Sentiment Analysis Where AI Meets Psychology (SAAIP)* (pp. 94–100).

14. Piryani, R., Madhavi, D., & Singh, V. K., (2017). Analytical mapping of opinion mining and sentiment analysis research during. *Information Processing and Management*, *53*(1), 122–150.

15. Pak, A., & Paroubek, (2010). Twitter as a Corpus for Sentiment Analysis and Opinion Mining, European Language Resources Association (ELRA), Proceedings of the Seventh International Conference on Language Resources and Evaluation (LREC'10), http://www.lrec-conf.org/proceedings/lrec2010/pdf/385_Paper.pdf (accessed 10 August 2021).

16. Brendan, O., Ramnath, B., Bryan, R. R., & Noah, A. S., (2010). From tweets to polls: Linking text sentiment to public opinion time series. Proceedings of the Fourth International AAAI Conference on Weblogs and Social Media., https://www.aaai.org/ocs/index.php/ICWSM/ICWSM10/paper/viewFile/1536/1842 (accessed 10 August 2021).

CHAPTER 18

Blockchain Makes Certain Truthfulness of Supervised Predictive Clinical Trial Data Analysis

D. ANAND KUMAR[1] and R. N. V. JAGAN MOHAN[2]

[1]*Research Scholar, GIET University, Gunupur, Odisha, India,
E-mail: anand.k.deva@gmail.com*

[2]*Associate Professor, SRKR Engineering College, Bhimavaram,
Andhra Pradesh, India, E-mail: mohanrnvj@gmail.com*

ABSTRACT

Blockchain skill is a make use of the aged computer science method recognized as hashing which creates an inimitable digital signature for each data block. The hashes build up serially, as new data is entered or altered and with every new block depending on the last. The ensuing blockchain creates an audit trial for valves that is easy to decode and validate even without looking at the actual data. Each time new clinical trial data is entered, the timestamp and the information containing the patient as well as his clinical reports and treatment are stored along with the hash of the previous block of data with respect to that patient on a new block with its own individual signature and this technology provides data exchanges that are more secure, more efficient and more transparent. In this chapter, by considering a blockchain of clinical trial data, supervised learning of machine learning (ML) methods are used for maximum likelihood estimation prediction of patients' diseases based on the observed symptoms. By combining both ML and blockchain technologies together, patients get the benefit of analyzing large amounts of data quickly with increased security. The experimental result estimates the blood pressure by the prediction method, least square regression and also finds the association between age and blood pressure using the correlation coefficient for clinical trial data.

18.1   INTRODUCTION

By way of the blockchain and artificial intelligence (AI), get together with the improved public even reaching above all trendy. There is in no way have been a better time to explore the intersection of both these skills. A blockchain can be defined as a distributed, decentralized, unchallengeable database (DB) for storing encrypted data. AI allows us to develop systems and models that are capable of doing tasks with more intelligence. Both the technology has their own individual level of complexity, but the merger of the two is beneficial to us in creating intelligent systems that can do better predictions. There is a possibility to join both Blockchain and AI together and use it for real world applications. In fact, over the past decade both technologies are developed and used in isolation, but now the situation was changed. AI revolutionized the progress being made in handling enormous amount of trajectory data accumulating from various sources which are to be protected from illegal and immoral activities. Blockchain and machine learning (ML) technologies can make a significant impact in the way the data is handled by providing more security and immutability [1]. Blockchain became the only feasible solution to store the evolving enormous amounts of data. Unlike cloud, it is a distributed storage in which data is stored in the form of blocks that are distributed across the entire network of computers like peer-to-peer. There is no central controlling authority and each computer or node stores a complete copy of the ledger which means that if one or two nodes in the network are compromised, the data remains unchanged and we will not lose it. Entire data stored in the Blockchain is encrypted and it cannot be tampered. So blockchain is the perfect storage method for sensitive or personal data which can be processed and analyzed with care by the use of various ML algorithms that can predict the occurrence of future events based on past experiences and take better decisions for clients. For example, consider healthcare industry where the generated trajectory clinical trial data from various patients from different locations is analyzed to detect, diagnose, and treat the diseases. In this chapter, we considered a private Blockchain for which public key cryptography; access control is provided during development of the chain, so that it is accessed only by the authorized parties who have appropriate access permissions. This is also known as a permission chain because only trusted and authorized parties will add new data blocks to the chain.

18.2 SUPERVISED LEARNING CLASSIFICATION FOR CLINICAL TRIAL DATA

Supervised learning akin to the name indicate the being there of a supervisor as a doctor. The basically supervised learning is a learning in which we diagnosis or predictive train the data analysis (DA) the machine using patient data which is well labeled that means some data is before now tagged with the right respond. After that, the machine is provided with a new set of clinical trial data so that supervised learning process analyses the healthcare training data and produces a correct outcome from considered data. The deferred repair to patients in critical and emergency cases may result in a huge cost and sometimes result in death. So after the prediction of a disease, it is important for hospital managers and doctors to advice the patient at right time to take necessary clinical treatment to overcome the severe affects of that disease with in stipulated time. To accomplish this, a common statistical method maximum likelihood estimation is used for estimating the parameters of a specific ML model, where we find the parameter values that maximize the likelihood of the observations for given parameters. For example, we are having the data of N patients where the probability of disease attack for a patient is 'p' and the probability of a not having a disease is '1–p.' If 'n' number of patients attacked with a disease and 'N–n' people fail to have that, then the likelihood is proportional to the product of the probabilities of successes and failures or (pn × (1 − p) (N − n)). Where n/N is the value of p that maximizes the likelihood of a disease for a particular patient based on his observed parameters.

18.3 CLINICAL TRAIL PREDICTIVE DATA ANALYTICS USING BLOCKCHAIN APPROACH

In supervised learning-based healthcare field, a private Blockchain skill is used to store the details of the patients and their treatment in blocks. This ensures that anyone who has access to this Blockchain can access the patients' clinical data. If we create a blockchain for storing the records of the patients in hospitals, it becomes very secure and no one can access the patients' details directly and make changes to it there by ensuring the anonymity of the patient. Only doctors are allowed to watch the details of the patient by using their details and public key with respect to the patient [2].

The present situation for many associations is that predictive analytics relies on a very small past data. This makes it difficult for data analytics to predict future affects effectively. A typical state of predictive analysis faces three key challenges namely as:

1. Lack of crucial history data to enable future predictive modeling.
2. Expensive data scientists that may or may not be effective due to limited history data.
3. Limited ability to scale.

With blockchain, these three problems can be overcome in one go. By appending all transactional data into the blockchain, the process of collecting and cleaning data for uniformity becomes simplified. Additionally, with the availability of better volume of data, the performance of predictive data models too is likely to go higher. The results of blockchain with predictive DA are evident. Blockchain shows the world a totally different view of how data can be stored and shared in a transparent, open source method. All three benefits of security, immutability, and transparency, offer significant ramifications for health care industry. With the data analytics domain, blockchain serves to facilitate better data availability in order to carry out powerful predictive modeling and statistical analysis. By using clinical trial data, predictive models can analyze the patient's condition and identify the potential risks caused due to the occurrence of several diseases. ML models can be trained to make accurate clinical predictions about the patient's condition by using history data. Blockchain technology uses distributed DB for data storage and sharing [3]. It also ensures that the data is valid and secured.

In peer to peer blockchain network, all nodes have access to the copy of distributed DB ledger which contains the records of every single patient connected to the blockchain. By retrieving, the data blocks from the blockchain DB we can apply ML methods to develop the models for analyzing the data in a better way for making accurate and efficient predictions fastly [4]. The nodes in the peer to peer network will authorize the patients' records. If the record gets verified, it will be attached as a new block with other valid records of the patient to form a chain. The new block is marked with previous block's hash and timestamp will be added to the existing blockchain. Once the block is added, it is permanent and unaltered. This method can also protect patient's privacy. With this technology latest treatment information, Clinical history of the patient, laboratory test results done by various physicians is also available. On the basis of Patients symptoms, clinical history, and examining the data of other similar types of patients suffering with same problems or symptoms the trained model can predict the disease and gives clinical suggestions in a cost-effective way (Figure 18.1).

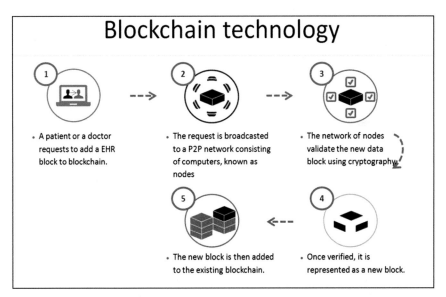

FIGURE 18.1   Patients clinical trial information storage using blockchain technology.

18.4   INDISPENSABLE PROCESS OF BLOCKCHAIN

Blockchain is a decentralized, shared public DB organized in the form of a linked list, with each block containing a hash of its preceding block. Proof-of-work algorithms are used to form each block, to attain consensus of this distributed system through the longest possible chain. Apart from the trustless distributed system bitcoin, blockchain is extendable in several ways by modifying the parameters of the chain. Data is collected from intended sources connected to the blockchain and processed to fit in a block by a process called mining. A cryptographic hash (also known as a digital fingerprint) is used to identify every block. The new block formed consists of a hash its previous block, so that all the blocks are linked in the form a chain from the genesis block to the newly formed block.

This technology is designed in such a way that the average time for generating a block remains fairly constant. The block generation time is controlled by using a difficulty value within the block. In the case of bitcoin, the blocks hash value is strictly smaller than a given value to be accepted. The given value depends on the total computation power of the Blockchain network. If the given value is smaller, it indicates that the Blockchain network is more powerful and it is more difficult to generate the block. The validated blocks are added to the blockchain in the order as determined by the consensus algorithm. Some nodes in a blockchain network are responsible for executing the consensus algorithms which validates the transactions and determines the order in which the new blocks are added to the blockchain. These nodes are called miners and this process is referred as mining. Once a miner receives a transaction proposal, it will proceed to verify whether the transaction is valid or not. Only valid transactions are integrated into a block. Later, the new block consisting of valid transactions is connected to the preceding blocks thereby creating a chain of blocks, known as blockchain [5]. Among all the nodes in the network, the blockchain is replicated, and every node has an identical DB of all the transactions in the network.

18.5 PROGRESSION OF BLOCKCHAIN WITH DATA AND SECURITY

Blockchain is an open, de-centralized, translucent catalog that records transactions between two parties, without the need for a third-party authentication. It makes sure that one piece of data can only have one owner, and that data can't be copied, only distributed. It is a linked collection of records called blocks. Every block contains a cryptographic hash of its previous block, a timestamp identifying the time and date of the transaction, and transaction data. The blockchain skill functions are based on the verification of a hash and digital signatures. Hashing is the process that the Blockchain uses to confirm its state. Each transaction requires one or more digital signatures which are obtained by combining public and private keys. Signatures ensure that the transaction is only made by the authorized parties. Further data blocks of the blockchain are secure as they are stored on different nodes which are not connected to the same processor. For this reason, the records can't be retroactively changed without changing all other blocks [6, 12].

In the case of specific diseases, in order to determine the effectiveness of particular medication, clinical trials are used. These tests may prove or disprove an offered hypothesis. During clinical trials, physicians will get a lot of valuable and sensitive information regarding statistics, laboratory test results, quality reports, demographics, etc. Criminals are interested in recording some of this sensitive data and misuse it for their personal benefit. To prevent this, the Blockchain technology provides an authentication method which allows users (patients and the physicians) to prove the authenticity of any document or test result registered in the system [7].

Proof-of-existence is provided in blockchain by adding data in the form of the transaction and validating it by all system nodes of the network. The data in the blocks of blockchain is immutable. This distinguishing characteristic allows storing the results of clinical trials in a secure way, making it impossible to alter the data. The proof-of-existence in the case of clinical trials is provided by comparing a unique data code, generated by the system using SHA256 method, with the original makes it possible to verify whether the data of clinical trials is modified or not. For each patient, the blockchain creates a hash for his health information block, together with a patient ID. Only authorized persons are allowed to retrieve necessary information without revealing a patient's identity. A patient can also decide to whom the access will be given and whether the access is full or partial [8]. The process of prediction from Clinical Trial Data Using Blockchain technology is as shown in Figure 18.2.

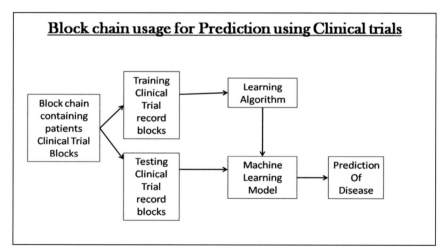

FIGURE 18.2   Predictions from clinical trial data using blockchain technology.

18.6   EXPERIMENTAL RESULTS

As part of our experimental results, the blood pressure of a woman is predicted from the following data, which shows the parameters age X, and systolic B.P.Y of 12 women and the two variables age X and B.P.Y correlated (Figure 18.3). The list of blood pressure of a woman is predicted clinical trial data is following two operations are (a) and (b) is below as follows:

Age (X)	B.P. (Y)
56	147
42	125
72	160
36	118
63	149
47	128
55	150
49	145
38	115
42	140
68	152
60	155

FIGURE 18.3   Age and blood pressure information of woman extracted from clinical trial data.

a. To estimate B.P., determine the prediction equation which is the least square regression equation of Y on X. We assume it to be Y = a0 + a1X

Its normal equations are:

$$\sum Y = Na0 + a1 \sum X$$

$$\sum XY = a0 \sum X + a1 \sum X^2$$

From the given data, $\sum X = 628$; $\sum Y = 1684$; $\sum X^2 = 34416$; $\sum Y^2 = 238822$; $\sum XY = 89894$; $N = 12$.

On substitution, we have:

$$1684 = 12a0 + 628\ a1$$

$$89{,}894 = 628a0 + 34416\ a1$$

From above two equations we get a0 = 80.77738, a1 = 1.138005.

So the prediction equation becomes Y = 80.777 + 1.138X.
Now the B.P. of a women with the age X = 45 is obtained as

Y (45) = 80.777+1.138(45) =131.987225 ≈ 132.

b. To find the association between parameters age and B.P, determine
the correlation coefficient 'r'

$$r = \frac{N\sum XY - \sum X \sum Y}{\sqrt{\left[N\sum X^2 - (\sum X)^2\right]\left[N\sum Y^2 - (\sum Y)^2\right]}}$$

$$r = \frac{12(89894) - (628)(1684)}{\sqrt{\left[(12)(34416)\right]\left[(12)(238822) - (1684)^2\right]}}$$

r = 0.8961

Age X and B.P. Y are strongly positively correlated.
To plot this Figure 18.4 namely, prediction of blood pressure of a woman
from clinical trial data we considered patient ages in one vector and their
corresponding blood pressure in second vector and we used parametric
likelihood-based estimation for predicting the clinical trial data.

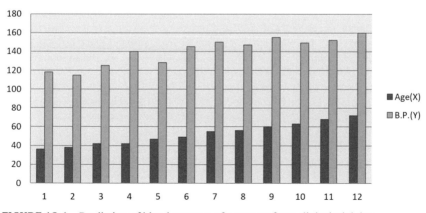

FIGURE 18.4   Prediction of blood pressure of a woman from clinical trial data.

18.7 CONCLUSION

Blockchain technology is emerging in the field of healthcare. Healthcare payers and providers are using this technology to handle clinical trials data and electronic medical records while maintaining immutability, security, and integrity. Better data sharing between healthcare providers means a higher likelihood of accurate diagnoses and more effective treatments. By considering, the clinical trial data retrieved from health care blockchain and applying ML models to that, we can predict the disease diagnosis for a given set of symptoms more accurately. Further, this can be extended to deep neural networks which allow using radiology images for better and more accurate diagnoses.

KEYWORDS

- **blockchain technology**
- **correlation coefficient**
- **digital signature**
- **hashing**
- **machine learning**

REFERENCES

1. Vyas, S., Gupta, M., & Yadav, R., (2019). Converging blockchain and machine learning for healthcare. *Amity International Conference on Artificial Intelligence (AICAI).*
2. Alhadhrami, Z., Alghfeli, S., Alghfeli, M., Abedlla, J. A., & Shuaib, K., (2017). Introducing blockchains for healthcare. *International Conference on Electrical and Computing Technologies and Applications (ICECTA).*
3. Zhang, G., Li, T., Li, Y., Hui, P., & Jin, D., (2018). Blockchain-based data sharing system for AI-powered network operations. *Journal of Communications and Information Networks, 3*(3).
4. Francisca, A. A., (2018). Big data, machine learning, and the blockchain technology: An overview. *International Journal of Computer Applications (0975-8887), 180*(20).
5. Prisco, G., (2016). *The Blockchain for Healthcare: Gem Launches Gem Health Network with Philips Blockchain Lab.* [online]. Available at: https://bitcoinmagazine.com/business/the-blockchain-for-heathcare-gem-launches-gem-health-network-with-philips-blockchain-lab-1461674938 (accessed 20 July 2020).
6. Nakamoto, S., (2008). *Bitcoin: A Peer-to-Peer Electronic Cash System,* https://bitcoin.org/bitcoin.pdf (accessed 20 July 2020).

7. Tasatanattakool, P., & Techapanupreeda, C., (2018). Blockchain: Challenges and applications. *International Conference on Information Networking (ICOIN)*.

8. Mettler, M., (2016). Blockchain technology in healthcare: The revolution starts here. *IEEE 18th International Conference on e-Health Networking*. Applications and Services (Healthcom).

9. Casino, F., Dasaklis, T. K., & Patsakis, C., (2019). A systematic literature review of blockchain-based applications: Current status, classification, and open issues. *Telematics Inform*.

10. Nichol, P. B., (2016). *Blockchain Applications for Healthcare*. [online]. Available: http://www.cio.com/article/3042603/innovation/blockchain-applications-for-healthcare.html (accessed 20 July 2020).

11. Novikov, S. P., Kazakov, O. D., Kulagina, N. A., & Azarenko, N. Y., (2018). Blockchain and smart contracts in a decentralized health infrastructure. *IEEE International Conference "Quality Management, Transport, and Information Security, Information Technologies."*

12. Singh, S., & Singh, N., (2016). Blockchain: Future of financial and cybersecurity. *2nd International Conference on Contemporary Computing and Informatics (IC3I)*.

CHAPTER 19

Blockchain Can Strengthen the Trustworthiness of Meta-Analysis

YUGANDHAR BOKKA[1] and R. N. V. JAGAN MOHAN[2]

[1]Research Scholar, GIET University, Gunupur, Odisha, India,
E-mail: yug.599@gmail.com

[2]Associate Professor, SRKR Engineering College, Bhimavaram,
Andhra Pradesh, India, E-mail: mohanrnvj@gmail.com

ABSTRACT

Blockchain expertise in arrears the secure dealings of cryptocurrencies similar to bitcoin can construct it easy to make transparent meta-analysis where reproducibility is raising concern. With the facility to form time-stamped and tamper-proof, records of measures that can reinforce the simplicity and reproducibility of meta-analysis. In this chapter, to construct a fog/cloud computing (CC) mechanism that should frequently record and track analysts' decisions like how studies are gathered and analyzed as part of a meta-analysis. In case, where meta-analysis current unpredictable outcomes. The blockchain can aid locate precisely how they chanced based on individual timestamps of logical decisions. The suggested mechanism would faultlessly form trustworthy patient clinical trial demographic profile documents without adding the difficult process of conducting a meta-analysis with the help of supervisor learning. The experimental result is on security based cloud access demographic clinical trial patient profiles from the cloud environment for classification purposes.

19.1 INTRODUCTION

In modernizations have the new technologies is altered day-by-day from present and future scenario. One can be the new technology is *cloud computing*

(CC), it changes the way you can use systems. CC plays a vital role that how we can store the data evidence and how it can run the applications instead of your programs and data on a separate system, and all hosted in the cloud. Leases can access data from CC and whole applications from anywhere across the globe, release you from the restrictions of the system, and make it easy for groups in various places to cooperate for access the cloud. Cloud is a cluster in enormous of unified computers, it can be individual computers, or servers they can be public or private. This cluster of cloud computers encompasses outside a single firm or enterprise. Any authoritative user can access these documents and applications from any computer over the internet.

Cloud computing is the distribution of not the same services over the net-world, cloud-based storage creates possible to maintain records to a remote database (DB) and extract them [1]. By using cloud storage, don't have to store the information on your system. Instead, you can access it from any location and download it. We can also alter files, simultaneously with other users. Cloud computing uses an internet skill and central for remote server is to support data and appliance [2]. It permits customers and trades to use appliances without mechanism and access their individual records at any computer with internet access. These cloud storage providers are in authority for custody the data obtainable and accessible, and the bodily milieu nearing extinction and consecutively.

19.2   WELL-BEING FOR DEMOGRAPHIC PROFILES DATA ATTRIBUTE

Demographic profiles-based attribute encryption (DPAE) contains a public-key (PK) based 1 to many other encryptions that allow the user to encrypted data and decrypted data based on features [3]. In that secret-key (SK) of a user and the cipher-text are based on features. In this type of model, the decryption of a cipher-text is probable only if all features of the user key contest the features of the cipher-text [4]. Decryption is desirable while the amount of identical is at minimum a constant value. Collusion-resistance is deciding safety features of demographic profiles data features encryption [5].

KP-DPAE includes the following steps:

- **Step 1 (Initial):** This procedure obtains input K (parameter) and income the value of PK as PK and a system master-secret key (MK). PK is used for encryption of message senders. MK is used for create SKs and is described ability.

- **Step 2 (Encrypted):** This procedure it obtains a message (M), the PK, and it contains of features as input. It outputs the cipher text (E).
- **Step 3 (Key Generate):** This procedure obtains as input to entrance a structure T and the MK. It generates an output as a SK that allows the user to decrypt a message encrypted under a set of features.
- **Step 4 (Decrypted):** It obtains as input the user SK for way in construction T and the cipher text (E), which encrypted under the feature set. This procedure outcomes the message (M), if, and only if the feature place gratify the user way in construction T.

The problem with KP-DPAE is the user can't decide who can decrypt the encrypted data. It selects only features for the data, it is improper in several appliances while a authority person has to faith the key issuer [6, 7] (Figure 19.1).

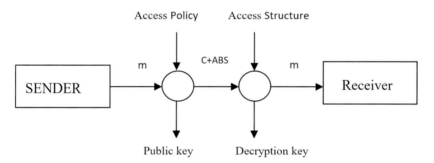

FIGURE 19.1   Demographic profiles data attributes security encryption and decryption.

In this regard, SHA-1 is technique for cryptography hash functions; generally, it is used to verify that a record does not change. Record verification using SHA-1 is skilled by comparing the checksums formed after apply the algorithm on the two records we want to compare. SHA-1 is the 2nd step of this cryptography hash function; replace the earlier SHA-0. The SHA-2 cryptography hash function also existing and SHA-3 is being developed [8].

19.3   META-ANALYSIS FOR PATIENT CLINICAL TRIAL DEMOGRAPHIC PROFILE

A Meta-analysis is an arithmetical analysis that coalesce the outcomes of several systematic learns. It is performed if there are many systematic reviews address the same question, by every patient revise reporting measurement that are estimated to have a few measure of error [9]. A meta-analysis can be used to make out this widespread effect. If the effect changed from one review to another review, then it is used to make out the cause for the difference [9]. Despite the make, decision on the service of the validity of a hypothesis can't be based on the results of a single review, while results are generally changed from one review to another review. A system is desired to produce clinical data across studies. Data description reviews are used for this purpose, but the detailed disease description review is mostly individual and becomes completely difficult when there are more than reviews involved [10]. Meta-analysis is help in study to estimate the evidence in areas as diagnosis data, medical history, sociology, sex differences, genetics, psychology, and others. The meta-analysis can help classify which questionnaires have been answered and which to be answered, which outcomes or people are most expected to give in important results, and which variant of the designed interference is to be most powerful. A demographic profile is generally defined by the following categories:

1. Gender, date of birth (DOB), occupation;
2. Contact details (name, address, phone);
3. Country, postal code, society, blood type;
4. Major diagnoses, major medical history, allergies;
5. Calamity details, insurance details.

19.4 META-ANALYSIS PROCEDURE FOR PATIENT CLINICAL TRIAL DEMOGRAPHIC PROFILE

Supervised learning based meta-analysis of clinical trial patient data by demographic profile is the being there of oversees the supervisor (doctor). Chiefly, supervisor learning is a study, which train the human intellectual using clinical trials data which is well labeled that means some data is already tagged with the correct prediction. Afterward, the human intellectual is provided with clinical trials, so that supervise learning meta-analysis procedure analysis and produces an accurate outcome from clinical trials training data. A meta-analysis is usually headed by a systematic review, it's allows identification and critical analysis of all the relevant proof [9]. The following steps in meta-analysis are as follows:

1. To do research issues diagnosis by the PICO model (population, intervention, comparison, and outcome);

2. Look for of patient history;
3. Assortment of learning:

- List of specific diagnosis reviews on a well-specified topic;
- Based on disease criteria;
- Make a decision whether unbiased reviews are built in to avoid review bias;
- Choose which longitudinal study on variables or events are allowable. For instance, when allowing for a meta-analysis of aggregate data:
  - Difference(discrete data);
  - Mean (continuous data);
  - Compute for continuous patients data, i.e., consistent sequentially to remove dissimilarity, but it incorporate an index of different groups:

$$\delta = \frac{u_t - u_c}{\sigma^2}$$

where, u_t is the treatment mean, u_c is the mean, σ^2 the variance.

- Choice of a meta-analysis model.
- Observe sources of between study heterogeneity of patients using sub-group examination or meta-regression.

19.5 DEMOGRAPHIC PROFILE BASED PATIENT CLINICAL TRIAL USING BLOCKCHAIN

The new-fangled facts of Blockchain to address these challenges and should be representing the completely medical research community. Blockchain bring the internet to its ultimate decentralization goalmouth [4]. Reproducibility, data sharing, private data privacy concern, and patient enrolment in medical tests are medical challenges for modern medical research. The aim of blockchain is that any service depend on secured third parties can be built in a translucent; decentralized, secure is the top of the blockchain. Hence, users have a high level of security, autonomy, trust of the data, and truthfulness on it. Blockchain allows for getting an important level of historical data and clarity of data for the whole document review in a medical test [11, 12] (Figure 19.2).

In general, blockchain is a collection of nodes. These nodes are connected with one other in a chain manner. In each node consists of different blocks, these are also known as chunks. Each block is connected to a preceding block. Each block consists of patient clinical data which can be used any of the reviewers. The blockchain protects and decentralize data store of sequence records. Majority of users can update the clinical data in the blocks, but data cannot be removed [4]. It is controlled by any of the users. The secure is encoded in set of protocols and maintained by user's community. The historicity and purity of data are the two major features in the data level. The blockchain ensures that actions are track in their correct in order, which mostly prevent a posterior rebuilding analysis [1]. The historicity and traceability of the data are among the functionalities of the technology, each operation with blockchain is time-stamped. This information is visible to the entire user, any user have a time-stamped data of proof. The chaining of blocks is achieved through another cryptography primitive, which use the hash functions [4]. All documents, the existence proof can be stored in blockchain. An important property of hash function is collision-resistant [1], i.e., no two different messages will generate the same hash output (Figure 19.3).

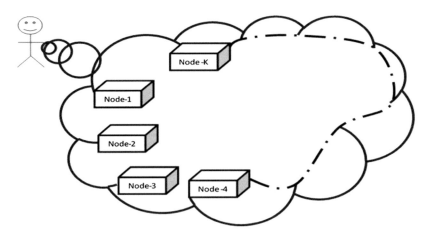

FIGURE 19.2   Decentralized blockchain system.

19.6   EXPERIMENTAL RESULTS

The experiment result demonstrated on demographic profile clinical records access from CC. To demographic profile clinical record DB using remote server needed in the same area as DB instance with the intention of it is able to connect to the outline wide-ranging the cloud network. To accomplish enhanced trustworthiness to compact with T^m and not hire it go down moreover considerable from 1.0. For the reason that the demographic profile records called m is probable, always we try to put T as nearer to 1.0 as probable. The wished entire trustworthiness of the system MSR is detained steady to 0.9 or 0.8 (Tables 19.1 and 19.2). Evoke that T is the least trustworthiness that we can accomplish for every DB access stage. For various values of T, the Tables 19.1 and 19.2 gauges (lowest) m such that T^m sinks below the desired whole trustworthiness MSR. This is completed by solving n in relation of T and MSR in the experiment of demographic patient profiles.

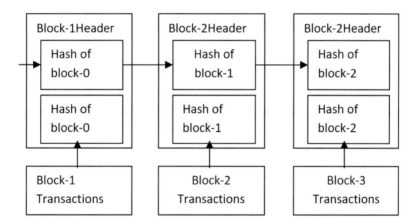

FIGURE 19.3 This pattern of the transactions in the previous block.

$$n = \log MSR/\log T.$$

TABLE 19.1 MSR = 0.9

T	0.9	0.99	0.999	0.9999	0.99999	0.999999
MSize	1000	1600	2700	3800	6500	10000

TABLE 19.2 MSR = 0.8

T	0.9	0.99	0.999	0.9999	0.99999	0.999999
MSize	1402	3010	5061	8062	15070	221300

By using attentive view of above tables, one can make sure to have high trustworthiness T of 0.999999 for each time accesses the records. We can run the trouble of decreasing complete trustworthiness MSR below 0.9 or even 0.8. Undeniably, the number of rights to use records is not anything then the execution of Demographic Clinical Trail Patient Records that is experiential in Table 19.2 as 221,300.

19.7 CONCLUSION

Attribute-based encryption methods are used like public key cryptography. The implemented process with the encryption procedure of key-policy attribute-based-encryption (ABE) in cloud appliance for patient data accessed. Blockchain can reinforce the trustworthiness of meta-analysis used safe time-stamping holds the key to reproducible meta-analysis. Blockchain system required sending and processing enormous data should allow some level of compromised security and centralization with a Blockchain-based protocol can be designed accordingly. Likewise, if user necessities are different such that increased adversarial conditions are required used the protocols can similarly be altered. Bitcoin cash, on the other hand, boosted scalability by raising block size limits, but the higher block size limits may compromise the security of the network. To make sure the scalability with appropriately chosen security modeling is the only way for blockchain facts to be accepted for automotive use cases.

KEYWORDS

- **attribute-based-encryption**
- **blockchain**
- **cloud environment**
- **demographic profile**
- **demographic profiles based attribute encryption**
- **meta-analysis**

REFERENCES

1. Xie, S., Dai, H., Zheng, Z., Chen, X., & Wang, H., (2017). An overview of blockchain technology-architecture, consensus, and future trends. *IEEE International Congress on Big Data.*

2. Miller, D., (2018). *Blockchain and the Internet of Things in the Industrial Sector IT Professional.*

3. Sahai, A., Bethencourt, J., & Waters, B., (2007). Cipher text policy attribute-based-encryption. *IEEE Symposium on Security and Privacy.*

4. Rudlang, M., (2017). *Comparative Analysis of Bitcoin and Ethereum.*

5. Li, M., Yu, S., Ren, K., & Lou, W., (2010). Securing personal health records in cloud computing: Patient-centric and fine-grained data access control in multi-owner settings. In: *Secure Comm.*

6. Newpor, C., & Cheung, L., (2007). *Provably Secure Cipher-Text Policy ABE ACM Conference.* Computer and Communications Security.

7. Chaowen, G., Junzuo, L., & Deng, J. W., (2013). Attribute-based-encryption with verifiable outsourced decryption. *Information Forensics and Security, IEEE.*

8. Waters, B., & Hohenberger, S., (2013). *Attribute-Based-Encryption with Fast Decryption Public Key Cryptography.*

9. Gauba, Z., Koticha, M., Uddin, M., & Mohavedi, A., (2018). *Meta-Analysis of Proposed Alternative Consensus Protocols for Blockchains.*

10. Laguillaumie, F., Attrapadung, N., Libert, B., & Herranz, J., (2012). Attribute-based-encryption schemes with constant size cipher texts. *Theory Computer Science.*

11. Aste, T., Tasca, P., & Di Matteo, T., (2017). *Blockchain Technologies: The Foreseeable Impact on Society and Industry.* Computer.

12. Pass, R., Seeman, L., & Shelat, A., (2016). *Analysis of the Blockchain Protocol in Asynchronous Networks.*

CHAPTER 20

Query Response Time in Blockchain Using Big Query Optimization

M. SRIKANTH[1] and R. N. V. JAGAN MOHAN[2]

[1]Research Scholar, GIET University, Gunupur, Odisha, India,
E-mail: srikanth.mandela@gmail.com

[2]Associate Professor, SRKR Engineering College, Bhimavaram – 534204,
Andhra Pradesh, India, E-mail: mohanrnvj@gmail.com

ABSTRACT

Blockchain is the core of computer technology. It is cryptographically endangered distributed and parallelized database (DB) knowledge for storing and transmitting facts. Each tuple in the DB broadly called a block and covers such as the query timestamp and a link to the previous block. Blockchain and Big data are two expertises in full swing then again they are also two complementary expertise. We can study the how the Blockchain executes big query analysis. Many pursuits query only require the traditional approach of ranking a set of matching results. However, some queries require a more complex approach, either because they are broad or ambiguous. It is important for a DB search engine to identify such queries along with to differentiate broad queries from ambiguous ones. In this chapter, we propose that there is a diversity of indications that the Hadoop-based DB search engine is used. This allows the Hadoop DB search engine to explore and improve the query as correct relevant. The experiment result is measured by the query response time (RT) for optimized queries and broad queries.

20.1   INTRODUCTION

Blockchain supports response time (RT) data processing, and the block-based consecutive access in Blockchain delays skillful query processing. The trustworthiness and unchangeable of blockchains combined with the support for decentralized, low faithfulness data processing offer new chances for cloud applications. In most queries are growth the RT for problematic logical queries, to their data consists of calculated results from the database (DB) tables that we can specify in become observable query table as of work out the phase a vital role that how we can store the data evidence and how it can run the applications as a substitute of your programs and data on a separate system, and everything is hosted in the DB. It is a vague accumulation of computers and servers accessed by the Client. Query DB in which user can access the tables data and whole DB application from any place across the sphere, releases user from the confined to the desktop and make it easy for groups in various places to cooperate for access the normal DB. It is an ever-growing a Client-Server model; it delivers the data allocation to computers system and other strategies on demand. Query DB is a one sort of model that available in all-inclusive and access to a shared pool of configurable computing resources to provide the facility and hastily with less management effort. It depends upon to sharing the resources and gets the consistency and economy scale, and it is more convenience in the interrelated network. It is mainly concentrated on data storage solutions to facilitate the users and enterprises with more capabilities to store and process the data owners from a distant place or across the globe.

As a matter of fact that, query processing is mainly focused on users spend the time and fare on computer infrastructure for organizations. To promoters are allows us to achieve the high-level based enterprise applications and allow IT teams too rapidly to meet fluctuating and unpredictable prior insist resources. A DB promoter as use is a typical model for various DBs that lead, when administrator does not to give up the much time. In the face of DB provides to ever-growing the hardware-virtualization, low-cost computers, and storage devices [1]. In addition, it is widespread in efficacy calculating and successful networks. It can measure the increasing and decreasing situations while computing necessarily and demand depreciates respectively. So it is a very much demanded service or helpfulness due to the advantages of the high computing power, high performance, low cost of services, scalability, ratification, and reliability.

20.2   ATTRIBUTED QUERY OPTIMIZATION

Attribute query optimization is relation between each query and its concern cost function that maps from multi-dimensional attributes value spaces to single dimensional cost value space. Which values are undefined at optimal time these query plans are matched with render to unique cost metric. How it is model cost of an individual query big data is nothing but quantity and scale only map job schedule input is a group of the attribute is then transform into another group of query analysis, here each send function have pairs as key-value [2]. For example, the broad query MasterLog include client name, Block, Prospect, Basin/SubBasic, State, WellType, Target Depth, Drilled Depth, Spuddinon, Rig_Type, Elevations, Surface, Coordinated, TDReachedon, HematicalTest, RigReleasedon. The disparity of amalgamation is vague for it could denote MasterlogTable.

20.3   MAP REDUCE APPROACH FOR IMAGE BIG DATA WITH BLOCKCHAIN

The notion of big query approach using MapReduce programming uses for query process in akin on numerous nodes. The amounts of MasterLog attributes data need to gather and managed the type of Big Data. It can also engage one or more of the subsequent features like volume, variety, velocity, and complexity. A MapReduce procedure holds two major errands as map tasks and reduces task. Reduce function processed the yield from map function supplied input and merged it query attribute key and value pair into a small group of query attributes [3]. The following query optimization steps are as follows:

1. **Input Point:** Now, we use a data reader that transmutes is every query as the input file and then transfer.
2. **Map Point:** Then Map is a user-defined job in which takes a series of queries key and pixel value pairs and processes then each one of them to make zero or more key and value pairs.
3. **Sort and Shuffle Point:** The specific key and value pairs are sorted using keys into a healthier block. The queries data list clusters the equal keys together then query values can be continual just in the reducer job. The reducer task starts with the shuffle and type period. Then it downloads the clustered key and value pairs onto the local appliance, so where the reducer is going on.

4. **Intermediary Keys Point:** In this mapper make key-value pairs is renowned as intermediary keys.
5. **Combiner Point:** It takes the intermediary query keys from the matched as input and puts on a user-defined code to increasing the pixel feature extracted values in a small possibility of one matched. A combiner is a type of local reducer in that gathering similar data from the map point into specific sets.
6. **Reducer Point:** The reducer gets the clustered query key-value paired query data as input and executes on a big query is broad then the DB search engine should propose modifications that query analyzer to guide the search towards more specific queries. These typically include category proposals, such as purifying from multi-dimensional attributes based results to specific result set. They may also include independent attributes modifications that propose valuable attributes to narrow the result set using Reducer function of Hadoop system for forecast the big query purposes.
7. **Output Point:** Each query process called a block and contains such as timestamp and a link to the previous block using a record writer. In this**,** output formatters that transform the final key and value pairs from the reducer function to store and conveying the query analysis information.

20.4   QUERY RESPONSE TIME (RT) WITH BLOCKCHAIN

The measure, RT inquiry is more effective tactic to improve MasterLog DB query performance. The finest implementations break down the time into discrete and independently measurable steps and find exactly the steps in which operations causes query analyze process delays. This works by focus overseer and developers on the most vital criterion reasons to wait and referred to as wait time inquiry and it allows number of queries to align their efforts with service level delivery for appliance. The below picture describes the RT monitoring process [4].

Each structural query language (SQL) query request passes through the DB illustration. With computing the time at each footstep, and the entire RT can be analyzed, before viewing Hadoop server query statistics, making guesses about their performance impact, wait, and RT methods measure the time taken to complete a preferred operation. The DB main query task is to respond with an outcome, and RT to make query DB performance decisions (Figure 20.1).

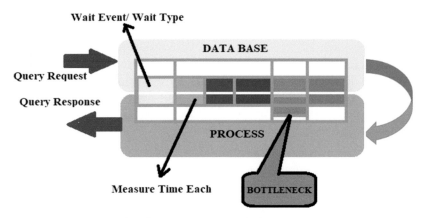

FIGURE 20.1   Response time for query process analysis.

Query RT for an exemplary Hadoop system application that comprises the subsequent series of actions. There need a certain total of time for each action. It takes for the user to think of and enter a query or appeal for the RT does not comprise the time. The DB server executes query optimization and the retrieves any user-defined routines and outdoor routines. For initially, then application is forwards a query to the DB server. So the DB server retrieves that add or update the proper records and performs input/output operations directly related to the query. Then the DB server performs any background input or output operations such as logging and the page dusting that befall during the periods in which the query is still pending. Then the Big Query-based Hadoop application appears then the data or issues a confirmation and then the issues a novel prompt to the user. So the DB server returns an outcome to the application.

20.5   EXPERIMENTAL RESULTS

The experiment result measured the query RT for optimized queries and Broad queries. The recital of MasterLog query RT has taken as an extent through DML (data manipulation language) assertions with different situations. Each assertion has repeated as a minimum 30 times and for query response, time is well-known and middling intended for all iterations. All the SQL queries were executed using electronic health records (EHRs) DB [3]. Then the effecting query results in the retrieval of data and the number of results made in each instance tabularized with the RT values. The "Slow

Miserable" arc with the service of gained RT values. It is gained in separating all the entry RTs with the early entry RT. Then the RT values of totally the data sizes such as 10,000, 20,000, 30,000, 60,000, and 80,000 entries in conventional DB is distributed by the initial entry RT of the outdated DB, i.e., 10,000 entries. Figure 20.2 gets by the values. The same way is continual for the cloud DBRT values. To display with the values and both the curves are plotted. These curves show the Slow Miserable as a relationship between the two. The main goal of this workout is to realize query elapsed time for a query that retrieves small thing number of rows from big table used by perusing the complete table (Table 20.1).

TABLE 20.1 Query RT (Response Time) Values of Unlike Entries for Outdated and Big Data DB (Database) in Milliseconds

RT (Response Time) for Data Entries	Outdated Database(Milliseconds)	Database (ms)
10,000	3	2
20,000	5	4
30,000	7	5.3
60,000	12	9
80,000	15.3	11
100,000	16.4	12.6
150,000	18	14.3

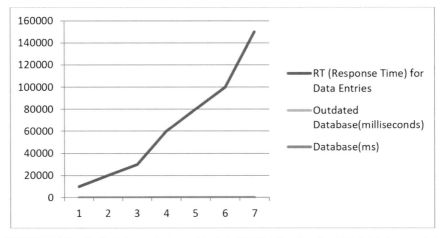

FIGURE 20.2 Radical change between big data DB (database) and outdated database recital while retrieving rows from tables.

From the above graph, there radical change between Hadoop DB and outdated DB recital while retrieving rows from tables. The experiment results demonstration that the outdated DB is accomplishment well for this query. At 10,000, the RT is almost in DB. At 20,000 entries, both the DBs have almost the same RT. At 30,000 entries, the DB has 6 RTs. At 60,000 entries, the Hadoop has 9 RTs in milliseconds. At 80,000 entries, the Hadoop DB is 11 times. At 1,00,000 entries, the Hadoop DB has 13 RT in milliseconds. At 1,50,000 entries, the Hadoop DB is 15 milliseconds.

20.6 CONCLUSION AND FUTURE VIEWPOINT

The performance of MasterLog from big data DB query RT has taken as coverage through Attributed based Query assertions with different conditions. The Hadoop based database search engine used for Query Optimization. To workout, the Hadoop database search engine to explore and improved the query as correct relevant. The query RT for optimized queries and Broad queries are in experimental.

KEYWORDS

- **big query analysis**
- **blockchain**
- **broad query**
- **data manipulation language**
- **Hadoop**
- **response time**
- **structural query language**

REFERENCES

1. Dileep, K. K., Jagan, M. R. N. V., & Srinivasa, R. M., (2016). Cluster optimization for similarity process using de-duplication. In: *IJSRD-International Journal for Scientific Research and Development* (Vol. 4, No. 06). ISSN: (online): 2321-0613.
2. Dileep, K. K., & Jagan, M. R. N. V., (2014). Optimizing the duplication of cluster data for similarity process. In: *ANU Journal of Physical Science* (Vol. 2). ISSN: 0976-0954.

3. Dileep, K. K., Jagan, M. R. N. V., & Vamsidhar, Y., (2012). Similarity based query optimization on map reduce using Euler angle oriented approach. In: *International Journal of Scientific and Engineering Research* (Vol. 3, No. 8). ISSN: 2229-5518.

4. Fariz, A. P., & Kusprasapta, M., (2018). Query support for data processing and analysis on ethereum blockchain. *International Symposium on Electronics and Smart Devices (ISESD)*.

5. Apache, H., & Hive, V. G., (2015). *Optimization of Multiple Queries for Big Data.* International Conference on Computational Intelligence and Communication Networks (CICN).

6. Foto, N. A., & Jeffrey, D. U., (2011). *Optimizing Multi Way Joins in a Map-Reduce Environment*. In IEEE.

7. Priyank, J., Manasi, G., & Nilay, K., (2019). *Enhanced Secured Map Reduce Layer for Big Data Privacy and Security*. In Springer.

8. Xiangming, D., & Brahim, B., (2016). Scheduling for response time in Hadoop map reduce. *IEEE International Conference on Communications (ICC) Conference Paper*. Publisher: IEEE.

9. Zhang, Z., Luo, B., & Cao, Z. (2011). The research on the query optimization on the distributed heterogeneous database based on the response time. *Proceedings of 2011 International Conference on Computer Science and Network Technology*.

CHAPTER 21

Transformative Blockchain Knacks for Bitcoin Cryptocurrency and Its Impacts

BHAVANI SANKAR PANDA[1] and CHANDAN KUMAR GIRI[2]

[1]*Assistant Professor, Computer Science and Engineering, GIET University, Gunupur, Odisha, India, E-mail: chintupanda@gmail.com*

[2]*Associate Professor, Computer Science and Engineering, CUTM, Paralakhemundi, Odisha, India, E-mail: chandankumargiri@gmail.com*

ABSTRACT

Every increasing use of digital or virtual currency in globe and its unpredictability, crypto-currencies are being adopted across world for various legal as well as illegal transactions. In returns received from crypto-currency, investments in recent times were enormous but there has always been an interrogation on their reality and reliability. At the present time, this technology signifies an innovative feature which will swaps available systems with distributed systems. This system supports different advance topographies like readiness, adjustment with the condition, error minimization, and decrease in price. This mechanism suited for many ongoing research and development also these technology usages cryptography techniques for security even though the current issues in digital crypto currencies. Bitcoin become success also it's rising conspicuousness since its launch, it has affected in a number of companies unveiling unconventional crypto-currencies. The study attempts to understand the transformation from traditional currency to digital crypto currencies as Bitcoin and others with respect to their precariousness, stability, and its impact in recent times and edify current time developments.

21.1 INTRODUCTION

Blockchain delivers a scattered record of transactions on a communication channel. It is mountable, protected, tamper-proof, and accessible by each patrician. It is shared transactions, distributed over a network of members, made up of series of data blocks; each by itself includes usual dealings. Different blocks are automatically bound with each other and added with tamper-proof cryptography mechanism. Also public transcribe of each operation was recognized. If more blocks there are, the less the probability that blocks can be altered. The well-known crypto currency for which blockchain technology was invented is the Bitcoin, invented by Satoshi Nakamoto in 2008. In clever way, he combines the previous cryptography technologies like asymmetric encryption, agreement, and Merkle tree to invent what is called bitcoin crypto currency. The first block (genesis) was initiated on 2009, afterwards the chain is cumulative every few minutes to reach around 52 k blocks on year 2018 with BTC price growing up to 8 k$ for each Bitcoin. The success of bitcoin triggered the technologists to think of its uncontrolled mechanism and begging with exploring about the theme. It is important to distinguish between bitcoin and blockchain, in which bitcoin is an automated crypto-currency which will be used to purchase different things. Its amenities based on incentivizing the applicant nodes like miners to confirm dealings and to condense the network as stable as maximum. Blockchain is the fundamental technology that helps the Bitcoin network to operate in an open, self-directed, and decentralized prototypical, where trust is enforced through asymmetric cryptography techniques.

21.2 BLOCKCHAIN PHENOMENON

Bitcoin involved most investors in the year 2017, because of maximum profit archived in investment and this mechanism for Bitcoin execution known as Blockchain. Its chunks of operation or transaction account details which shared among different users publicly using protected cryptography methods. Each chunk holds the previous operational info, timestamp, and new operational data through cryptographic hash function. Again, this technology permitted the availability of digital currency through scrutinizes the bitcoin crypto-currency request. For safekeeping reasons, the technical facts of digital coins are willful to close safekeeping fleabags. Bitcoin supports Blockchain which keeps the records of all valid user transactions. In this technology, the customers' open digital wallet through private/public key sets and then trade required coin through the digital wallet. The safety mechanism of the user's digital wallet depends on the security level of user's private key. Some individuals hide bitcoin as a private key and others hide the secrete key into a plain text. The information security technique is the key to the development of synchronous Internet technology. The distributed mechanism, scripted mechanism, password mechanism and decentralized mechanism of the blockchain present a perfectly new viewpoint for the development of Internet information security technology.

Various people still obscure with blockchain and bitcoin even if they are not the same. Bitcoin is one of the application usages blockchain technologies to upload data into a cloud server and sometimes packing in to an atomic site, as well as breaking everything into small pieces and allots those transversely to the whole communication networks. In Blockchain, many flaws available which create vulnerable issues by certain security and procedural challenges which includes accessible, secrecy leak, self-interested, etc., and creates different obstructs in the wide application of blockchain. Three essential components of bitcoin are like trades and scripts of it, Agreement, and withdrawal and P2P communication network. It expresses how trades are stimulated also provides answers as to how can we eliminate the banking system and apply P2P trades. This is related to many who are related to the banking system. In this most of backers keeps, a copy of the data and it is the accountability of the bank. This called distributed ledger, which includes all past and present trades' information due to which source and destination users totally free from different third party controls. Afterwards, they are given to a new communication controller called agreement and receive or decline their operations based on ledger data. Applying agreement mechanism, elects net users for trades to be approved or unapproved. This mechanism is a scattered point to point scheme leads no additional server or single point of control system. Bitcoin are created through a procedure entitled as mining, which was involved in different findings for the solution to complex problem (Figure 21.1).

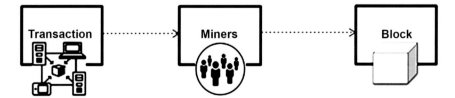

FIGURE 21.1   Relationships of miners with block and transaction.

Conferring to the above figure almost every sponsor in the bitcoin architecture can act as a miner who using the computing power in order to solve the complex problem. The sponsor's nodes that are traversing across different areas through mesh network. These are basically two types are mining nodes where users are participate in the establishment of blocks and are encourage with bitcoin to agreement of their occurrence in the network and non-miners benefits from the bitcoin system without take part in block creation. Another main part of this motivation is to allot bitcoin currency. The operation is typically based on public-private key set technology and cryptographic hash functions, where trades are sign up and circulated on a public network. Asymmetric cryptography, are mathematically confined and can't be swapped. The public and private key functions used for transactions and then grouped into blocks, exchanged, and authorized by a network of different nodes with an agreement that determines which blocks are recognized (Figure 21.2).

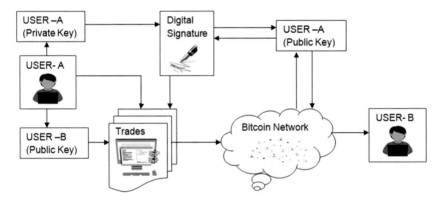

FIGURE 21.2   Anatomy of a bitcoin transaction.

21.3   NECESSITY OF BITCOIN TECHNOLOGY

After mounted a bitcoin digital wallet on our personal computer, mobile, and other supportive digital device, it will create our first Bitcoin address statement and we can make more when we need one. Blockchain technology is an open transcribe that, entire scattered point-to-point bitcoin network depends on. Each definite trade is involved in the Blockchain so that the Bitcoin digital wallet can able to compute the available balance and new trades may be confirmed. Trade is a transmission of cost between bitcoin addresses that get involved for digitally sign dealings to deliver a mathematical resistant that they are lawful. Bitcoin digital wallets keep a undisclosed piece of data denoted as private key which is used to sign agreements to provide a mathematical resistant that they derived from the owner of the digital wallet. The digital signature scheme protects the transaction from being transformed by anyone once it was assigned with digital signature. Every transaction is transmission between users and should be validated and guaranteed by the bitcoin communication network within a few minutes or mostly within an hour (Figure 21.3).

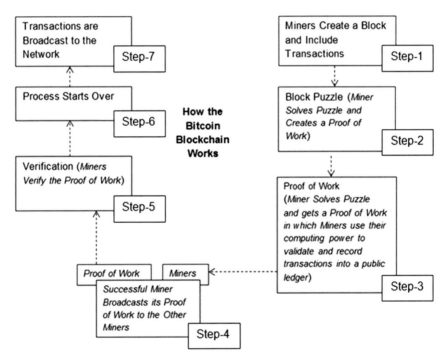

FIGURE 21.3   Bitcoin blockchain workflow.

21.3.1 BITCOIN

These are generated as rewards for fee processing work in which users allocate their computing service to confirm and noted payments into a public record. This action is called mining and miners are remunerated with transaction payments and newly created Bitcoins. This can be swap over for other cash, resources, and facilities (Table 21.1).

TABLE 21.1   Bitcoin Statistics

Characteristics	Description
Coin limit	21 million
Algorithm	SHA-256
Mean block time	10 minutes
Difficulty retarget	2016 block
Block reward details	Halved every 210,000 blocks
Initial reward	50 BTC
Current block reward	25 BTC
Block explorer	blockchain.info
Created by	Satoshi Nakamoto
Creation date	January 3rd, 2009

21.3.2   HISTORY OF BITCOIN

Bitcoin cryptocurrency, an electronic currency uses decentralized mechanism without a banking system or using solo admin that sends from user to another user on the P2P Bitcoin communication network. It was developed by an anonymous group of users as shareware and released in the year 2009.

21.3.3   BITCOIN WALLET

This software enables users to receive, store, and send bitcoins by empowering users to create and manage a collection of private keys. Private keys are the numbers users keep secret using a Bitcoin wallet that enable them to move Bitcoins by solving a math problem that the Bitcoin network can validate, for example, https://www.weusecoins.com/wallet.

21.3.3.1   CATEGORIES OF BITCOIN WALLETS

It is basically four types like online bitcoin wallets gain access from the web through any internet linked device. Hardware Bitcoin wallets is a physical device which was designed to protect Bitcoins from vulnerable use. Software Bitcoin wallets are applications moved or download to a mobile phone, PC or tablet PC, etc. Paper Bitcoin wallets are private keys made from an offline PC that has never been connected to the internet service (Figure 21.4).

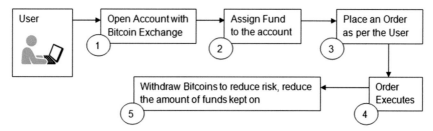

FIGURE 21.4 Step wise process to buy cryptocurrency online.

Blockchain public records used for bitcoin transaction which uses sequence of blocks and contain a hash value of the previous treated as unique one. Bitcoin grid execute its software by observance blockchain by exploit a blockchain and exploit sequence of network bonds legalize dealings. Afterwards, add those to their replica of the record and then broadcast all transcribe add to the other bonds. For archive atomic identification to the series of proprietorship individual network bonds stocks own replica of blockchain. Every time a new group of recognized transactions called a block, is formed, attached to the Blockchain, and rapidly available to all knots, without needing central mistake. It permits bitcoin software to decide when a specific bitcoin was consumed, which is needed to prevent multiple spending.

21.3.4   BITCOIN NETWORK REGULATOR

This Bitcoin network technology no one heads as well as owns which uses internet technology for communication and distribution. Dealings are tested by miners who have whole trade of bitcoin cloud mining as choices. Even if designers are improving the software and they can't force a change in the Bitcoin protocol because all consumers are free to select what software and version they wants to use. For benefit in order to stay responsive with each other, all consumers need to use software obeying with the similar rules. This only can work properly with a whole agreement among all consumers. All consumers and designers have a robust incentive to guard this agreement.

21.3.5   WORKING PROCESS OF BITCOIN

According to the consumer viewpoint, Bitcoin is a mobile app or a computer program that delivers an individual bitcoin digital wallet and allows a consumer to send and receive digital coins called bitcoins. In this bitcoin communication network sharing a massive public record called blockchain. This record holds every transaction which ever managed and enables a consumer's PC to verify the authority of each transaction. The genuineness of each transaction is secured by digital signatures corresponding to the sending addresses therefore allowing all consumers to have total control over sending bitcoins. Due to which in this scheme no scam, no chargeback's and no recognizing of information that could be compromising integrity (Figures 21.5–21.7).

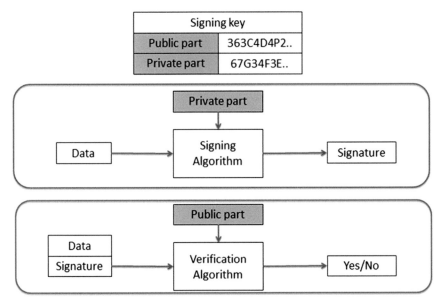

FIGURE 21.5   Integrating private and public key for signing and verification.

21.3.6   SOLVING PROBLEMS THROUGH BITCOIN

Recent in the globe digital currency bitcoin is currently the most widespread well-known cryptocurrency. This technology is trying to solve an important problem faced by various people across the globe. The huge amount of cash in circulation is mainly controlled by traditional bankers and they employ the controls of inflation and currency debasing as technique of growing and supervisory product fees all over the economy. Atomic and official control of currency is the problem which bitcoin is trying to resolve. Like bitcoin, many other digital cryptocurrencies, gives influence to the people using the cryptocurrency as they do not need any specialists in a atomic centralized system for doing trade and make procurements. Initially it actuality P2P and non-atomic in nature which enable people trade without having the fixed centralized rules of the transaction dictated by third parties. Digital cryptocurrency enables consumers to act secretly and this eliminates authority conventionally detained by financial bodies.

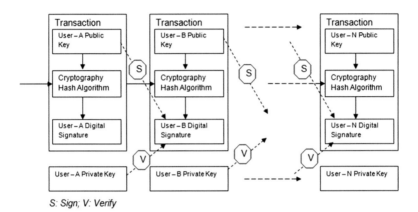

S: Sign; V: Verify

FIGURE 21.6   Transaction process between different users.

21.3.7   CRITICISM

21.3.7.1   ENERGY DEPLETION

Bitcoin technique was carped due to the quantity of electricity used during mining. According to 2015, many economists anticipated about the recent abilities used by all miners, by which collective electricity intake becomes 166.7 megawatts. By end of the year 2017, worldwide bitcoin mining action was established to ingest between 1 to 4 GW of electricity. The high end approximation's bitcoins total intake levels amount approximately 6% of total power used up by worldwide banking networks. Day by day bitcoin intake increased 100 times even if of today's levels. Intake bitcoins amount up to 2% of worldwide electricity intake.

21.3.7.2   COST

For minimizing the price, bitcoin miners will setup at cooling places where geothermal energy intake is less cold air is free like arctic region. In this, miners use hydro electronic energy in different regions like Himalayas regions, Austria, Washington, etc., for minimizing electric price. Subsequently, miners paying attention to different vendors like Hydro Quebec shows they have energy oversupplies. As per the University Cambridge research, it says much about bitcoin mining which was done in china uses subsidized electricity and which was provided by the local Govt.

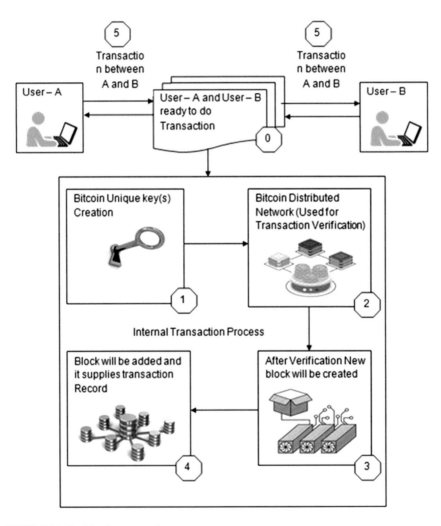

FIGURE 21.7   Bitcoin transaction process.

21.3.7.3 HUSTLE SYSTEM AND MONOLITH SYSTEM ANXIETIES

According to different journalists, economists' included central banking systems stated different anxieties that the bitcoin is hustle. In the year 2013, a law expert at the University of Chicago told about real hustle systems leads to scam. As of the report in July 2014, the World Bank decided that crypto-currency bitcoin was not a thoughtful hustle system. The Swiss Federal Council inspected the anxieties that bitcoin might be a pyramid scheme in the year June 2014; meanwhile, the instance of bitcoin, distinctive possibilities of incomes are absent, it's difficult to assume that this technique becomes a pyramid system.

21.3.7.4 SECURITY ISSUES AND ILLEGAL TRANSACTIONS

Like other computational application, digital crypto-currency also supports CIA scheme that is confidentiality, integrity, and availability. Authentication uses public key crypto as digital signatures, integrity uses digital signatures and cryptographic hash function; availability will be satisfied with broadcast messages to the P2P network and confidentiality with false name. In this scheme integrating digital signature by create a message digest using a cryptographic hash and then, encrypt the message digest with your private key. In this from a given message we will create hash value using hashing algorithm then digital signature algorithm add private key and generate digital signature. Once signature created attached it to the input message and produce signed message through digital signature scheme (Figure 21.8).

Bitcoin having different security issues like how can we protect against Sybil attacks needs bonds to resolution crypto mysteries for promise dealings to blockchain. If any splits in the blockchain then find out the mechanism to slowdown it. This chooses longest chain of blocks to wins. How much of the communication network must authorize a transaction for it to be considered as accepted. In this, only one bond wants to resolve the crypto mystery to create required new block. However, a single block may not be decent enough for authorization. Suppose that one block is part of a fork then it may get replaced in the near future. Bitcoin trusts on the fact that no single unit can control most of the processing power on the required bitcoin grid for some important span of period. By this scheme, it's able to outspread undergrowth of blocks faster than the rest of the network which would allow choosing which transactions to appear in the blockchain.

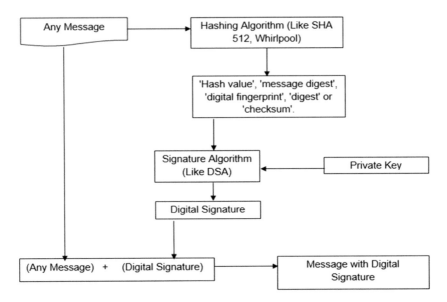

FIGURE 21.8   Process for creating digital signed message.

Due to the unknown nature of cryptocurrency, Bitcoin is frequently used for unethical and illegal activities online. The working gloomy net marketplace is examples where gloomy net markets carry on to expand as traders and contractor are able to remain unidentified even though improved government scrutiny observation. Bitcoins are mostly used by criminals who paying attention to the financial controllers, law-making bodies, law implementation, and the electronic media. We have much abidance in several news outlets have declared that the acceptance of bitcoins centers habit to acquisition unlawful belongings. By changeable nature and the abuses, anybody can control it out of reality. It happens because of the abuses worldwide.

21.3.8   LEGALITY OF BITCOIN

Bitcoin is never legally acceptable as a substitute for much country's lawful tender. On 1 February 2018, Government of India detailed that it will do everything to terminate usage and types of virtual moneys due to unlawful practices. Country like India doesn't consider crypto-currency and legal as well as these in financing illegitimate activities as part of the payment system. These crypto resources are used in funding illegal doings or as part of the payments system in India, that's why it does not recognize as legal tender and a substitute inspire blockchain technology in payment systems. RBI declared a ban on the auction or purchase of cryptocurrency for entities controlled by RBI. Similarly, the United States has taken a usually encouraging stance toward Bitcoin, however some government organizations work to stop or reduce Bitcoin usage for unlawful transactions. Bitcoin are generally greeted in utmost parts of the world. However, some nations have truly banned them or their usage. Generally, prohibitions were forced to decentralized environment, the threat to their current monetary system or else just because suitable guidelines are yet to be permitted, by March 2019 nine countries according to the "Coin.dance," which decided to do so are Afghanistan, Pakistan, Algeria, Bolivia, Bangladesh, The Republic of Macedonia, Saudi Arabia, Vanuatu, and Vietnam. Again, there are some countries where Bitcoin is somewhat controlled and cannot be merchandised or used for any payment. In more extreme cases, the nations like China, India, Ecuador, Indonesia, Morocco, Zambia, Nepal, Egypt, American Samoa, and Qatar have even forbidden crypto exchanges. Even though the prohibitions and restrictions these regulations incapable to abolished crypto-currency trades. As it supports decentralized mechanism, it is simply impossible to exclusion them and many people in those countries usage webs like https://localbitcoins.com, https://paxful.com, https://bisq.network, https://coinmarketcap.com, etc., even forbidden to trade it with others.

In India, there are thoughtful anxiety leads to following facts for dimensions case against crypto-currencies: (a) All crypto-currencies are private creativities by non-sovereigns. (b) No unique fundamental value for all the attributes of crypto-currencies. (c) No secure insignificant cost of these private crypto-currencies. (d) Unproven risky rise and fall of prices. (e) Can't help the purpose because of inconsistent currency. (f) As per worldwide review it is not been accepted as a legal fond in any authority. (g) Committees approval should be forbidden as of trade with crypto-currencies.

21.4 CONCLUSION

This explained about application zones linked to crypto-currencies where we drew the fundamentals of cryptocurrencies for non-professional spectators. Second, we illustrated different benefits portrayed by bitcoin and scrutinized problems with setting out. Then we bend back into the details and controlling challenges obtainable by bitcoin and crypto-currencies. Bitcoin is a mysterious and spread out expense system could certainly transform the economy and helps to exit the uneven power of banking schemes and democratize financial exchange. An anonymous malicious decoder may not be the procedure for the future but real honesty is required for the subsequent test to be successful. It is an indication that opportunities the blockchain, proof of bitcoin transaction log, policy for creating a smart contract for decentralized platform. We express about blockchain role in future where bitcoin being the head application of an open record but the present enactment suffers from permissive economic policy and serious faults, such as large entity with access to large computing facilities to operate the open records.

The broader investigation relates to the upcoming fiat on cryptocurrency and the opportunity of public construction and distribution built on blockchains as the foundation for the record interchange. Blockchain technology groups with social construction propose may succeed progressively be suspicious of self-governing currencies. Our investigation further limited to a serious investigation of the usage of the first broadly adopted non-proprietorial virtual crypto-currency like bitcoin. As a spot of confrontation to open market belief, virtual cryptocurrencies may be limited, but as a consolidating norm for supportive sharing together with the self-governing agreement currency for commercial marketplace. It may have a robust habitation in future, just as supportive arrangements gained subsistence with mass customer entrepreneurship. This chapter tells a new form of accommodating recreational area online may be permitted to blockchain, also it will not to be constructed on bitcoin.

KEYWORDS

- **bitcoin**
- **blockchain**
- **cryptocurrency**
- **cryptography**
- **hashing**
- **ledger technologies**

REFERENCES

1. Andresen, G., (2011). *BIP 11: M-of-N Standard Transactions*. https://github.com/bitcoin/bips/blob/master/bip-0011.mediawiki (accessed 20 July 2020).

2. Andresen, G., (2012). *BIP 16: Pay to Script Hash.* [online]. https://github.com/bitcoin/bips/blob/master/bip-0016.mediawiki (accessed 20 July 2020).

3. Chohan, Usman W., (2017). Assessing the Differences in Bitcoin & Other Cryptocurrency Legality Across National Jurisdictions (September 20, 2017), http://dx.doi.org/10.2139/ssrn.3042248.

4. Back, A., (2002). *Hashcash-A Denial of Service Counter-Measure.* http://www.hashcash.org/papers/hashcash.pdf (accessed 20 July 2020).

5. Bamert, T., Decker, C., Wattenhofer, R., & Welten, S., (2014). Blue wallet: The secure Bitcoin wallet. In: *Proc. 10ᵗʰ Int. Workshop Secur. Trust Manage* (pp. 65–80).

6. Bos, J. W., Halderman, J. A., Heninger, N., Moore, J., Naehrig, M., & Wustrow, E., (2014). Elliptic curve cryptography in practice. In: *Proc. 18ᵗʰ Int. Conf. Financial Cryptogr. Data Secur. (FC'14),* 157–175.

7. Courtois, N. T., (2014). On the longest chain rule and programmed self destruction of crypto currencies. *Computing Research Repository, Tech. Rep.* abs/1405.0534.

8. Douceur, J., (2002). The Sybil attack. In: *Proc. 1ˢᵗ Int. Workshop Peer-Peer Syst.* (pp. 251–260).

9. Finney, H., (2011). *Best Practice for Fast Transaction Acceptance: How High is the Risk.* https://bitcointalk.org/index.php?topic=3441 (accessed 20 July 2020).

10. Fleder, M., Kester, M., & Pillai, S. (2015). Bitcoin transaction graph analysis. Massachusetts Institute of Technology (MIT). *Computer Systems Security.*

11. Goldfeder, S., Bonneau, J., Felten, E. W., Kroll, J. A., & Narayanan, A., (2014). Securing bitcoin wallets via threshold signatures. *Tech. Rep.*

12. Goldfeder, S., et al., (2015). Securing bitcoin wallets via a new DSA/ECDSA threshold signature scheme. *Tech. Rep.*

13. Kroll, J. A., Davey, I. C., & Felten, E. W. (2013). *The Economics of Bitcoin Mining, or Bitcoin in the Presence of Adversaries.* In: Proc.

14. Lamport, L., Shostak, R., & Pease, M. (1982). The Byzantine generals problem. *ACM Trans. Program. Lang. Syst., 4*(3), 382–401.

15. Leslie, L., (1998). The part-time parliament. A*CM Transactions on Computer Systems (TOCS), 16*(2), 133–169.

16. Malkhi, D., & Reiter, M., (1998). Byzantine quorum systems. *Distrib. Comput., 11*(4), 203–213.

17. Massias, H., Avila, X. S., & Quisquater, J. J., (1999). Design of a secure time stamping service with minimal trust requirement. In: *Proc. 20ᵗʰSymp. Inf. Theory Benelux (SITB'99).*

18. McConaghy, T., Marques, R., & Muller, A., (2016). *Bigchain DB: A Scalable Blockchain Database (DRAFT).* Berlin.

19. Merkle, R. C., (1980). Protocols for public key cryptosystems. In: *Proc. 1980 Symposium on Security and Privacy, IEEE Computer Society.*

20. Merkle, R. C., (1987). A digital signature based on a conventional encryption function. In: *Proc. 7ᵗʰ Conf. Adv. Cryptol. (CRYPTO'87).*

21. Miller, V. S., (1985). Use of elliptic curves in cryptography. In: *Proc. 5ᵗʰ Conf. Adv. Cryptol.* (pp. 417–426).

22. Nakamoto, S., (2008). *Re: Bitcoin P2P e-Cash Paper* [online]. http://www.bitcoin.org/bitcoin.pdf (accessed 10 August 2021).

23. Reid, M. H. F., (2011). An analysis of anonymity in the bitcoin system. In: *2011 IEEE International Conference on Privacy, Security, Risk, and Trust, and IEEE International Conference on Social Computing.*

24. Rivest, R. L., (2004). Peppercoin micropayments. In: *Financial Cryptography.*
25. Ron, D., & Shamir, A., (2013). Quantitative analysis of the full bitcoin transaction graph. In: *Proc. 17th Int. Conf. Financial Cryptogr. Data Secur. (FC'13)* (pp. 6–24).
26. Schoenmakers, B., (1998). Security aspects of the E-cash™ payment system. *State of the Art in Applied Cryptography*.
27. Szabo, N., (1998). *Secure Property Titles with Owner Authority*. https://nakamotoinstitute. org/secure-property-titles/ (accessed 20 July 2020).
28. Turek, J., & Shasha, D., (1992). The many faces of consensus in distributed systems. *IEEE Comput., 25*(6), 8–17.
29. Wright, Ping. *Distributed Ledgers Are the Future of Identity Security*. Tech target. https:// searchcloudsecurity.techtarget.com/news/450303520/Ping-Distributed-ledgers-are-the-future-of-identity-security (accessed 10 August 2021).

CHAPTER 22

Blockchain-Based Smart Contract for Transportation of Blood Bank System

NIHAR RANJAN PRADHAN,[1] D. ANIL KUMAR,[2] and
AKHILENDRA PRATAP SINGH[1]

*[1]Department of Computer Science, National Institute of Technology Meghalaya, Shillong – 793003, Meghalaya, India,
E-mail: niharpradhan@nitm.ac.in (N. R. Pradhan)*

[2]GIET University, Gunpur, Rayagada, Odisha, India

ABSTRACT

Tracking the shipment of blood is difficult, and violation of maintaining critical parameters leads to wastage and spoilage. IoT sensors can be used to monitor temperature, pressure, humidity, exposed to light, broken seal, etc. In our work, we implemented and tested an Ethereum based smart contract in solidity platform. Although several research works have been carried out in the blood bank inventory system, few of them focus on its security, transparency, and traceability. Health record sharing without modification is essential. Blockchain-based blood bank systems can resolve the need for individuals to access, trace, manage, and share their health and blood-related information which is immutable. Blood traceability can increase the availability systems to blood bank system and determines the number of quantities to transport from cities to a rural area. Delay in availability, results in delay in surgery or may lead to death. The blockchain-based system is a confidential environment which acts as a communication hub between donors, doctors, testing labs, and recipients or patients. To address these issues, we have designed a blockchain-based solution for the blood bank system.

22.1   INTRODUCTION

Blood products and shipping can be created using RFID with blockchain technology. The lifetime of blood bags is 21 days. It can't be produced like manufacturing other products because of donor willingness and need a gap of three months if donated before. Blockchain implementation has advantages like:

1. Open (everyone can join);
2. Tokens (rewards to blood donor);
3. Transparency (anybody can see blood availability);
4. Traceability (reduce the delay in transportation, improve monitoring of blood bags);
5. Distributed (nobody owns the network);
6. Anonymization (ensures privacy);
7. Immutable (donor/receivers can see all their previous records) [2].

There must be some decision strategies to provide the demand-supply of blood for patients that appear in regular and irregular way like an accident. Many tools, such as regression models or parametric model, are used to predict the future demand for blood components. Good estimation can reduce the excess or stock out of stocks in blood centers. Some popular models T-weekly (monthly) moving average model. According to WHO in every 15 minutes, 901 people need blood and 202 of them is a woman giving birth and districts may find a shortage of blood packets but in other over 65. Blood is perishable raw material, and a mixture of districts may have stock. Blockchain-based blood traceability various cells like plasma, nutrients, antibodies, clotting agents, and coordination among hospitals can reduce the delay and proteins, salts, hormones, and waste products. As it can save some precision life. It can reduce human error, improve somebody's life major precautions can be taken for blood monitoring of blood bags, improve the safety of patients and demand and supply mismatch, decrease in blood availability hospital staffs, improve the efficiency of the management. Storage cost is high; Blood is obsolescence after 21 days, quick distribution during an emergency. Blood componentizing separates raw blood to RBC, platelets, plasma, and cryoprecipitate. Blood transfusions of RBC are required for major surgeries like liver transplant, open heart surgery, accidents, kidney failure, and sickle cell anemia, trauma-due to burns and accidents and leukemia. Transfusion of platelets is required in bleeding disorders, cancer therapy and opens heart surgery whereas plasma transfusion is very risky

[1]. Blood traceability can increase the availability system to integrate blood bank system.

22.2   TRADITIONAL BLOOD BANK SYSTEM

22.2.1   NECESSITY OF HUMAN BLOOD BANKS

Blood bank system is related to some or any person's life; it must be handled carefully. Pathology department in every hospital is most important. It processes the blood which will be given to patients. Blood undergoes several tests to confirm the blood can be delivered to patients or not, and the receiver should not be affected by any disease or viruses [3] (Figure 22.1).

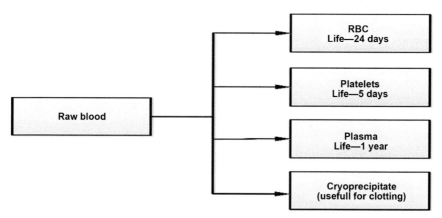

FIGURE 22.1   Various blood components.

22.2.2   RISKS INVOLVED IN TRADITIONAL BLOOD BANKS

1. Effective decision making policy as various blood groups-8 blood groups, components-4 major components, blood group, and its component substitutability.
2. No interaction medium between hospital and public for blood donation schedules or events.
3. Public afraid to donate blood.
4. Blood transfusion data is not secure.
5. The public is unaware of the availability of blood stocks in hospitals.

6. No facility and rewards for donors. Blood screening result is not known to the donors. Even they don't know how many times they have donated their blood. Physical ledgers are maintained but not distributed.
7. Donor health condition is not given from the blood test report.
8. Different hospitals use different software or platform which is not secure and easily modifiable.

22.3 PROPOSED SYSTEM-BLOCKCHAIN-BASED BLOOD BANK

Blood banks can be connected distributive and decentralized way. Information related to donors, patient, and hospitals can be collected in blocks. The blockchain-based blood bank can efficiently regulate the blood. Excess stock of blood can be easily transported to the nearest blood bank having a shortage of blood. In the proposed model, private blockchain is used for storing, the user medical records while consortium Blockchain for secure indexes for users. Private blockchain of blood bank contains the encrypted medical reports of users. It is a storage layer whereas the consortium blockchain gives services to end-users by searching the indexes. Figure 22.2 describes the supply chain architecture of the blockchain-based blood bank system [4].

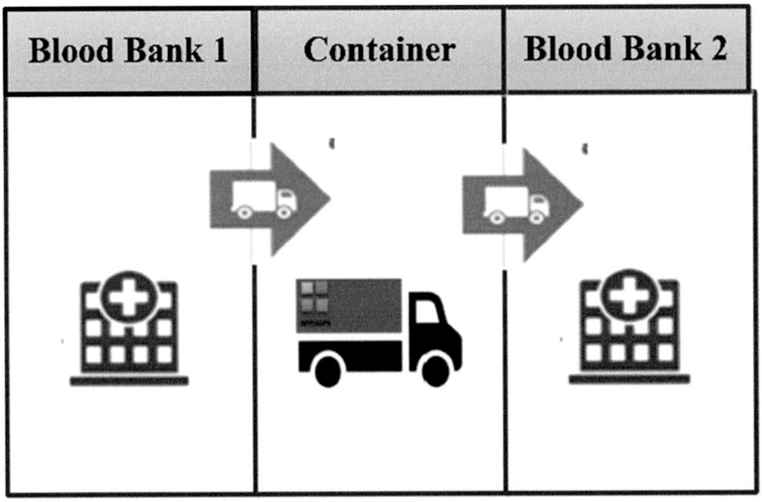

FIGURE 22.2 Blood supply chain.

22.4   IMPLEMENTATION AND TESTING

The smart contract written in solidity is implemented using Remix IDE: http://remix.ethereum a web browser. We have considered a single-echelon that is a single sender and single receiver Blood bank. We consider three entities sender, receiver, and transport container. Each entity has an Ethereum address. The smart contract is designed to solve the Blood supply chain tracing and transparency [5]. It takes all the factors such as temperature, route, and payment terms. The smart contract generates the passphrase at sender side before shipping, and at the receiver side, the same passphrase is entered for delivery. This approach can be extended to multi-echelon. The smart contract calls the function at a certain time. Modifiers are used so that only sender can call some specific function. Similarly for transport container and receiver (Figure 22.3). A snapshot in Figure 22.4 is given for violation of rules.

```solidity
Home      bloodtrace.sol ✕

1   pragma solidity ^0.4.0;//version 0.4 or higher
2
3▾  contract BloodTransport{
4
5       //participating entities with Ethereum addresses
6       address container;
7       address public sender_owner;
8       address public receiver_hospital;
9       string public content;//description of container content
10      bytes32 public passphrase; //recived passphrase when money is deposited
11      string public receivedCode; //recived code to be hashed
12▾     enum packageState {
13          NotReady, PackageContainerReadyforSelfCheck, ReadyforShipment,
14          RequestDeposited, StartShipment,WaitingforPassphrase, ReceiverAuthentiated,
15          WaitingForCorrectPasscode, ShipmentReceived,
16          AuthenticationFailureAborted,Aborted }
17      packageState public state;
18      uint startTime;
19      uint daysAfter;
20      uint shipmentPrice;
21      //sensors
22      enum violationType { None, Temp, Open, Route, Jerk}
23      violationType public violation;
24      int selfcheck_result;//1 or 0 indicating the self check result of Bloodtransport veichle
25      int tempertaure; //track the tempertaure between 1 degree to 6 degree centigrade
26      int open; //if the container opens 1 , 0
27      int onTrack; //to track the route 1 , 0
28      int jerk;//sudden jerk 1, 0
29
```

FIGURE 22.3   Smart contract for blood bank transportation.

The proposed system utilizes distributed ledger, which makes patients data immutable. The following are the stages involved in blood bank process.

```
103
164 ▾    function violationOccurred(string msg, violationType v, int value) OnlyContainer{
165          require(state == packageState.StartShippment);
166          violation = v;
167          state = packageState.Aborted;
168 ▾       if(violation == violationType.Temp){
169              tempertaure = value;
170 ▾           if (tempertaure <= 1 or tempertaure >= 6){
171                  TempertaureViolation( msg ,true, tempertaure);
172              }
173          }
174 ▾       else if(violation == violationType.Jerk){
175              jerk = value;
176              SuddenJerk(msg, true, jerk);
177          }
178 ▾       else if(violation == violationType.Open){
179              open = value;
180              SuddenContainerOpening(msg, true, open);
181          }
182 ▾       else if(violation == violationType.Route){
183              onTrack = value;
184              OutofRoute(msg , true, onTrack);
185          }
186          Refund();
187      }
188
189
190  }
```

FIGURE 22.4   Smart contract functions in case of violation.

22.4.1   DONOR REGISTRATION

In this process donor, preliminary details like name, gender, blood type, mobile no, email, city, last time of donation, age, and weight will be filled. Figure 22.5 explains the steps.

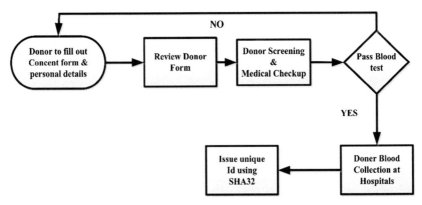

FIGURE 22.5   Donor registration.

The eligibility test of donor's age should be in between 18 and 60. Hemoglobin should be normal. Blood pressure and body temperature should be normal. Donors must not have taken medicine in last 48 hours and should not have diseases like HIV, jaundice or syphilis, etc. The donor should not be addicted to any drugs [9].

22.4.2   DONOR BLOOD COLLECTION

It maintains a record of donors, date of donation, and details of blood unit collected. The donors are informed about test results (Figure 22.6).

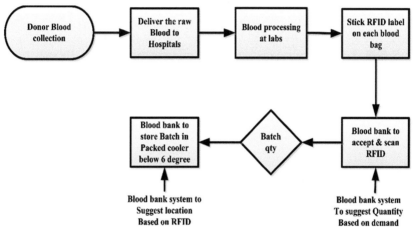

FIGURE 22.6   Donor blood collection.

22.4.3   BLOOD PROCESSING

Raw Blood is batched into standard units and is labeled with RFID labels after being processed. During blood processing, the temperature is measured and monitored. Each bag contains donor number, date of collection, blood group type, component type, and date of expiry. All these information are recorded in blockchain. Following are the stages:

1.   Testing the raw blood for infectious disease.
2.   Blood component preparation: This test is performed to find either plasma or serum. Plasma is the liquid that contains blood and is separated from the blood spinning procedure. Then it appears at the top and blood cells at the bottom. The serum is like plasma which

is allowed to clot. Here blood collected with no anticoagulant. It is collected the same way as plasma.

3. Donor patients compatible assessment.
4. Transfusion on the patients.
5. Complication control-after transfusion it evaluates the complication of transfusion.

22.4.4 RECIPIENT REGISTRATION

The Doctors/specialist or patients who need blood for medical operations are the recipient. Details like hospital name, patient's age, DOB, blood groups, genders, and others will be entered during registration. After registration, blood can be supplied from the same hospitals where the patient is admitted or if not available, then immediate supply from the nearest blood bank. The recipient's relatives or doctors can verify the donor details, lab test reports, blood temperature, and other details by RFID code (Figure 22.7).

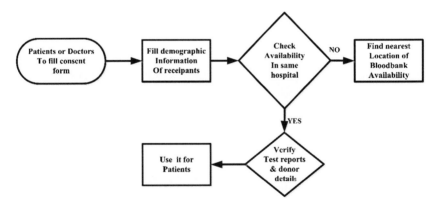

FIGURE 22.7   Recipient registration.

22.4.5 AN APPLICATION FLOW

The application flow describes how systematically the data moves from donors, blood processing unit, and inventory management, blood distribution, and finally to recipients. Figures 22.8 and 22.9 differentiates between traditional and blockchain-based blood bank system.

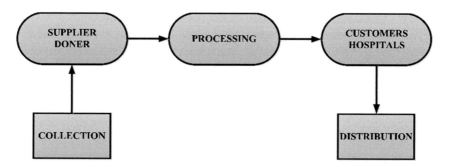

FIGURE 22.8   Traditional blood bank structure.

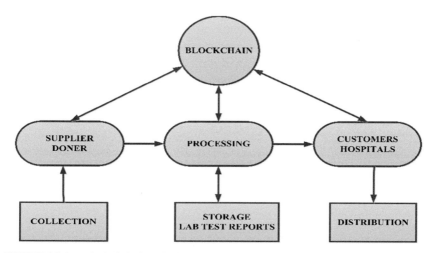

FIGURE 22.9   Blockchain-based blood bank structure.

22.4.6   *BENEFITS OF THIS APPLICATION*

It integrates the sharing of information among all stakeholders, improves the existing blood information system by traceability and transparency. This application reduces operational cost, human errors and increases information sharing, scalable, and traceable process [10–12].

Access to a distributed ledger for transparently sharing medical data can guarantee data security. The blood bags with RFID code contain unique donor ID, testing information, product code, and registration no, license number, and expiry date. This information can be viewed by collection, production, and storage and distribution staffs. It monitors the quality of blood starting from suppliers, storage, to transportation. The monitored data throughout the process is visible to the customers. The RFID reader reads the blood package with RFID and sensor. The package information is like block which contains RFID address, manufacturer, timestamp, sensor type, and sensor data. RFID (Radio-Frequency Identification) is an automatic identification communication technology which identifies multiple, high speed moving objects through radio frequency signal. The block is interconnected with cryptographic function so that it creates an immutable digital database (DB) of blood packets at each instance of time. Before blockchain implementation, we could able to see only the expiry date and manufacturer. Implementation of Blockchain in blood supply avoids blood contamination, blood wastages, and losses due to spoilage. Figure 22.10 shows how RFID is used in blood supply chain management [13, 14].

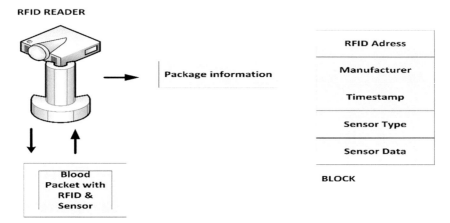

FIGURE 22.10   Blockchain-based RFID blood traceability system.

22.4.7   *TRANSACTION FLOW*(FIGURE 22.11)

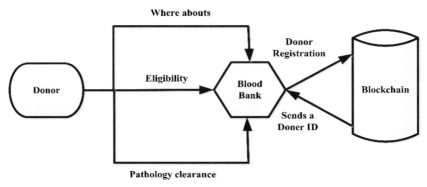

FIGURE 22.11   Donor blockchain.

22.4.7.1   DONOR BLOCKCHAIN

- **Donor Registration:** For every donor, a genesis block is created with the following data [15]:
  - Unique ID;
  - Donor name;
- **KYD Document:** Document is stored in a file, and a hash of the document is stored in a block. At the time of accessing document is verified against hash like in Merkle tree.
  - DOB;
  - Email address;
  - Contact number.

22.4.7.2   BLOOD BANK BLOCKCHAIN

Blood bank blockchain captures the data in the following steps:

1. Raw blood collection;
2. Batch mixing;
3. Transfusion.

22.4.7.3   RECIPIENTS TRANSACTIONS BLOCKCHAIN

- **Recipients Registration:** This Blockchain contains the data like:

- Recipients ID;
- Name;
- Identity;
- Delivery date;
- Location;
- Email IDs.

22.5   CONCLUSION

This blockchain-based blood bank system considers all factors such as donor test reports, transfusion, and testing reports, storage of blood information, expiry reports and brings more transparency and traceability by taking the critical data at all stages. The primary objective is the availability of blood and its component, minimizing the waste of blood, controlling the demand-supply chain. Demand supply model may not be uniform for all hospitals. It is different for cities and rural areas. Rural hospitals sometimes cannot afford their blood banks. Since it deals with somebody's life, delay in transport and safety cannot be compromised. RFID code on every blood bag can be used to trace and view the entire history of the blood, which brings transparency and gives confidence in distributing to a location where there is a shortage of blood.

KEYWORDS

- **blood bank system**
- **demand-supply chain**
- **radio-frequency identification**

REFERENCES

1. Lowalekar, H., & Ravichandran, N., (2014). Blood bank inventory management in India. *Opsearch, 51*(3), 376–399.
2. Vanany, I., Maryani, A., Amaliah, B., Rinaldy, F., & Muhammad, F., (2015). Blood traceability system for the Indonesian blood supply chain. *Procedia Manufacturing, 4,* 535–542.
3. Selvamani, K., & Rai, A. K., (2015). A novel technique for online blood bank management. *Procedia Computer Science, 48,* 568–573.
4. Saha, A., Amin, R., Kunal, S., Vollala, S., & Dwivedi, S. K. (2019). Review on blockchain technology based medical healthcare system with privacy issues. *Security and Privacy*, e83.
5. Hasan, H., Al Hadhrami, E., Al Dhaheri, A., Salah, K., & Jayaraman, R., (2019). Smart contract-based approach for efficient shipment management. *Computers and Industrial Engineering.*
6. Ali, M. S., Vecchio, M., Pincheira, M., Dolui, K., Antonelli, F., & Rehmani, M. H., (2018). Applications of blockchains in the internet of things: A comprehensive survey. *IEEE Communications Surveys and Tutorials.*
7. Buterin, V., (2015). *On Public and Private Blockchains.* [online]. Available at: https://blog.ethereum.org/2015/08/07/on-public-and-private-blockchains/ (accessed 20 July 2020).
8. Dinh, T. T. A., Wang, J., Chen, G., Liu, R., Ooi, B. C., & Tan, K. L., (2017). Block bench: A framework for analyzing private blockchains. In: *Proceedings of the 2017 ACM International Conference on Management of Data* (pp. 1085–1100).ACM

9. Yue, X., Wang, H., Jin, D., Li, M., & Jiang, W., (2016). Healthcare data gateways: Found healthcare intelligence on blockchain with novel privacy risk control. *Journal of Medical Systems, 40*(10), 218.

10. Silva, F. O. S., Cezarino, W., & Salviano, G. R., (2012). A decision-making tool for demand forecasting of blood components. *IFAC Proceedings Volumes, 45*(6), 1499–1504.

11. Abbas, A., & Khan, S. U., (2014). A review on the state-of-the-art privacy preserving approaches in the e-health clouds. *IEEE Journal of Biomedical and Health Informatics, 18*(4), 1431–1441.

12. Dorri, A., Steger, M., Kanhere, S. S., & Jurdak, R., (2017). Blockchain: A distributed solution to automotive security and privacy. *IEEE Communications Magazine, 55*(12), 119–125.

13. Tian, F., (2016). An agri-food supply chain traceability system for china based on RFID and blockchain technology. In: *2016 13ᵗʰ International Conference on Service Systems and Service Management (ICSSSM)* (pp. 1–6) IEEE.

14. Novo, O., (2018). Blockchain meets IoT: An architecture for scalable access management in IoT. *IEEE Internet of Things Journal, 5*(2), 1184–1195.

15. Hazzazi, N., Wijesekera, D., & Hindawi, S., (2014). Formalizing and verifying workflows used in blood banks. *Procedia Technology, 16*, 1271–1280.

16. Mettler, M., (2016). Blockchain technology in healthcare: The revolution starts here. In: *2016 IEEE 18ᵗʰ International Conference One-Health Networking, Applications, and Services (Healthcom)* (pp. 1–3). IEEE.

17. Yue, X., Huiju, W., Dawei, J., Mingqiang, L., & Wei, J., (2016). Healthcare data gateways: Found healthcare intelligence on blockchain with novel privacy risk control. *Journal of Medical Systems, 40*(10), 218.

CHAPTER 23

Blockchain Traversal: Its Working Aspects and a Brief Discussion on Its Security

ANAND KUMAR JHA,[1] RITIKA RAJ,[1] ANJANA MISHRA,[1] and BROJO KISHORE MISHRA[2]

[1]*Department of Information Technology, CVRCE, CVRCE, Bhubaneswar, India, E-mail: anjanamishra2184@gmail.com (A. Mishra)*

[2]*Department of Computer Science and Engineering GIET, Gunupur, Odisha, India*

ABSTRACT

Blockchain has been a topic of great interest and it has caught the eyes of researchers all over the world for last about a decade. Blockchains are referred to as the records that are constantly increasing and are referred to as blocks, and they are connected to each other with the help of cryptography. Blockchain provides us a much more secure and resilient transaction. It is one of such inventions that have contributed a lot in maintaining security and privacy during transactions. With the help of blockchain and its components, we have been able to maintain transparency in transaction system. So, in this chapter, we will be dealing mainly with the bitcoins and other cryptocurrencies that has contributed in evolution of blockchain. We will study about its various prospects its origin and some of its advantages and disadvantages.

23.1 INTRODUCTION

23.1.1 BITCOIN

Bitcoin is the cryptocurrency which runs on blockchain. It is a peer to peer network [9]. It is a decentralized system which allows you to do a value transfer transaction at a lower transaction where no third party are involved between the sender and the receiver. Bitcoin avoids the double spending through the basic structure of blockchain which involve verification transaction. It reduces the tremendous cost of the transaction. It reduces the value as low as $1 as compared to $5 which the bank was charging. Earlier most of the people were having no idea how the money is transferred from one bank to another. But because of bitcoin transection from the beginning time all the transactions are written in the public ledger, where everyone is watching this ledger and having their own copies of it. When the Bitcoin wallet is generated, it gives us two parts. One is the Public part, where an address is stored about where to send money and the next that is the secret part is a Key. Key is part where cryptography works. It is authenticated by the best cryptography algorithm available which is very difficult to hack. Grab your reader's attention with a great quote from the document or use this space to emphasize a key point. To place this text box anywhere on the page, just drag it (Figure 23.1).

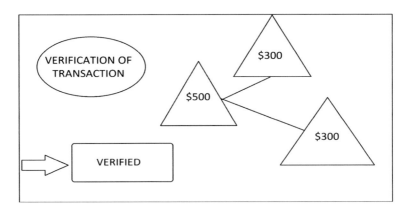

FIGURE 23.1   Working of bitcoin.

23.1.2   *BLOCKCHAIN*

Blockchain is the digital ledger that is used to record all bitcoin transactions across many computers. Blockchain is the era of new technology [10]. Blockchain as the name suggests are long chains or sequences of blocks that store information about the transactions in a digital form. The concept of a single universal ledge is the heart of blockchain. It is something as fundamental and parallel to the internet. Earlier the concept of bitcoin as cryptocurrency has gained many regards. It helps us to carry out the transactions without the involvement of any third party. Blockchain stores the digital signatures of both the parties and has new parameters in the security paramount. It stores the transaction details like date, time, and dollar amount. The Blocks are added with a unique code called the Hash code. Blockchain operates in a peer to peer network with the development in the field of cryptocurrency the concept of Blockchain is elevating to new heights.

Blockchain is the technology that power bitcoin. Blockchain is suite of distributed ledge technologies that is programmed to record values. It stores information in batches called blocks which is linked together in form of a chain. To make a change to the particular data recorded in a particular block, you need to go through all the blocks so the new transaction is added to the existing block (Figure 23.2).

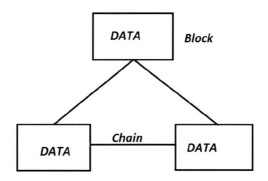

FIGURE 23.2   Blockchain.

23.2   GROWTH OF BLOCKCHAIN TECHNOLOGY

In 1991, Stuart Haber and Scott Stornetta envisioned through which many people got the information about the cryptographically secured chain of blocks.

In 1992, this system was more enhanced and was upgraded to incorporate Merkle Trees that increased the efficiency and was able to collect a greater number of documents on a single block. In 2008, Santoshi Nakamoto introduced whitepaper Bitcoin.

In 2009, the Bitcoin was offered up to open source community. Here Blockchain provides the trust to the community because it records all the important information related to the transaction.

In 2014, after being acquainted with all the technology, many understood the difference between the cryptocurrency and blockchain. They started investing in and thumbing their knowledge about it. As the blockchain doesn't involve the third party between the transaction, it reduces the cost. According to a recent survey 15% of financial institutions are currently using this technology [11].

In 2013, Ethereum was invented by Vitalik Buterin. In this amazing concept of EVM "Ethereum virtual machine (VM) was introduced. Ethereum is a platform where we can create a decentralized application which is DAPPS. It also has its own cryptocurrency known as ETHER. The main objective of Ethereum's programming language is to allow developers to write more programs in which blockchain transactions could get automate and govern outcomes.

23.3   WORKING METHODOLOGY

23.3.1   ADDITION OF NEW BLOCKS

In the case of blockchains, every new block is added at the end of the chain. And is mapped with a unique code called hash code. For a Block to add in the chain the following major steps are to be followed. For a block to be added there must be an occurrence of a transaction. The transaction details must be verified with the available public records, this is made liable for ensuring proper and secure Block. The transaction details are then stored in the block with the digital signatures of both the sides are preserved and stored. Each block is assigned with a unique code and this code is called hash code. Then the Block is added to the end of the chain and this is how the sequence is added and the process continues (Figure 23.3).

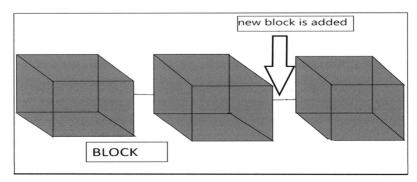

FIGURE 23.3   Working of Blockchain.

23.4   PROS AND CONS OF BLOCKCHAIN

23.4.1   ADVANTAGE

- In Blockchain the values are maintained and are kept unchanged.
- The transaction is verified by a huge peer to peer network.
- Cryptocurrencies like Bitcoin are not stuck during some financial issues.

23.4.2   DISADVANTAGES

- Cryptocurrencies like Bitcoin can lead to a growth of an economy which is beyond the control of the government.
- Illegal transactions will be hard to be traced.
- Transactions will be fixed that is once a transaction is made it cannot be reversed.

23.5   SECURITY

Blockchain are often questioned for their security breach and for their reliability deficit. The blocks are stored at the end of the chain each assigned with a unique hash code, and to alter the contents of any block it is very difficult. The code of each block is formulated with certain mathematical functions and if the content of any of the block is modified then a new digital code is assigned to the block. Suppose if anyone tries to edit the contents of these blocks then any change made to the block then it gets a new Hash code but the block next to edited block has the previous hash code only. And anyone who is trying to hack the blockchain has to modify the hash code of all the following blocks which will require a very strong computation and an efficient algorithm to achieve it. There are basically two type of keys private and public keys.

23.5.1   PRIVATE KEY

Private key always remains confidential. This key is used for data decryption. Private key is also called secret key. These keys are faster compared to public keys. Private keys are the symmetrical key, i.e., there is only one key and the other is only the copy of it. There are only available with the two communication parties.

23.5.2 PUBLIC KEY

Public key is a key which is made available to everyone that is it is publicly accessible. This is used for the data encryption. These keys are relatively slow to encrypt and are a symmetrical. These keys are for public. In this, every transaction is mentioned in the public ledger so it is halfway around the globe. It is the founder and core of network security [12].

23.6   CONCLUSION

Blockchain is an integrated technology to ensure keeping records and process transactions in a decentralized manner. Because of this technology, Global transactions are recorded and shared among the systems. Blockchain and cryptocurrencies are great inventions and can really create a great impact on the financial market. People will have an altogether new platform to invest and carry out their transactions. But however, it has not gained the trust of many people in common, because of some of its above-mentioned feature. It is a new mode and few of modifications like involvement of government will really increase the trust of people. The technology gap has to be bridged and people must understand what Bitcoin and Blockchain are how they work and also, we need to ensure transparency and security has to be full-proof as well. So that in future this great technology can we widely used and can reach to the grass-root level.

KEYWORDS

- **bitcoin**
- **blockchain**
- **cryptocurrencies**
- **digital signature**
- **hash code**
- **private key**
- **public key**
- **security**
- **transactions**

REFERENCES

1. Ahram, T., Sargolzaei, A., Sargolzaei, S., Daniels, J., & Amaba, B., (2017). Blockchain technology innovations. *2017 IEEE Technology and Engineering Management Conference (TEMSCON)*.
2. Chen, G., Xu, B., Lu, M., & Chen, N., (2018). Exploring blockchain technology and its potential applications for education. *Smart Learning Environments, 5*(1).
3. Dabbagh, M., Sookhak, M., & Safa, N., (2019). The evolution of blockchain: A bibliometric study. *IEEE Access, 7*, 19212–19221.
4. Dinh, T., Liu, R., Zhang, M., Chen, G., Ooi, B., & Wang, J., (2018). Untangling Blockchain: A data processing view of blockchain systems. *IEEE Transactions on Knowledge and Data Engineering, 30*(7), 1366–1385.
5. https://blockgeeks.com/guides/what-is-blockchain-technology (accessed 20 July 2020).
6. https://hackernoon.com/the-evolution-of-blockchain-where-are-we-f0043b2d0cd0 (accessed 20 July 2020).
7. Memon, M., Hussain, S., Bajwa, U., & Ikhlas, A., (2018). Blockchain beyond bitcoin: Blockchain technology challenges and real-world applications. *2018 International Conference on Computing, Electronics, and Communications Engineering (ICCECE)*.
8. Wang, R., He, J., Liu, C., Li, Q., Tsai, W., & Deng, E., (2018). A privacy-aware PKI System based on permissioned blockchains. *2018 IEEE 9th International Conference on Software Engineering and Service Science (ICSESS)*.
9. Van Der Horst, Choo, K. R., & Le-Khac, N. (2017). "Process Memory Investigation of the Bitcoin Clients Electrum and Bitcoin Core," in *IEEE Access, 5*, 22385–22398, doi: 10.1109/ACCESS.2017.2759766. .
10. Sapra & Dhaliwal, P. (2018). "Blockchain: The new era of Technology," 2018 Fifth International Conference on Parallel, Distributed and Grid Computing (PDGC), pp. 495–499, doi: 10.1109/PDGC.2018.8745811.
11. Masoud, M. Z., Jaradat, Y., Jannoud, I., & Zaidan, D. (2019). "CarChain: A Novel Public Blockchain-based Used Motor Vehicle History Reporting System," *2019 IEEE Jordan International Joint Conference on Electrical Engineering and Information Technology (JEEIT)*, pp. 683–688, doi: 10.1109/JEEIT.2019.8717495.
12. Wang, R., He, J., Liu, C., Li, Q., Tsai, W., & Deng, E. (2018). "A Privacy-Aware PKI System Based on Permissioned Blockchains," *2018 IEEE 9th International Conference on Software Engineering and Service Science (ICSESS)*, pp. 928–931, doi: 10.1109/ ICSESS.2018.8663738.

CHAPTER 24

Improved Framework for E-Healthcare Based on Blockchain and Fog Computing

K. LAKSHMI NARAYANA,[1] S. PRIYANGA,[2] and K. SATHIYAMURTHY[3]

[1]PhD Scholar, Department of CSE, Pondicherry Engineering College, Puducherry, India, E-mail: kodavali.lakshmi@gmail.com

[2]MTech, Department of CSE, Pondicherry Engineering College, Puducherry, India

[3]Associate Professor, Department of CSE, Pondicherry Engineering College, Puducherry, India

ABSTRACT

In the present day's e-Healthcare applications, activity recognition system (ARS) is the most significant undertaking in remote checking of patients experiencing physical medical issues for taking quick action. The cloud is far away from data and it is not capable to handle huge bandwidth data due to network latency. The goal of the Fog computing is to decrease the data that needs to be transferred to the cloud for data processing, and to increase the efficiency. Fog computing improves the QoS and also reduces network bandwidth. All machine learning (ML) algorithm performances depend on the quality of the training data. If the training data is inadequate or it is modified by attackers, then ML algorithm would miss predict and give invalid results. As Blockchain is a decentralized model, it has attracted both industry and research nowadays. Blockchain data structure is practically difficult to forge. Hence, in this chapter, Fog is used to reduce the network bandwidth and blockchain is used to store training data to avoid forging.

24.1 INTRODUCTION

A blockchain is a growing chain of blocks, which are linked using cryptography in a decentralized environment. Each block contains timestamp, transaction data, and cryptographic hash of the previous block as shown in Figure 24.1 [15]. Blockchain is unsusceptible to change of data by its design. In a blockchain, transactions among two parties are recorded in an efficient, verifiable, and permanent way [16]. Such a blockchain can present an innovative solution for long standing problems of security related to data storage in centralized systems. Blockchain can be considered as new face of cloud computing (CC), and is expected to reshape the organizational and individual behavioral models.

Distributed database (DB) is among the salient features of blockchain. This sort of DB exists in numerous replicas across different nodes forming a peer-to-peer network, means that no centralized server or DB exists [17].

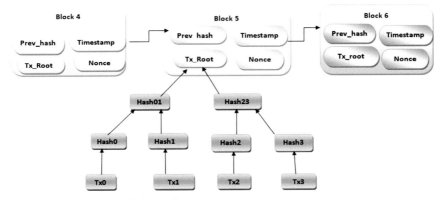

FIGURE 24.1   Blockchain internal structure.

Every computer in the network is called a node and every node in the network receives a duplicate copy of Blockchain that automatically gets downloaded. Transactions are digitally signed with a public key that uses two mathematically related keys. Due to the complexity of mathematical formula used, it is practically difficult to guess the key, and crack the transaction. The public key is used to sign and encrypt a message to be sent and the corresponding recipient can decrypt the message using their private key. To maintain the Blockchain DB as a "World Wide Ledger" data with respect to all, new transactions are propagated to all nodes [17].

Smart contracts are the lines of code which are stored on a Blockchain and execute automatically when predefined conditions and terms are checked and met. Smart contracts works on the following simple "if/when… then…" statements that are written into code on a Blockchain. A network of computers executes the actions when predefined conditions and terms are checked, and met. The benefits of smart contracts are more significant in business alliances, where they are used to impose some agreement so that all participating parties can be certain of the outcome without an intermediate party involvement (cointelegraph.com).

A consensus algorithm is a procedure through which all the peers of the blockchain network arrive at a common understanding regarding the present state of the distributed ledger. Consensus algorithms establish reliability and trust among unknown peers in a distributed environment. A transaction is added to the blockchain only after it has been validated using a consensus protocol, ensuring that it is the only correct version, and each record is also encrypted to provide an extra security (cointelegraph.com). Popular consensus algorithms are: PoW, PoS, and PoET. Others algorithms like Proof of Importance, Proof of Activity, Proof of Weight, Leased Proof of Stake, etc., can also be used. The types of blockchain are public, private, and consortium.

1. **Public Blockchain:** It does not need any permission for access. Anyone can become a part of it, and can also be a validator and participate in consensus. Generally, these type of networks offer economic incentives for those who secure and use some sort of Proof of Stake or Proof of Work algorithm. One of the very popular public Blockchain is the bitcoin Blockchain.

2. **Private Blockchain:** It needs access permissions. No one could join unless they are invited by the network administrators. This type of Blockchains is being considered as a middle-ground for organizations when they are not happy with the degree of control given by public networks. A Consortium Blockchain is said to be semi-decentralized.

3. **Cloud Computing (CC):** It is anon demand availability of computer system resources, especially computing power and data storage, without any direct active management by the user. The term is normally used to describe data centers (DCs) available to many users over the Internet. It offers different services to the users. CC is a blend of number of concepts such as service oriented architecture (SOA), Virtualization, etc., [1]. Because of this, many organizations are switching to CC. In cloud, computing it is easy to access, manage,

and compute user data, but it has security risks. The conventional security techniques are not enough, so the proposed system has introduced Blockchain to store training data set instead of using cloud [2].

4. **Fog Computing:** It was presented by the Cisco systems as new model to ease wireless transfer of data to distributed devices in the internet of things (IoT) network model [2]. The goal of the Fog computing is to decrease the data that needs to be transferred to the cloud for data processing, and to increase the efficiency. Fog computing improves the QoS and also reduces network bandwidth. Fog is used to deliver data and place it in a node closer to the user who is at the edge of the network. Here 'edge' means edge of the network to which the user is connected, also known as *'edge computing.'* In the proposed system, Fog computing is very much helpful, since IoT devices are having low computing capability and less storage. In daily life, normally abnormal activities have been done unusually and they need complex computations to classify. Hence, all complex computations will be performed in cloud and routine activities will be classified in Fog only. The devices such as Raspberry Pi, routers, smartphones, personal computers, etc., are used for implementing fog. There are many fog applications which involve machine-to-machine or human-to-machine interactions such as zooming the video camera, locking a door, applying brakes on a train, or sending an alert to a technician regarding preventive repair.

5. **Machine Learning (ML) Algorithm:** These have been introduced earlier. These algorithms are classified based on either learning style or similarities [18].

   ML algorithms are classified by learning style consists of supervised learning, unsupervised learning, semi-supervised learning and reinforcement learning (RL), etc. In supervised learning, input or training data has a predefined label. A classifier is designed and trained to predict the label of test data. The classifier is suitably tuned to get a suitable level of accuracy. In unsupervised learning, input or training data is not labeled. A classifier is designed to group unsorted information based on similarities and differences. In RL, the algorithm is trained to map action to situation so that the reward or feedback signal is maximized.

6. **Ensemble Algorithm (web.engr.oregonstate.edu):** Its main purpose is to utilize different learning algorithms to get good predictions, than those that are obtained by using single learning algorithm. These algorithms combine several ML techniques into one model so as to reduce bagging, boosting, stacking, etc.

24.2   LIMITATIONS OF TRADITIONAL ARS

Many existing works [1, 3] have been made on Activity Recognition System by using either Blockchain or fog computing but not both. In this chapter [1], both fog and Blockchain for ARS is used but the detailed usage of Blockchain is not mentioned. The goal of the Fog computing is to decrease the data that needs to be transferred to the cloud for data processing, and to increase the efficiency. All ML algorithm performances depend on the immutability of the training data. Blockchain data structure is practically difficult to forge. Hence, in the proposed work, Fog is used to reduce the network bandwidth and Blockchain is used to store training data to avoid forging.

24.3   PROPOSED SYSTEM

Initially, videos are captured by cameras which are fixed in Hospitals or where there is a need to monitor and those videos is given as input for Fog devices (Ex: Raspberry IoT Device). Fog device divides the video into frames and perform preprocessing operations on it to extract features that contain significant human activity. Fog searches with these features in blockchain for getting video classification results. In the proposed system, training data has been maintaining in Blockchain to avoid modifications by either active or passive attackers in future.

Fog provides classification results directly if it matches frame features in blockchain. If the computations are complex and they cannot be computed in fog network, only those video frames are uploaded through internet to CC data centers, then in cloud different ML algorithms (Ensemble algorithm) are applied on it for classifying the given video or frames. Cloud may interact with blockchain for training data if it requires for classification purpose. After classification, results will be sending back to fog by cloud. Then fog add this new record (consists of frame features, date, time, location, and classification result) into the blockchain training data and provide same results as output. The overview of the framework of ARS has shown in Figure 24.3. Fog computing decreases the delay of uploading large volumes of data to cloud, decreases bandwidth consumption, and associated costs with it. Any ARS performs the two basic tasks, i.e., detection of human action and recognizes the class of action. For many human activities, some features or properties are common. In our proposal, the feature descriptors are acquired from the keyframes of the training data [1] (Figure 24.2).

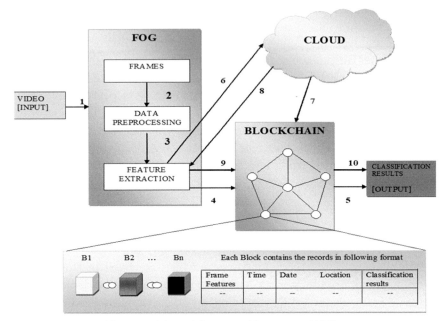

FIGURE 24.2   Architecture of the activity recognition system (ARS).

The algorithm for the proposed activity recognition system is as follows:

Algorithm: ARS _with_ Fog_ and_ Blockchain (video)

Input: Video
Output: Activity Recognition
Data structures: Blockchain
Extract frames from video, let total number of frames be 100 (#frames)
//Preprocessing the extracted frames (removing redundant frames)
i=0, new_F_list[ ], dup=0, k=0, frames[1.100]
while(i< #frames) { j=0
while(j < #frames) {
if(frames[i] == frames[j]) dup=1
j++ }
if(dup == 0) { new_F_list[k] = frame[i]; k++; }
}
//Feature Extraction from frames
i=0, features[ ], c= count(new_F_list[])
while(i<c) {

features[i]= new_F_list[i].features
i++ }
Blockchain Database consist of two columns which consoles [feature list, labeled activity]
//Matching features with Blockchain database by fog
i=0
if(features[i] in Blockchain database)
print (Blockchain features[i].activity)
else
//Cloud will recognize the activity with help of blockchain training data and classifies using //Ensemble algorithm
Activity = cloud (features[i])
Fog updates Blockchain database with new record, i.e., [features[i], activity]
print(activity)

So as to decrease computation complexity in the classification of features, the extracted feature descriptors are separated into subsets of action classes or clusters constituting the vocabulary of key features [1]. In this classification, a multiclass SVM is used by treating the vocabulary of key features as a feature vector to determine the action class of video frames [1]. The action vocabulary is an abstraction of the feature vector that represents a particular action class which is used for classification (Figure 24.3).

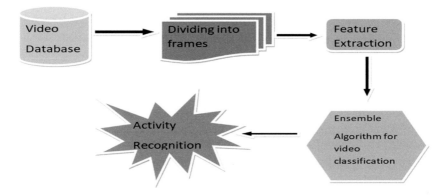

FIGURE 24.3   Framework of activity recognition system (ARS).

24.4 APPLICATIONS OF ARS (ACTIVITY RECOGNITION SYSTEM)

ARS can be used to monitor patient's activities time to time and to take appropriate actions immediately after detecting irregular activities. For example, if patients suddenly face breathing problems or heart problems, at that time patients body movements are something different than regular activities. These irregular movements can be recognized by proposed system and give signal to take necessary further action.

ARS can also be used in traffic monitoring systems, where vehicle traffic flow videos are closely observed, detects if any irregular traffic flow (traffic jam) and sends signal to control rooms to take appropriate actions immediately.

Another application of ARS can be used in houses or hospitals where old age people are taking rest alone in their rooms. Take caring persons are busy with their works in other rooms. If camera is placed in old age people room with ARS then activities of old age people are monitored, detects irregular activities of old age people if any and give signal or alarm immediately, so that take caring person or nurses could react to protect them.

Last but not least, interesting application of ARS can be used in all important public places like airports, railway stations, bus stations, traffic monitoring places, parks, malls, and theaters. These areas were already fixed with cameras. If ARS is used with these cameras, then the ARS system could react within the time for catching thieves instead of tracing them later after incident has happened.

24.5 CONCLUSION

This chapter provides improvised framework for activity recognition system in e-healthcare applications based on blockchain and fog computing with ensemble ML model. ML algorithm performance will be increased because proposed work is storing training data in blockchain; it can't be modified or altered once entered into Blockchain. The implementation for above improved ARS framework need to do practically to test how well it works in practical and real-time situations in coming days. tensor flow is an open-source platform for ML. It has many libraries, tools, and community resources that allow developers and researchers to build and deploy ML-powered applications easily (tensorflow.org). Blockchain applications can be implemented using Hyperledger Fabric and Composer. Hyperledger Explorer, which is one of the Blockchain modules, can be used to access the Blockchain contents easily through online (hyperledger.org).

ACKNOWLEDGMENT

This research work is supported by the UGC Minor Project Grant for the work titled "Design of Dynamic Heath Risk Prediction Model through IoT" in 2016–2017.

KEYWORDS

- **activity recognition system**
- **blockchain**
- **cloud computing**
- **consensus**
- **e-healthcare**
- **fog computing**
- **smart-contract**

REFERENCES

1. Islam, N., Faheem, Y., Din, I. U., Talha, M., Guizani, M., & Khalil, M., (2019). A blockchain-based fog computing framework for activity recognition as an application to e-healthcare services. *Future Generation Computer Systems*, *100*, 569–578.
2. Shabnam, K., & Surender, S., (2017). Fog computing: Characteristics and challenges. *International Journal of Emerging Trends and Technology in Computer Science (IJETTCS)* (Vol. 6).
3. McGhin, T., Choo, K. K. R., Liu, C. Z., & He, D., (2019). Blockchain in healthcare applications: Research challenges and opportunities. *Journal of Network and Computer Applications*.
4. Bruyn, S. A., (2017). *Blockchain an Introduction* (pp. 1–43). University of Amsterdam.
5. Das, K., & Behera, R. N., (2017). A survey on machine learning: Concept, algorithms, and applications. *International Journal of Innovative Research in Computer and Communication Engineering*, *5*(2), 1301–1309.
6. http://web.engr.oregonstate.edu/~tgd/publications/mcs-ensembles.pdf (accessed 20 July 2020).
7. https://cointelegraph.com/bitcoin-for-beginners/how-blockchain-technology-works-guide-for-beginners#where-can-blockchain-be-used (accessed 20 July 2020).

8. https://www.hyperledger.org/projects/fabric (accessed 20 July 2020).

9. https://www.tensorflow.org/ (accessed 20 July 2020).

10. Joshi, A. P., Han, M., & Wang, Y., (2018). A survey on security and privacy issues of blockchain technology. *Mathematical Foundations of Computing*, *1*(2), 121–147.

11. Li, X., Pang, T., Liu, W., & Wang, T., (2017). Fall detection for elderly person care using convolutional neural networks. In: *2017 10th International Congress on Image and Signal Processing, Biomedical Engineering and Informatics (CISP-BMEI)* (pp. 1–6). IEEE.

12. Wang, W., Hoang, D. T., Hu, P., Xiong, Z., Niyato, D., Wang, P., Wen, Y., & Kim, D. I., (2019). A survey on consensus mechanisms and mining strategy management in blockchain networks. *IEEE Access*, *7*, 22328–22370.

13. Yuan, Y., & Wang, F. Y., (2018). Blockchain and crypto currencies: Model, techniques, and applications. *IEEE Transactions on Systems, Man, and Cybernetics: Systems*, *48*(9), 142–1428.

14. Zheng, Z., Xie, S., Dai, H. N., Chen, X., & Wang, H., (2018). Blockchain challenges and opportunities: A survey. *International Journal of Web and Grid Services*, *14*(4), 352–375.

15. Wenbo Mao, (2013). "The role and effectiveness of cryptography in network virtualization: a position paper", ASIA CCS '13: *Proceedings of the 8th ACM SIGSAC Symposium on Information, Computer and Communications Security*. pp. 179–182. https://doi.org/10.1145/2484313.2484337.

16. Bruyn, A. Shanti. (2017). *Blockchain An Introduction*. University Amsterdam, 1–43.

17. Joshi, A. P., Han, M., & Wang, Y. (2018). A survey on security and privacy issues of blockchain technology. *Mathematical Foundations of Computing, 1*(2), 121–147.

18. Das, K., & Behera, R. N. (2017). A survey on machine learning: concept, algorithms and applications. International Journal of Innovative Research in Computer and Communication Engineering, 5(2), 1301–1309.

CHAPTER 25

An IoT and Blockchain-Based System for Acute Security Check and Analysis

ADARSH PRATIK,[1] ABHISHEK BHATTACHARJEE,[1]
ROJALINA PRIYADARSHINI,[1] and SUBHAM DIVAKAR[2]

*[1]Department of IT, C.V. Raman College of Engineering, Bhubaneswar,
India, E-mails: adarshpratik.2010@gmail.com (A. Pratik),
abhishekb496@gmail.com (A. Bhattacharjee),
priyadarhini.rojalina@gmail.com (R. Priyadarshini)*

*[2]Department of ETC, C.V. Raman College of Engineering, Bhubaneswar,
India, E-mail: shubham.divakar@gmail.com*

ABSTRACT

This chapter aims at addressing the security concerns found by the people for their homes, by the colleges for their labs with hi-tech technologies, by the corporate for their offices and various other places. Mainly in the places, where there is a presence of expensive items with limited or no security enforced, mainly due to the lack of controlled access mechanisms. In this chapter, we have provided a security system based on IoT (internet of things) architecture paired with IT architecture and firebase as database (DB) as a service model of cloud computing (CC). Our model checks for activity in any premise in two phases. Phase 1 (Access Details)-during active hours it keeps a check on number of persons coming in and going out of the premises. Phase 2-during inactive hours it checks for unauthorized intrusions as well as same of phase 1. The mechanism is such that in phase 1 (during active hours) the program keeps a track of incoming and outgoing persons (by counting mechanism) with their facial identity and provides a notification on app and website with the date and time. In phase 2 (during inactive hours) there is a check of no (zero) person being present in the premises and if there is any intrusion then the end-user is instantly notified over the app with a live stream of video playback. The devices security is handled using Blockchain to avoid any unauthorized access of any data of the user and if tried to enter into the network then the IP address of the intruder is also put into the blockchain to help report about the device impersonation.

25.1   INTRODUCTION

Security is a major concern in any city. Thefts or unauthorized access of any place are getting very frequent even in this modern era due to lack of proper security. It is difficult to always keep track on individuals accessing your place physically. Hence, a system is required which can automatically keep track on every activity going on inside a particular area [1]. Such a system can be developed using IoT technology with proper use of sensors and then make them work together to complete a certain goal. This chapter shows the work that has been done using this technology and the entire system was designed to work automatically with proper internet connection and power backup. This system is designed to keep a track of count of persons that are coming in and going out of our concerned area along with their facial image being captured and stored for any future analysis and investigation. Also, it keeps a check of the concerned area when it is closed and there is no one present in the premise. It can then automatically detect any kind of motion and record the required evidence and also store it for any kind of future investigation. This makes it very easy for the user to always keep a check on any kind of activity going on in their place. The user can easily tap into their device over an application to analyze their data [2].

Authentication and authorization are also the most important factors that led to the development of technologies providing high security to the users. When it comes to security and trust in the digital world lots of care is needed to be taken care of while dealing with online crypto-currency, online purchases, or online storing of data. This need for high security and ease of tracking down the previous records led to the development of blockchain that was used in the cryptocurrency bitcoin by a person or a team using the name Satoshi Nakamoto in 2008. Since then blockchain has emerged as a tool that has applications in diverse areas. It exists on a peer to peer network. Blockchain is actually lists of records that keeps growing and which are linked using Cryptography. Here the device details (such as IP address and MAC address) can be stored into a blockchain. This will ensure the security of the device as well. The device will be tamper-proof from any outer interference.

If there is a network change in the device then a set of new credentials will be updated in the blockchain, and hence the user will be alerted over it.

25.2   RELATED WORKS

IoT is used in many applications like security, automation, smart grid, connected health (tele-medicines/digital health), farming, and many more. Also, blockchain is used in finance, payments, IoT, personal identification, and other security purposes [3]. Here various methods of security and automation are discussed which can be obtained with the use of IoT and blockchain. Ravi Kishore Kodali et al. [4] implemented wireless home security and home automation as a dual aspect of their project. The system sends voice calls to the user for any movement near entrance and also triggers alarm, lights, fan, and air-conditioner based on user's discretion. Pavithra and Ranjith [5] in their paper exploited Raspberry Pi and IoT technology for real-time home safety and monitoring. They described the connection of IR (infrared), PIR (passive infrared), and fire sensors to GPIO ports of Raspberry pi and then used the website as an interface for users. Jasmeet and Punit [6] used Intel Galileo 2nd generation development board for real-time tracking of the devices at home. The system also has a voice control system. So that appliances can be triggered with voice or through android application. Quadri and Sathish [7] proposed a system which is designed using ARM-11 architecture with Raspberry Pi-3 board, USB camera and DC motor which enables remote controlling of door locking system. They also developed a website as user interface. Prakash and Venkatram [8] in their chapter used IoT with fingerprint recognition technique; they also deployed other hardware modules, i.e., raspberry pi, Wi-Fi router, gas sensor, fire sensor, and door fringe motor sensor for security purpose and automation.

25.3   PROPOSED WORK

This chapter shows the security concern of smart cities and a system is developed following such concerns to provide proper security to various indoor places in the cities. To come up with the solution, this system was equipped with IoT technology to provide acute security check and to keep the system secure we have implemented blockchain on it. The entire system was divided into two modules (Module 1 and Module 2). Separation of the modules was done to make the system run under two circumstances. Both the modules depend on each other to function. The first module has IR sensors paired with the camera to collect data that is count of persons along with their images being captured and the second module has PIR-red sensor paired with the camera to collect data from the heat signature movement along with video recording (see Figure 25.1).

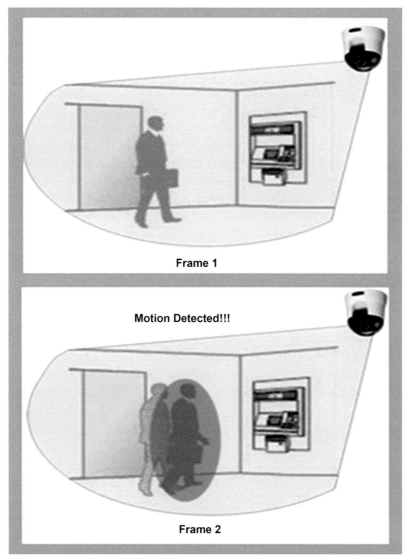

FIGURE 25.1   PIR sensor used to detect movement.

Source: https://nascompares.com/best-home-and-office-ip-cameras-for-nas/.

25.3.1   MODULE I

This module works during the active sessions in the workplace. Generally, at the time of entry into any indoor place, which is always through the main door, the door has the IR sensors working continuously to detect any object. In this time, the PIR also remains active. Both devices being active together makes the main security check. Suppose there is an office that opens at 8 am. In general, there should be no one present inside the office before 8 am as everyone went home from the office at night. So initially, the count of persons inside the office is zero. In such situation when the count of persons in the premises is zero, both the sensors remain active. When there is an entry in the premises the count of person in the program increases, hence leading to the switching off of the PIR. This is done to avoid the passive infra-red to unnecessary detect movement and record videos. Now the person count is continuously checked for both incoming and outgoing persons. The entire counting process is handled by the IR sensors. Along with the counting there is image capture mechanism taking place only during the entry. Reason behind this is image of people accessing the place is to be taken at the entry only after that, the activity of the person would always be under the CCTV surveillance. The entire data is stored for any future security check (see Figure 25.2).

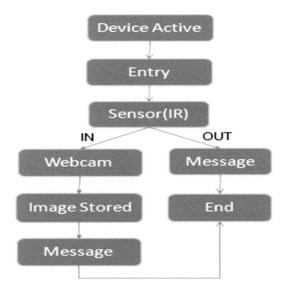

FIGURE 25.2   Module I, when people go in and come out.

This module completely works over the IR during the active hours. When again the count becomes zero then both the sensors work in parallel. So, Module I shows the counting process that only happens when there is activity in the premises.

25.3.2 MODULE II

This module works for the inactive hours or when there is no one in the premises. If there is a legit entry through the main door, where the IR sensors are working then this module is turned off. If not then the PIR remains on to detect movement inside the premises. Suppose there is a break in the premises not from the main door but from any other point then this module (Back door entry) using the PIR it could detect the movement and record the video at the time of the movement (see Figure 25.3).

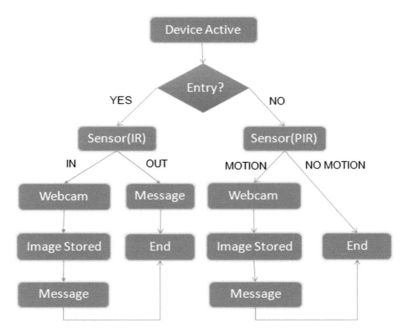

FIGURE 25.3   Module II-when there is unauthorized access.

This was the security of the indoor premises. Now for the security of the device all the IoT devices are connected with the wireless networks so that they can run. Thus it becomes vulnerable to attacks from the hackers and the most important aspect is to keep a track of the IP address that gets connected to the wireless networks so that if the unknown user attempts to make a connection then its IP gets recorded in the block thus it can be easily found out who is trying to get unauthorized access to the system. The power of Blockchain in terms of security can be considered from the fact that altering even a single block can be tracked easily which makes it highly secure for storing medical records, transaction details. Thus, this helps us in keeping our system secure. Here also every network connection are stored in the blockchain and any changes being made in the network can be detected can alerted.

25.4   IMPLEMENTATION

25.4.1   *IOT INFRASTRUCTURE*

Components used to build this system are discussed in subsections.

25.4.1.1   *RASPBERRY PI3*

It is a tiny computer that works under Linux distribution. Here it is used as a device that can control all the sensors using the applied program. All the sensors are connected to this device using jumper wires. The device is connected to internet and to firebase to send all the data to the cloud for easy access and data analysis (DA) of the files in future. The entire control of the system is depended on the functioning of this device.

25.4.1.2   *INFRARED (IR) SENSORS*

These sensors are used here to detect the object through the doors and keep the count of incoming and outgoing persons. The sensors are programmed in such a way that it would easily understand when a person is going in and when it is going out.

25.4.1.3   *PASSIVE INFRARED (PIR) SENSOR*

This sensor is used to detect motion (see Module 2). It uses the heat radiated from the object to give the movement detection. This device solely concentrates on movement detection and does not do any part of counting the number of persons.

25.4.1.4   CAMERA

It is used to capture images during the entry of a person and record videos during the movement detection.

25.4.2   FIREBASE

The entire data is collected from the sensors are stored in Google's firebase that provides backend-as-a-service. As the data is required in real time, that is why firebase is used here to get the application and the website transfer data in real-time and receive quick response from the user.

25.4.3   ANDROID AND WEB APPLICATION

As for every little development, to showcase the security system an android application and website is delivered in which a user is provided with the user identification code for their device. The user can access the data of the device from the identification code. Such an application makes it easy for the user to tap into their security feeds anytime and always stay updated with real time security alerts.

25.4.4   BLOCKCHAIN

Putting the network details in the blockchain can make the network secure from outside attackers. If tried to change any section of the network or trying to access the device from another network then the new details get stored in the blockchain which cannot be removed and can be tracked down to find the attackers. Also, another aspect of implementing blockchain in this system is that the faces of people that are store in the database (DB) have to be kept under high security. Blockchain helps in keeping the data secure by allowing only one device to access the DB and that to the one whose MAC address is stored in the blocks of blockchain. Thus, anonymous users would not be able to gain access to the DB.

25.5 RESULTS

This system provides counting of person both ways (in and out) and stores the data in the DB which in turn provides real-time data transfer to the user for every action. Users would be given the authorization to control the notification and alerts as per their need. This would make the system flexible for the user. For every entry inside the in counter gets raised by one and for every exit, the out counter is raised by one. Both of them together gives the total traffic count coming in and going out of the area. If both subtracted to each other give the result of zero then it shows that the area is empty. In such situation, the PIR starts doing in job.

PIR detects motion during inactive hours or when the area under surveillance is empty, it helps in reporting any unauthorized access or intrusion by streaming of the video to the user in real-time as well. This could help in faster reporting of the crime in the nearest Police Station by the user with adequate evidence. The evidence is also secured and stored in the cloud. This helps the user to gain access to their files at any point of time.

To only provide security to the premises was not enough and the security of the device was also necessary. Hence, the devices security was controlled with the help of blockchain which would keep the network of the system in check from unauthorized control and any private data of the device secure from malware and unchecked access.

25.6 CONCLUSION

This chapter shows how both IoT and blockchain together can be used as a security system that can give security to both an indoor area and the device that is doing the security check. It is very common nowadays among people using IoT environments for various types of functionality like home automation and smart system. This makes it easier for a human to accomplish a task but the security of the device also matters the most. It is very useful to add the IoT device to the blockchain as it can be only accessed by the controlling party and no unauthorized access is permitted. Hence, Safety of the premises can be left to the IoT infrastructure and safety of the device to the blockchain.

KEYWORDS

- **blockchain**
- **cloud**
- **controlled access**
- **infrared**
- **internet of things**
- **passive infrared**
- **security**

REFERENCES

1. Madakam, S., & Ramaswamy, R., (2014). Smart homes (conceptual views). In: *2014 2nd International Symposium on Computational and Business Intelligence* (pp. 63–66). IEEE.

2. Lee, I., & Lee, K., (2015). The internet of things (IoT): Applications, investments, and challenges for enterprises. *Business Horizons, 58*(4), 431–440.

3. Swan, M., (2015). *Blockchain: Blueprint for a New Economy*. O'Reilly Media, Inc.

4. Kodali, R. K., Jain, V., Bose, S., & Boppana, L., (2016). IoT based smart security and home automation system. In: *2016 International Conference on Computing, Communication, and Automation (ICCCA)* (pp. 1286–1289). IEEE.

5. Pavithra, D., & Balakrishnan, R., (2015). IoT based monitoring and control system for home automation. In: *2015 Global Conference on Communication Technologies (GCCT)* (pp. 169–173). IEEE.

6. Gupta, P., & Chhabra, J., (2016). IoT based smart home design using power and security management. In: *2016 International Conference on Innovation and Challenges in Cyber Security (ICICCS-INBUSH)* (pp. 6–10). IEEE.

7. Quadri, S. A. I., & Sathish, P. (2017). "IoT based home automation and surveillance system," *2017 International Conference on Intelligent Computing and Control Systems (ICICCS),* pp. 861–866, doi: 10.1109/ICCONS.2017.8250586.

8. Prakash, N. S., & Venkatram, N., (2016). Establishing an efficient security scheme in home IoT devices through biometric fingerprint technique. *Indian Journal of Science and Technology, 9*(17), 1–8.

9. Best Home and Office IP Cameras For NAS (2019). https://nascompares.com/best-home-and-office-ip-cameras-for-nas/ (accessed 20 July 2020).

10. Dorri, A., Kanhere, S. S., Jurdak, R., & Gauravaram, P., (2017). Blockchain for IoT security and privacy: The case study of a smart home. In: *2017 IEEE International Conference on Pervasive Computing and Communications Workshops (PerCom workshops)* (pp. 618–623). IEEE.

CHAPTER 26

Limitations of Facial Emotion Recognition Using Deep Learning for Intelligent Human-Machine Interfaces

SABYASACHI TRIBEDI and RANJIT KUMAR BARAI

Department of Electrical Engineering, Jadavpur University, Kolkata, West Bengal, India, E-mails: stribedi15@gmail.com (S. Tribedi), ranjit.k.barai@gmail.com (R. K. Barai)

ABSTRACT

Emotion recognition is an integral part of any human-machine interaction (HMI) system. Proper emotion recognition allows for HMI systems to choose the successive appropriate responses, given the context and the emotion expressed by human(s). The advent of deep learning using deep neural networks (DNN) has made incredible strides in achieving and even exceeding human accuracy in image classification and face detection. However, the adoption of such models for HMI systems is severely limited due to the constraints imposed by the formal definition of classification. Convolutional neural networks (CNN) have become the de-facto DL variant in the last decade for image-based classification tasks because they combine the feature extraction and classification steps into one mathematical model. But we still have not obtained a CNN-based framework which is flexible enough for the goal of a user-friendly adaptive HMI system. In our research pursuits, we have experimented with pre-trained CNN for FER and attempted to extend it towards video data in order to allow for HMI systems to obtain a continuous input feed of the user's emotions. We believe that this is the best way to develop HMI systems which can respond in a befitting intelligent manner, but the approach is beset with problems such as mutual exclusivity of emotion classes, prediction noise, etc. This chapter sheds into these issues which HMI systems suffer from while using CNN for

FER-related tasks with video data in Section 26.3. To illustrate where and how they arise, we have included one of our relevant experiments in Section 26.2 where we use two pre-trained CNN models to perform FER on a video of a group of people conversing. In Section 26.4, we discuss about potential solutions to overcome and circumvent the issues while still using the same CNN models as before.

26.1   INTRODUCTION

Emotions are an integral and essential component of human life. Emotions can be by themselves a very strong medium of communication expressed through facial muscles, hands, body gestures, and modulations of the voice. This chapter focuses specifically on visual perception of emotions through a person's facial expressions [1]. Mentions that facial expressions contribute to 55% to the effect of a speaker's message on the listener, followed by vocal modulations (38%) and message content (7%). Expressed emotions also provide information regarding the current cognitive state of the person, like boredom, interest, stress, and confusion. Understanding and interpreting facial expressions contribute to the depth and dimensionality of information that can be collected from a person in various situations and contexts, apart from his or her spoken words and messages. Automatic recognition of facial expressions is becoming increasingly important in designing HMI systems that are responsive or sensitive to an interacting human's current emotional state. Mower, Mataric, and Narayanan [2] have proposed an emotional classification paradigm based on emotional profiles (EPs). Majority of the publications on the topic of emotion recognition from images are due to the famous facial emotion recognition (FER) in the wild (EmotiW) challenge. The challenge consists of three separate competitions which are group-level emotion recognition, "engagement in the wild" and the audio-video sub-challenge. In the context of the chapter, the objective of group-level emotion recognition is to classify an image of a group of front-facing people's perceived emotion. Gupta et al. [3] put forward a novel approach for classifying global emotions for group-level images using which the authors were able to achieve a final test accuracy of 64.83% with a ranking 4th among all challenge participants in EmotiW 2018. The approach uses two different techniques together as an ensemble predictor: CNN for obtaining a global representation of the image while an attention mechanism produced local representation by extracting individual facial features and merging them. Rassadin, Gruzdev, and Savchenko [4] describe an approach to the EmotiW 2017 Challenge by using a pre-trained convolutional neural network (CNN) for facial feature extraction and an ensemble of random forest (RF) classifiers at a validation accuracy of 75.4%. Although this chapter doesn't refer to the first EmotiW competition for a comparative analysis, the problem statement addressed is directly derived from it. More specifically, our proposed approach has obtained inspiration from Tan, Lianzhi et al.'s publication [5] where the authors use two CNN models for learning from individual faces as well as from the whole group-level image as a whole. This approach was able to win EmotiW 2017 with 83.9% and 80.9% on the validation and testing datasets respectively.

This chapter puts forward gentle critical observation at the research and development of DNNs. DNNs as a Universal Function Approximator [6] has performed remarkable in small problem domains, but leaves much room for going beyond the scope of what they could be trained for. In the EmotiW challenge, the classes defined were "positive," "neutral" and "negative." We believe that this could be too restrictive in the attempt to showcase our observations. Therefore, in Section 26.2, we discuss about one of our own experimentations where we use with pre-trained CNNs with 7 emotion classes instead: "surprise," "sad," "fear," "distress," "angry," "disgust," and "happy" to effectively highlight the observed limitations of FER in current academic literature and industrial applications. Section 26.3 discusses the observed limitations in the context of our experimental setup so that it can be easily understood in a pragmatic way. We conclude our chapter with proposals of alternative ways to overcome the discussed limitations.

26.2   EXPERIMENTATION

In this section, we talk about one of our experiments which will later serve as an example of how DNNs for FER prove to be limited in their application in HMI systems. Subsection 26.2.1 will discuss our neural network pipeline which detects human faces from a video, catalogs, and tags them along with the duration of the video and then performs FER on all of them. Subsections 26.2.2 and 26.2.3 will contain information about the two pre-trained CNNs and how are they made available to the general public for use.

26.2.1   PIPELINE

The pipeline (as shown in Figure 26.1) accepts a video in a standard MPEG-supported format and converts it into an array of frame images.

Dlib's [7] face detection model API allows us to pass each frame to the pre-trained CNN, which detects and crops the faces of human subjects in the frame. The API also has an internal implementation of face identification based on similarity. We used this implementation in conjunction with a dictionary database (DB): every face detected by the CNN was matched with every other face stored in the DB. If they are labeled as more than 75% similarity, they were labeled with same list of faces in the dictionary and were given a unique identification number. Once all the frames from the video were read, the dictionary of cataloged crops of human facial images were fed into the pre-trained CNN from Octivaio et al. [8] which returned a column vector of softmax probabilities for seven emotion classes: {happy, sad, neutral, angry, fear, disgust, and surprise}. The emotion classification vectors for every face of every distinct person detected in the video were accumulated together to form a time series matrix for each person. Figure 26.2 illustrates the idea of the time-series matrix. Every column was an emotion class softmax prediction vector for a person with IDi from a frame at time stamp t_i where that person's face was detected and identified.

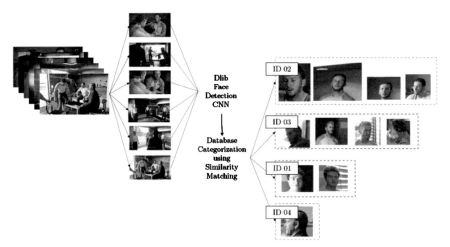

FIGURE 26.1 Experimental pipeline used for video-based FER.

26.2.2 FACE DETECTION

Dlib [7] has provided an open-sourced DNN model and Python API on GitHub, to perform face detection and identification. This model allows us to load some facial images of people, whom we want to detect and identify in real-time, and then use the pre-trained CNN to identify them from a new unseen image.

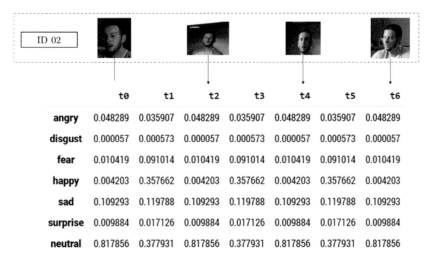

	t0	t1	t2	t3	t4	t5	t6
angry	0.048289	0.035907	0.048289	0.035907	0.048289	0.035907	0.048289
disgust	0.000057	0.000573	0.000057	0.000573	0.000057	0.000573	0.000057
fear	0.010419	0.091014	0.010419	0.091014	0.010419	0.091014	0.010419
happy	0.004203	0.357662	0.004203	0.357662	0.004203	0.357662	0.004203
sad	0.109293	0.119788	0.109293	0.119788	0.109293	0.119788	0.109293
surprise	0.009884	0.017126	0.009884	0.017126	0.009884	0.017126	0.009884
neutral	0.817856	0.377931	0.817856	0.377931	0.817856	0.377931	0.817856

FIGURE 26.2 The concept of a time series matrix of emotion class predictions for a person with ID 01.

26.2.3 EMOTION CLASSIFICATION

The model for FER has been provided with an open-source license on Github by the authors Octivaio et al. [8] who have also published a paper on the same model, which described their experiments in FER and gender recognition tasks, performed using the same model. The model has been trained on the FER 2013 [9] dataset for FER and the IMDB gender dataset [10]. The architecture is novel and unique in the fact that there isn't any fully connected layer (FCL) anywhere before the Softmax classification layer at the end. The chapter explains their decision to do away with the FCL is based on their observations with the VGG16 [11], the inception V3 [12], and the xception [13] architectures where 90% of their learnable parameters (neuron weights and biases) are in the FCL layers. In order to reduce the size of the model, and increase training speed while maintaining a decent level of classification accuracy, they proposed the architecture as shown in Figure 26.1. This was perfectly satisfied with our expected requirements of our experimentations because we wanted to work with light-weight versions of

CNN which can be natively deployed to execute on and integrated into HMI systems. They were able to achieve a combined accuracy of 96% accuracy on Gender Classification and 66% accuracy for FER with the same model (Figure 26.3).

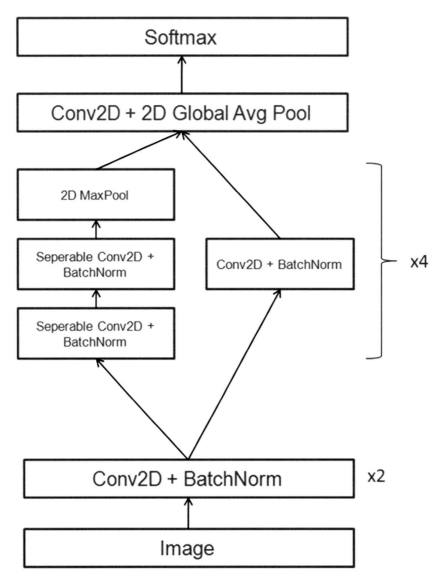

FIGURE 26.3   CNN architecture for [8].

26.2.4 EXPERIMENTAL RESULTS AND DISCUSSIONS

We plotted the time series matrix introduced in Sub-Section 26.2.1 and illus-trated in Figure 26.2 in seven area plots for each of the seven emotion classes. The video used to create the emotion plots discussed in his sub-section can be downloaded from this special URL [14]. Each emotion plot has four overlapping areas for the four persons identified in the video, showing the probability of emotion experienced along the duration of the events as they unfold in the video.

To set the context and evaluate the inferences implied from the plots, we need to first understand the clip scene. The person with ID 04 owns a private military company where a specialized assault team of three soldiers, detected, and cataloged with IDs 01, 02 and 03 are employed. They were recently deployed to assault a terrorist stronghold and rescue a high-profile government official held hostage for ransom. After a triumphant rescue operation, they relax at an army camp room, discussing the operation and its consequences for which they believe they may be called upon again to report for a new mission. The owner of the company, ID 04 enters during their discussion and shares a shot of whiskey with them, congratulating, and appreciating their actions.

ID 01 is very excited throughout the whole scene and has a visible cheer. ID 02 is a more serious-looking fellow which contemplates about the mission going wrong numerous times and how the terrorists will react to them assaulting in their home. ID 03 is more subtle and professional in his facial expressions, and enjoys a couple of jokes at the expense of ID 01 and ID 02. ID 04 maintains a more or less happy face when he appears unannounced amongst ID 01, ID 02 and ID 03 in the second half of the clip.

Figure 26.4 shows the probability variations of the emotion "happy." ID 02 and ID 03 manage to show a happy face at times in the first half of the video scene. ID 02 isn't much into his colleagues' jokes after the first few seconds of the video. ID 04, the superior to the three others persons, arrives midway in the scene, where he offers to share alcohol with the other three. They share a hearty happy moment for a few seconds (indicated by the sharp spike in the middle of Figure 26.4) before turning to serious talks. ID 01 is implied to be a jolly person who has traits of happy expressions throughout the whole video.

Figure 26.5 shows the probability variations for the "surprise" emotion. It is clearly visible that the ID 03 has a prominent expression of surprise right when ID 04 enters the room. ID 04 exhibits some amount of surprise as well at that time. ID 02 is shown to express almost negligible amount of surprise in the entire scene: this does match with his moody and quiet behavior as seen in the video. ID 01 has been shown to have a surprise face more or less the entire scene. This is open for a more subtle interpretation: ID 01 being the most joyful person among the 4 subjects in the video, can the expressions of happiness and surprise have a close relation between each other such that the CNN can detect both of them in some quantity.

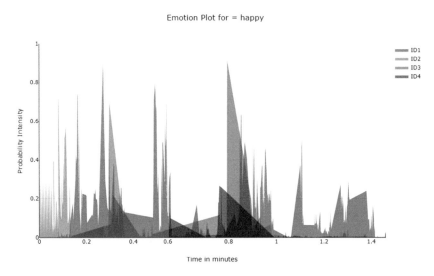

FIGURE 26.4 "Happy" emotion area plot.

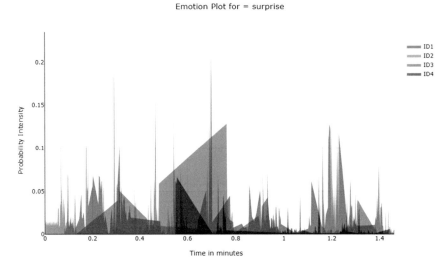

FIGURE 26.5 "Surprise" emotion area plot.

Figure 26.6 shows the probability variations for the "neutral" emotion. Neutral traits are a major consistent feature of the scene. ID 03, known to be the most professional and subtle in behavior among the four subjects, has the most prominent neutral facial traits which drops as the video goes along, with the introduction of ID 04 in the scene. ID 02 remains a constant level of neutral expressions throughout. ID 04, being superior by professional rank, maintains neutral traits when he comes into the scene.

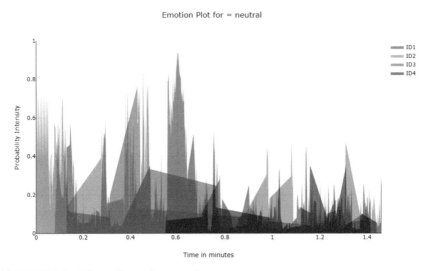

FIGURE 26.6   "Neutral" emotion area plot.

Figure 26.7 shows the probability variations for the "Fear" emotion. ID 02 is shown to have a large response for the fear trait before the appearance of ID 04. During this time, ID 02 was discussions about the potential political consequences that they have to address due to their recently completed mission. The graph hints at the possibility that ID 02 is very thoughtful about the issue which can explain the moody and non-response behavior for the previous three emotions. He maintains the fear trait in his expressions even after ID 04 is introduced, which shows he is occupied with the thought heavily. ID 03 and ID 04 are shown to exhibit a low but a consistent appearance of fear in their facial expressions: this can be interpreted in their more serious intents and facial dispositions as individuals. ID 01 is almost non-responsible for this class.

Figure 26.8 shows the probability variations for the "angry" emotion and it has a very active response for ID 01, ID 03, and ID 04. Now the interpretation can be debatable in account of the convincing proofs we have for the emotion responses from the previous 4 plots. On inspecting the video, one may not find a proper understanding why the CNN detected traits of anger among the subjects. An easy answer can be that the facial structure of the hypothetical subjects is rendered in such a way that the CNN misreads them. The CNN has been trained on real human faces, and hence may fall prey to graphical rendering issues of hypothetical video game character faces.

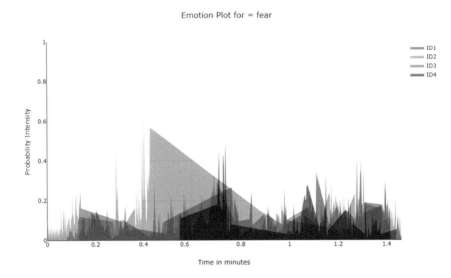

FIGURE 26.7   "Fear" emotion area plot.

Figure 26.9 shows the probability variations for the "Sad" emotion and it has a very high response for all the subjects in the video, particularly for ID 03, ID 02, and ID 04. ID 01 does have some time periods of "sad" response but it is not active throughout. This prediction can be interpreted to along the lines of the "neutral" face response. Given the context of the video scene is tragic and serious in nature; a serious face is expected from all the subjects. On inspecting the video, one may find that time segments where a subject had a serious solemn face is where the model found a considerable amount of the "sadness" trait in it.

Figure 26.10 shows the probability variations for the "disgust" emotion. This emotion is non-existent in every context of the video and hence the plot reflects so. The probability can be seen to be less than 0.1.

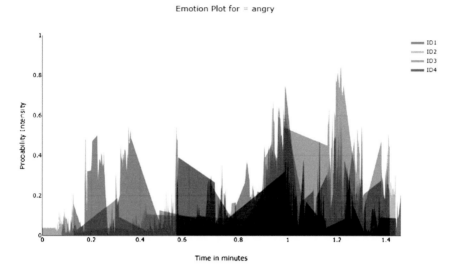

FIGURE 26.8   "Anger" emotion area plot.

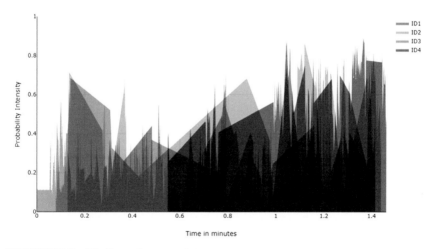

FIGURE 26.9   "Sad" emotion area plot.

26.3   DISCUSSION ON INHERENT LIMITATIONS

DNNs by themselves are extremely narrow in the domain and flexibility of the tasks they excel at. In our experiment, we find that the pretrained CNN for FER can only work with three predefined emotion classes on which the CNN had been trained for. Moreover, the softmax classification layer at the end can only produce probability values which imply that the classes are mutually exclusive to each other. For an HMI system to exhibit intelligent understanding of the interacting user, this is a very narrow and strictive piece of information that doesn't allow for complex inferences later on. For example, if the user is being indecisive or unclear in his/her facial expression, the CNN may classify the expression as being one of amongst the three emotion classes, which would be a bad prediction to base future inferences on. It doesn't have to capability to answer in, i.e., to interpolate between the emotion classes it has been trained to predict for. Emotions are not mutually exclusive of each other, and one may express certain expressions which have been triggered by multiple emotions (for example, of anger and sadness).

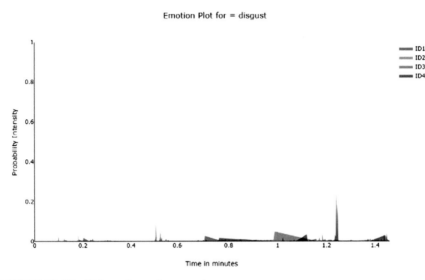

FIGURE 26.10   "Disgust" emotion area plot.

Many video-based FER systems use the frame-based approach discussed in our experiment. This simplifies the engineering aspect and allows for modularity in development. The two CNNs, one for face detection and FER are trained and evaluated separately independently of each other and then brought other to fit as individual components of a larger pipeline. However, this isn't a very optimal way to perform classification on sequential data where the previous data sample is related to the next. Recurrent neural networks and its variants like LSTM and GRU were developed for the sole purpose of performing deep learning with sequential data. However, the features and the prediction outputs are not at all well-defined in the context of the aim of our experiment. There has been no literature which shows RNN variants can perform face detection or FER. RNN variants are supervised algorithms and require relevant annotations which define what it needs to learn from the sequential data.

Using our pipeline of two CNNs, we were able to easily implement a video-data based FER system. Performing FER on a frame-by-frame basis means that for every frame, we obtain a set of emotion class prediction vectors for every face detected. Generally speaking, videos contain 24 to 30 frames per second. The sample video used in our experiment had 25 frames per second. For a 2-minute clip, there were 3000 frames in total. ID 01, 02 and 03 were available and detected almost throughout the entire duration of the 3000 frames, and hence we obtained three prediction vectors of length 7 for every one of those 3000 frames. Every prediction vector corresponded to a facial expression of a subject captured at 1/25 of a second or 0.04 seconds. Fayolle and Droit-Volet [15] tells us that this duration of capture is too small for a human facial expression to fully register. What we end up with is a lot of variance in the predictions collected for a single subject. CNNs are not completely accurate in their predictions and may not correctly classify an expression. So, we obtained an array of predictions of facial expressions taken at intervals of 0.04 seconds which may not actually correspond to the true facial expressions of the subject at all.

26.4   CONCLUSION

Our chapter puts forward two main constraints of approaching FER using DNNs which restricts the potential for a more human-like inferring capability of other humans' expressions and emotional state to HMI systems. We have discussed the problem of assumed mutual exclusivity of classes in classification DL models. Such a model cannot go extrapolate or interpolate between the classes for which they have been defined and trained on. Combating the issue of high variance in frame-based predictions is a complex problem. This is because training a FER-CNN with an RNN variant has not been performed and shown to be successful in any contemporary literature.

The nature of the proposed solutions to the limitations discussed in the previous section is based on an understanding of the mathematical model used behind DNNs in general. Tweaking the structure to build more novel models to suit the requirements of experimental aims can result in much-needed breakthroughs in the field of HMI systems. HMI systems when equipped with the power of understanding and monitoring changes in a person's perceived emotional states with time in a social group setting offers a promising goal in using machine learning (ML) for psychological, psychiatric, and human resource management studies.

KEYWORDS

- **convolutional neural networks**
- **emotion classification**
- **emotion noise**
- **face detection**
- **facial emotion recognition**
- **multi-label classification**

REFERENCES

1. Albert Mehrabian, *"Silent Messages: Implicit Communication of Emotions and Attitudes,"* Wadsworth Publishing Company (1 July 1972), ISBN-13; 978–0534000592.
2. Mower, E., Mataric, M. J., & Narayanan, S., (2011). A framework for automatic human emotion classification using emotion profiles. *IEEE Transactions on Audio, Speech, and Language Processing, 19*(5), 1057–1070.
3. Gupta, A., Agrawal, D., Chauhan, H., Dolz, J., & Pedersoli, M., (2018). An attention model for group-level emotion recognition. In: *Proceedings of the 2018 on International Conference on Multimodal Interaction* (pp. 611–615). ACM.
4. Rassadin, A., Gruzdev, A., & Savchenko, A., (2017). Group-level emotion recognition using transfer learning from face identification. In: *Proceedings of the 19th ACM International Conference on Multimodal Interaction* (pp. 544–548).
5. Tan, L., Zhang, K., Wang, K., Zeng, X., Peng, X., & Qiao, Y., (2017). Group emotion recognition with individual facial emotion CNNs and global image based CNNs. In: *Proceedings of the 19th ACM International Conference on Multimodal Interaction* (pp. 549–552). ACM.
6. Csáji, B. C., (2001). *Approximation with Artificial Neural Networks* (Vol. 24, p. 48). Faculty of Sciences, EtvsLornd University, Hungary.
7. *Github Repository.* https://github.com/ageitgey/face_recognition (accessed 20 July 2020).

8. Arriaga, O., Valdenegro-Toro, M., & Plöger, P., (2017). *Real-time Convolutional Neural Networks for Emotion and Gender Classification*. arXiv preprint arXiv:1710.07557.

9. *FER2013 Kaggle Dataset*. https://www.kaggle.com/c/challenges-in-representation-learning-facial-expression-recognition-challenge/data (accessed 20 July 2020).

10. *IMBD Gender and Age Dataset*. https://data.vision.ee.ethz.ch/cvl/rrothe/imdb-wiki/ (accessed 20 July 2020).

11. Simonyan, K., & Zisserman, A., (2014). *Very Deep Convolutional Networks for Large-Scale Image Recognition*. arXiv preprint arXiv:1409.1556.

12. Szegedy, C., Vanhoucke, V., Ioffe, S., Shlens, J., & Wojna, Z., (2016). Rethinking the inception architecture for computer vision. In: *Proceedings of the IEEE Conference on Computer Vision and Pattern Recognition* (pp. 2818–2826).

13. Chollet, F., (2017). Xception: Deep learning with depth wise separable convolutions. In: *Proceedings of the IEEE Conference on Computer Vision and Pattern Recognition* (pp. 1251–1258).

14. https://bit.ly/2ZRadm3 (accessed 20 July 2020).

15. Fayolle, S. L., & Droit-Volet, S., (2014). Time perception and dynamics of facial expressions of emotions. *PLoS One,9*(5), e97944.

16. Fu, J., & Rui, Y., (2017). Advances in deep learning approaches for image tagging. *APSIPA Transactions on Signal and Information Processing*, 6.

Part III
Application and Case Studies

State of the Art of Machine Learning Techniques: A Case Study on Voting Prediction

GOURI SANKAR NAYAK,[1] RASMITA PANIGRAHI,[2] and
NEELAMADHAB PADHY[3]

[1]*MTech Research Scholar, School of Computer Engineering (CSE),
GIET University Gunupur, Odisha, India, E-mail: mr.gouri4u@gmail.com*

[2]*PhD Research Scholar, School of Computer Engineering (CSE),
GIET University Gunupur, Odisha, India, E-mail: rasmita@giet.edu*

[3]*Associate Professor, School of Computer Engineering (CSE), GIET
University Gunupur, Odisha, India, E-mail: dr.neelamadhab@giet.edu*

ABSTRACT

Voting methods are the primary methods of evolutionary techniques where machine learning (ML) plays a vital role. Ensemble methods help to create multiple models then combined them into to accurate resultant set. It is one type of classification technique used for voting prediction model to reduce the difference (bagging/(boosting) or improves forecast (stacking). Linear regression techniques are used to construct the optimization model as well as implemented to predict the training data set, accessed from a Kaggle site. The prime substance of this chapter is to produce the basic concept of ensemble learning by using the voting technique. For a better understanding of this concern, we are describing the advanced algorithms by using hands-on training on a real-life problem.

27.1   INTRODUCTION

The voting for full forecasting in concert can be achieved and adopting the most appropriate analysis. To collate an implementation between a quiet bulk and the prospective unclean voting enforced individually one and the other voting approach to bootstrap aggregate and explored the work of data set. However, the two simplified ensemble approach, Boosting, and bagging have accepted abundant consideration. The technique mentioned earlier need resample or reevaluation of instruction sets from the model data. Previously different algorithms frequently implemented as specific resample training data set. Boosting was an advanced approach for developing the achievement of any anemic literature algorithm, which lacks being somewhat more than an arbitrary guess. To associate the allocate in an ensemble, boosting takes a carry bulk vote of their forecasting, and the bagging algorithm needs a bootstrap case to frame the allocation in an ensemble. Subsequently, random forest (RF) depends upon the decision tree and individual level of trees build upon on the character of appearance case separately. Because of this substance, we can say that without the help of the desired focal point on the sequence level, a confidential outcome can be identified. It has been proved that bagging is one kind of better ensemble learning technique, used for experimental data for classification [6]. The author designed a voting ensemble classifier with multi-types features to identify wafer map defect patterns in semiconductor manufacturing [5]. Ensemble learning is a machine learning (ML) technique where more than one learner implemented the data sets to find a solution of the same task by extracting a set of predictions then grouped into single composite prediction. Here a collection of models are used to obtain their result individually. The main character abides by a specific ensemble approach accepting quiet bulk voting under bootstrap aggregate is essential bagging. Difficult-to-analyze instances are determined the particular nearby class boundaries. For example, support vectors in SVM (support vector machine) are the difficult-to-analyze cases (Figure 27.1).

27.2   RELATED WORK

Bauer and Kohavi [1] have illustrated an experimental identification of the Voting class design: bagging, boosting, and versions ML. Agarwal et al. [2] proposed a specific SVM, which is well-established as a leading classifier with maximal certainty and gives margin root mean square error (RMSE). The act of classification gets right of entry can be determined out in phrases of the accuracy of the rule by taking a look at facts set. Clustering get admission is installed on unsupervised to know about the above because there are no predefined lessons. In this get, access to facts can be grouped collectively as a cluster. Amrieh et al. [3] have completed an act of pupil's predictive model is determined through the stated different classifiers, like as ANN (artificial neural network), NBDT (Naïve Bayesian and decision tree). Further, we implemented an ensemble technique toward increase an act of the above classifiers. Mythili et al. [4] has analyzed the efficiency of ML algorithms inconclusive with the consequence of result, paternal literacy, masculine, recession, and the area within the exercise and consider of school students achievement. It is identified that the RF act is perfect than that of various algorithms engaged in the study. The literature review outcome is given in Table 27.1.

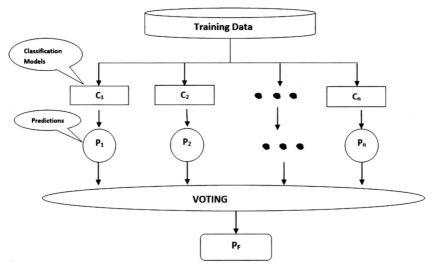

FIGURE 27.1 Voting model.

27.4 VOTING ALGORITHMS

The various voting algorithms are carried out for beneath descriptions. Every algorithm precedes an action and training data set as absorption and runs an action more than one instance through growing the handling of training data set detail. The resultant classifiers are blended to build very last classifiers this helps to distribute set.

TABLE 27.1 Literature Review Outcome

Machine Learning Technique	Merits	Demerits
Bayesian Networking	The graphical representation offers it a bonus of breaking complex issues into specific smaller fashions.	Slow in classifying facts units with many functions.
	Use a technique which is inspired by means of convolution biological system.	Gets a stack in nearby optima (overfitting)
Genetic Algorithm	It has the functionality of solving optimization problems for the duration of classification.	Selection of kernel characteristic is not straight forward.
	The algorithm is easy to analyze mathematically.	
Support Vector Machine	All computations are done in area the use of kernels giving it apart for use nearly.	Slow in training and requires extra memory space.
	Easy to implement and can deal with multi-class issues.	Slow in training and calls for huge reminiscence space.
K-Nearest Neighbor		It is computationally complex because to classify a check sample involves the consideration of all training case.
		If trees aren't pruned back it causes over becoming.
Decision Tree	It has a unique structure therefore smooth to interpret.	Type of records must be considered while building tree (i.e., Categorical or numerical)
	It has no trouble in handling high dimensional data set.	
	It is based totally on human reasoning concepts which are not particular.	Its construction has an excessive stage of generality there by means of excessive consumption of resource.
Fuzzy Logic	It offers a representation of Uncertainty.	
	Simple to enforce and effective.	The final results of Clustering depend on how cluster facilities are initialized to specify k value.
K-Means Algorithm	Simple to enforce and effective.	The final results of Clustering depend on how cluster facilities are initialized to specify k value.
		The set of rules works for the simplest numerical information.

27.4.1 BAGGING ALGORITHM

Bagging/Bootstrap set of rules is motivated through consistently examined *m* mentioned in distinction to the instruction data set with a difference. B bootstrap samples x_1, x_2, ..., xB is developed, and a classifier CL_i is constructed from every bootstrap sample X_i. In very last classifier, CL_i is constructed from CL_1, CL_2, ..., CL_B whose results are expected most easily by way of its sub-classifiers, with ties damaged arbitrarily (Table 27.2).

TABLE 27.2   The Bagging Algorithm [8]

Input: Training set *T*, Inducer *I*, Integer *B* (number of bootstrap samples).
1. For i = 1 to B 2. {
3. T' = bootstrap sample from T.
4. $CL_i = I (T')$
5. }
6. $$CL^*(x) = \frac{\arg\max}{y \varepsilon Y} \sum_{i:CL_i(x)=y} 1 \quad \text{(The conclude label y)}$$
Output: classifier *CL**

For an inclined bootstrap, the pattern is an example in the training set has possibility, $1 - \{(1-1/m)\,m\}$ of reality is decided on the least as soon as inside m times are randomly decided on from the training set. For large m, this is about $\{1 - 1/e\} = 63.20\%$, that means that every bootstrap pattern includes most effective approximately 63.20% precise detail from the instruction set. These sickness purposes exclusive classifiers to be constructed if the prompt is ambiguous (e.g., neural networks, decision trees) and the overall performance can enhance if the precipitated classifiers are excellent and now not correspond, yet bagging might also barely corrupt the act of robust algorithms (e.g., k-nearest neighbor) being efficaciously Smaller instruction parts are used for training set in each classifier. The authors analyzed the influence of several voting methods on the performance of two classification algorithms [7].

27.4.2 ADABOOST ALGORITHM

Like Bagging, the AdaBoost algorithms bring about a hard and fast of classifiers and vote them. Behind it, there are two algorithms regulate basically.

The AdaBoost algorithms, proven in Table 27.3, make the classifiers basically, even as Bagging can make them in lateral. AdaBoost also modify the weights of the training instances maintain as entering to each inducer well-traditional up on classifiers had been previously assembled

TABLE 27.3   The AdaBoost Algorithm [8]

Input: Training set T of size m, Inducer I, Integer B (wide variety of trials).

2.   T' = T with instance weights assigned to be 1.

3.   For i = 1 to B 3. {

4.   CLi = I (T')

5. $$\epsilon_i = \frac{1}{m} \sum_{x_j \in T' : CL_i(x_j) \neq y_j} weight(x)$$

6.   If $\epsilon_i > \frac{1}{2}$, the training set T' to a bootstrap sample from T with weight 1 for every instance and go to step 3(this step is up to 25 times after which we exit the loop).

7. $\beta i = \epsilon_i \mid (1 - \epsilon_i)$

8.   For-each $x_j \in T'$, if $CL_i(x_j) = y_j$ then weight $(x_j) = $ weight $(x_j). \beta_i$.

9.   Normalize the weights of instances so the total weight of T' is m.

10.   }

11. $$CL*(x) = \frac{\arg\max}{y \varepsilon Y} \sum_{i : CL'i'(x)=y} \log \frac{1}{\beta_i}$$

Output: classifier **CL***

The objective is to effort the inducer to lessen everyday errors over diverse enter distributions 1. Given an integer B decide the range of trials, B weighted training sets T1, T,2, …, TB are set up in sequence and B classifiers CL1, CL2, …, CLB are built. A final classifier CL* is making using a weighted voting scheme: the substance of each classifier relies upon on its activities at the educational sets used to construct it.

27.5   PROPOSED METHODOLOGY

The below model specifies that classification models are implemented on the preferred data set to predict individual result related to make the voting ensemble. The voting ensemble will give you the final prediction (PF). During the training classification, the three classifiers are used (RFC, ML, SVM). For better understanding refers the Figure 27.2.

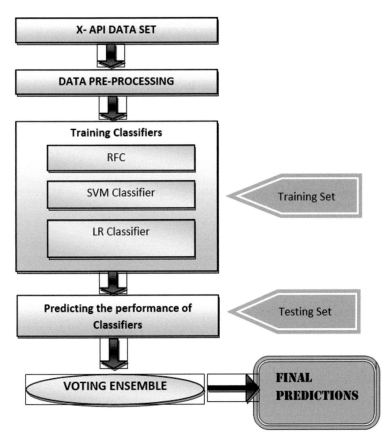

FIGURE 27.2 Proposed ensemble structure, where; RFC: random forest classifiers; SVM: support vector machine; LR: logistic regression.

- **Phase 1:** This is one of the data set for student academics, which we get from www.kaggle.com web portal. This set deals 16 features of 400 to 500 students. These appearances are classified into four parts, which is shown in Table 27.4.
- **Phase 2: Preprocessing the Data:** The facts set are uniformly implemented in the proposed work. The regulated of statistics is achieved via 0's and 1's normalization. The data Processing includes: (a) data of cleaning and (b) feature of selection.

  1. **Data of Cleaning:** It is a crucial convert task that assists to ignore beside the point facts component and lost values. In these analysis paintings, the dataset consists of 4% of lacking values;

consequently, the statistics with lost profits are detached from the dataset. Finally, the dataset is decreased to 481 considerations subsequently, which acts data cleaning.

2. **Feature of Selection:** In statistics, processing characteristic choice is a crucial step, a motive of this approach is to describe a set of functions that could anciently define the input records, lessen the size of the feature area, and eliminate the unnecessary as well as wrong facts. In this research painting, the International capabilities had been considered for research inside the field of learning analytics and data mining as international functions assist in figuring out the nature of the intellectual.

TABLE 27.4   X-API Dataset [9]

Features	Categories
Demographical Features	Citizenship
·	Gender-specification
·	Native place
·	Parent responsible for student
Intellectual Features	learning stages
·	Stage
·	Standard levels
·	Selection identification
·	Student nonattendance days
Parental Features	Parent acknowledges survey
·	Parent educate
·	Achievements
Noticeable Features	Groups arguments
·	Inspect resources
·	Practical class
·	consider advertisement

*   **Phase 3: Classification Algorithms:**

    1. **Random Forest Algorithm (RF):** RF is a technique to get to understand a set of regulations that reviews the aggregated outcome of distinct classification methods. For one and the other, arbitrary forest, rules are recycled and analyzed. Numerous analytical trees are being built in RF, and the corresponding tree is being qualified

in a bootstrap sample of the genuine facts of education. Every tree in the RF is going to forge a ballot for some people to join X. The result of the RF could then be chosen based on the bushes majority vote. RF Area is an ensemble classification technique

Consisting of a large number of decision trees, the outcome of the random collection of laws in the forest is the magnificence which is the output class mode with the help of the character tree and with few information sets, RF overfit.

2. **Support Vector Machine (SVM):** It is a gadget which, by knowing or understanding, is a controlled device studying methodology. It is used in records to explore and realize styles. To each classification and ranking, SVM is used in particular. SVM is a binary regular expression and picks each of the two classes for each of the education stats. From the SVM method, a new version is formed which is then used to categorize a brand new item into either one or a magnificence. Since SVM supports kernel characteristics, with less computational complexity, it could study various mixtures of the assigned features. SVM avoids the concern of overfitting.

3. **Logistic Regression:** It's a volatile structured, dichotomous. Determinants of both the established variable are all in unbiased factors in this model and can be measured on a nominal, ordinal, interval, or ratio scale. The affiliation between the impartial and the base variable is nonlinear in logistic transformation.

- **Phase 4: Voting Ensemble:** In the majority vote casting, certain classifiers produce a forecast for each instance within the testing statistics. The PF for every case is the only which receives the maximum votes and it more significant than half of the votes. The pseudo-code of the Voting used inside the proposed work is as follows in Table 27.5;

TABLE 27.5   Pseudocode for Ensemble Classifier Based on Voting Prediction [9]

Pseudo Code
Ensemble Classifier Based on Voting Prediction
Step 1: Implement three classifiers (random forest, support vector machine, and logistic regression) under the training data set.
Step 2: Find the performance of the three classifiers individually and differentiate them.
Step 3: Performing voting for every observation.
Step 4: Differentiate the performance of the voting with three classifiers, (i.e., random forest, support vector machine, and logistic regression).

27.6 CONCLUSION

Initially, this proposal is assumed that ensemble by using ML for which trains the diverse as well as accurate classification. Here diversity is achieved for a verity of architecture, hyper-parameter settings, and training techniques. However, ensemble methods are very successful in record performance settings behalf challenging datasets and are among the top awarded of Kaggle data science. Logistic regression and RF have found that majority voting ensemble method help us to arrive at better performance accuracy.

KEYWORDS

- **artificial neural network**
- **ensemble methods**
- **evolutionary techniques**
- **machine learning**
- **Naïve Bayesian and decision tree**
- **root mean square error**

REFERENCES

1. Bauer, E., & Kohavi, R., (1999). An empirical comparison of voting classification algorithms: Bagging, boosting, and variants. *Machine Learning, 36*, 105–139.
2. Agarwal, S., Pandey, G. N., & Tiwari, M. D., (2012). Data mining in Ed education: Data classification and decision tree approach. *International Journal of e-Education, e-Business, e-Management, and e-Learning, 2*(2).
3. Elaf, A. A., Thair, H., & Ibrahim, A., (2016). Mining educational data to predict student's academic performance using ensemble methods. *International Journal of Database Theory and Application, 9*(8).
4. Mythili, M. S., & Shanavas, A. R. M., (2014). An analysis of student's performance using classification. *IOSR Journal of Computer Engineering (IOSR-JCE)* (Vol. 16, No. 1, Ver. III, pp. 63–69). e-ISSN: 2278-0661, p-ISSN: 2278-8727.
5. Muhammad, S., Bilguun, J., & Jong, Y. L. (2019). *A Voting Ensemble Classifier for Wafer Map Defect Patterns Identification in Semiconductor Manufacturing.*
6. Emine Yaman, Abdulhamit Subasi (2019). Comparison of bagging and voting ensemble machine learning algorithm is a classifier. *International Journals of Advanced Research in Computer Science and Software Engineering* (Vol. 9, No. 3). ISSN: 2277–128X.
7. Leon, F., Floria, S., & Bădică, C. (2017). Evaluating the effect of voting methods on ensemble-based classification. *Proceedings of the 2017 IEEE International Conference on Innovations in Intelligent Systems and Applications, INISTA 2017.* Gdynia, Poland. doi: 10.1109/INISTA.2017.8001122.

8. Kim, H., Kim, H., Moon, Hi, & Ahn, H. A weight-adjusted voting algorithm for ensembles of classifiers, Journal of the Korean Statistical Society, Volume 40, Issue 4, 2011, 437-449, https://doi.org/10.1016/j.jkss.2011.03.002.

9. Salini, A., & Jeyapriya, U., A Majority Vote Based Ensemble Classifier for Predicting Students Academic Performance, International Journal of Pure and Applied Mathematics Volume 118 No. 24 2018 ISSN: 1314-3395 (on-line version) url: http://www.acadpubl. eu/hub/

CHAPTER 28

Climate Prediction by Using a Machine Learning Approach: A Case Study

K. SIVA KRISHNA[1] and NEELAMADHAB PADHY[2]

[1]*Research Scholar, School of Computer Engineering (CSE), GIET University Gunupur, Odisha, India, E-mail: ksivakrishna@giet.edu*

[2]*School of Computer Engineering (CSE), GIET University Gunupur, Odisha, India*

ABSTRACT

Agriculture is the most popular and crucial sector in the Indian economy. Globally by 2050, the agricultural products will be decreased, but the population will be increased by 1.68 billion. There are many factors which reduce the productivity of agriculture. One of the significant effects of agriculture is climate change (CC); this can affect agriculture in several ways, like average temperature change, rainfall, and extremes in climate. In this research, the article author is trying to explore the relationship between climate and agriculture. In this research chapter, the authors concentrated on predicting the results by collected the data from Govt. of Odisha for Jeypore region. For this secondary data applied the machine learning (ML) models like linear regression and multiple linear regression. Compared the predicted results of these two ML models and concluded that the results obtained by MLR are best as compared to LR.

28.1 INTRODUCTION

In agriculture, climate change (CC) is the essential factor that can create an environmental threat that adversely affects. Agriculture is interrelated to CC, globally these two took place, and their relationship between world population and world food is imbalanced which is important in increasing of the production.

In this chapter, the authors are trying to predict CC and humidity level for the year 2020 by considering the secondary data as a dataset for the region of Jeypore, Odisha. For this, researchers have used the machine learning (ML) algorithms which are applied to the agricultural data, which is already gathered by the IoT devices.

28.1.1 INTERNET OF THINGS (IOT)

The objects which are linked together through the internet, collects the data through sensors and process that data to perform specific data. In IoT-based smart farming, the crop field is supervising by a system which is having sensors. These sensors can control the temperature, humidity, light, and soil moisture.

28.1.2 IMPORTANCE OF IOT IN CLIMATE CHANGE (CC)

For better farming, the farmer's needs weather condition. For slowdown the destructive course of CC IoT, solutions can be useful. IoT devices with international standards are properly aligned for climate protection and which is taking an active role by global innovators. Nowadays, IoT is booming because it is connected with number of devices which are expected to go beyond the 60 billion thresholds in 2016, and the IoT market is forecast to produce $14.4 trillion in improved revenues and lower costs by 2022.

28.1.3 MACHINE LEARNING (ML)

The one of the applications of artificial intelligence (AI) is ML, which provides a system which has learning ability and improves from experience without explicit program. ML functions broadly divided into two categories, namely, supervised, and unsupervised learning.

28.1.4 AGRICULTURE

Agriculture is the backbone of our economic system. In the Indian economy, it plays a crucial role. Agriculture provides food and raw material along with this it also provides employment opportunities to a considerable proportion of the population. A global annual emission in 2010 is about 20% to 25% where forestry and land-use change contributed in agriculture. Both non-agricultural land it is converted and greenhouse gases which are helping the agriculture from CC.

28.1.5   IMPACT OF CLIMATE CHANGE (CC) ON AGRICULTURE

Global environmental issues threaten our ability to meet the basic human needs which induced the CC. In CC, greenhouse gases are increasing which is likely to affect crops differently from region to region. In this regard, there are several ways of agriculture like productivity, profitability, and prices which are affected by the climatic change. Change of climate is not only a significant problem, but it creates changes like snow, evaporation, global rainfall, streamflow, and other factors affect quality and supply of water.

28.2   LITERATURE REVIEW

In Agriculture, the researchers have enhanced the various aspects, which improve the quality and quantity of agriculture productivity. The work is done by researchers on many different parameters like soil testing, weather conditions, and crop management. Forgiving the best results in agriculture Climate is also one of the significant parts.

So in this research article, we are concentrating on CC in agriculture. For this research, with the help of secondary data and it was collected from Govt. of Odisha for the Jaipur region and predict the CC. Before going for actual implementation, the authors reviewed different articles which will give exact relation between the climate and agriculture. In one of the chapters, the researchers were predicted and given the CC. For example, a dataset created for temperature by the University of Delaware and by Dell, Jones, and Olken (2012, 2014) and IMF (2017) its precipitation is collected. Actual measurement points are far fewer than Indian data and it is dependent.

From the following graph, with the help of 45 weather stations from India are gridded by the database (DB) of Delaware temperature, and 210 weather stations are gridded by IMD data. The precipitation of Delaware DB depends on Indian rainfall data compared to an actual one (Figure 28.1).

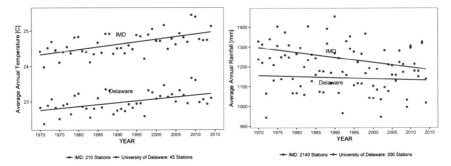

FIGURE 28.1   Temperature and rainfall comparison of Indian and International Data. a) Average annual temperature; b) average annual rainfall.

Source: Data collected from IMD and the University of Delaware for survey.

In Ref. [1], the researchers explained the relationship between climate and agriculture. They concluded that the relationship between environment and agriculture plays a significant role in developing countries because most of the agricultural activities depend on climatic conditions. In this article, the researchers noted down that agriculture contributes to 58% of total N_2O emission, which creates nitrous oxide of 4.5 million tons per year. The researchers also found that CC showed its impact on developing countries due to lack of technology. With the help of smart farming, we can give some solutions to CC. In the end, the authors concluded that climate affects the agriculture through the emission of greenhouse gases like CO_2, methane, and nitrous oxide for this the agriculture is only the solution.

Khajuria et al. [2] found that agriculture is facing economic and ecological challenges and the significant impacts like change in climate, change in crop yield, which are the impacts on agriculture system. The issue was explained that influences and accessed the vulnerability of the agriculture sector. The researchers concluded that improving the agriculture management with adaptation strategies on the CC, which enhances the capacity for a socio-ecological system by creating the climate-resilient pathways.

In Ref. [3], the authors focused on smallholder farmers in West Africa and how the farmers used climate forecasting for crop management decisions and whether it gives any benefits. For this, the authors conducted workshops in Senegal and West Africa which are two agro-ecological zones for farmers, which used a participatory approach. In the end, authors concluded that in Bacfassagal, the overall limited effect of the adaptation is relatively insufficient for the estimated yields.

A logic model for structuring the appraisal of a program on climate services is developed, which exhibit in a case study of the Caribbean Agro-meteorological Initiative (CAMI). Vogel et al. [4] developed a logical model which improved social and economic outcomes by linking the weather data, agriculture, and climate with decision making. Finally, the authors found logic model framework for climate, logical sequence of activities are grouping of the components of climate services, which are necessary for success. By applying the logical model to CAMI, it finds the one step area of progress. Significant progress was done by CAMI in a short time and has set in action on several critical mechanisms for a thriving climate services for agriculture.

Soares et al. [5] done the research for the climate information in Europe towards a conscious of growing fragmentation of accessible information and the desire amongst end-users and it can able to centralize and coordinate climate data. The authors also highlighted, in order to enhance the function of climate and uptake some of the modern factors that to be addressed effectively so that it is easy for decision-making for economic sectors throughout Europe. At the end, the authors discussed in the survey sample that the decadal climate predictions are appeared. In particular, for decadal climate predictions there is a clear interest, these remain understood poorly and restricted to research-based applications.

28.2.1   CLIMATE CHANGE (CC) BY MONTH IN THE YEAR 2018 AT JEYPORE

From Ref. [11], for the year 2018 the monthly CC in Jeypore as per the records was 36.2°C (97.2°F) is the maximum temperature, which is recorded in the month of May, minimum temperature is 19.3°C (66.74°F) for the month December and 24.7 °C (76.46°F) was the annual average temperature.

28.3   EXPERIMENTAL WORK

28.3.1   ARTIFICIAL INTELLIGENCE (AI) IN THE AGRICULTURAL INDUSTRY

For an agricultural revolution, AI plays a significant role and with the help of fewer resources world must produce more food. The most popular applications of AI in agriculture have three major categories like agricultural robots, predictive analytics, and crop and soil monitoring.

28.3.2   ML ALGORITHM-IMPLEMENTATION OF LINEAR REGRESSION

The implementation of LR approach is done on the dataset (secondary data) which contains the attributes date, minimum temperature, maximum temperature, humidity levels both min and max, precipitation, and many more instances for this dataset calculated the mean and stander deviation but the difference is more that means the data is noisy data, cleaned the dataset with the help of some methods. The authors chose the humidity and precipitation for plotting the graph because these two instances are positively correlated to each other as compared to other instances. After that plotted a graph between the humidity levels and precipitation the following depicts the same. For this graph, calculated the mean value of humidity, and maximum humidity levels are showing zero precipitation. At the end, it has been given the accuracy of 84.67% (Figure 28.2).

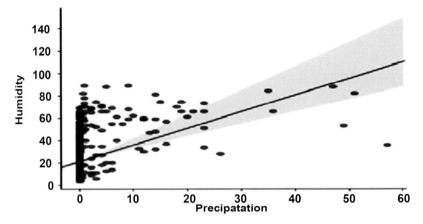

FIGURE 28.2   Weather change in agriculture.

28.3.3   ML ALGORITHM-IMPLEMENTATION OF MULTIPLE LINEAR REGRESSION

For this approach implementation initially created model, along with model created the training set and test set, with the help of training set and test set calculated the target value which will be the value of change in temperature. The total of 36 instances in this dataset and compared target value to each and every individual instance and plotted the graph. Figure 28.4 depicts how the changes are done for target value and each and every individual instance. In this, the authors consider the x-axis is target value which constant for all graphs, it may vary in the values and y-axis is each and every individual instance in the graph. At the end got the accuracy of 92.25% (Figure 28.3).

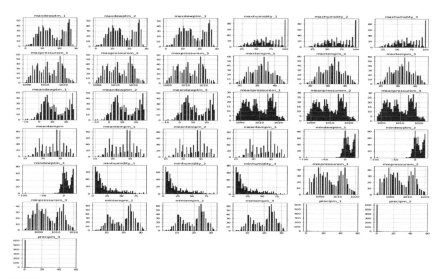

FIGURE 28.3   Comparison of target (change in temperature) values with each and every individual instances.

28.3.4   COMPARISON OF ACCURACY BETWEEN LR AND MLR

Figure 28.4 depicts the summarized accuracy comparison between the two algorithms LR and MLR, here in this the *x* label is represented the algorithms used and *y* label is represented the percentage.

28.4   CONCLUSION

For Indian economy, agriculture is the crucial sector but it is affected due to CC, for this chapter, the authors done a research on CC by reviewing the different articles and implementation is done on secondary data which is collected from Govt. of Odisha for Jeypore region with the help of LR and

MLR algorithms. Finally, got 84.67% accuracy by applying the LR model and whereas for MLR got 92.25% accuracy so that the authors concluded that MLR approach given better results than LR.

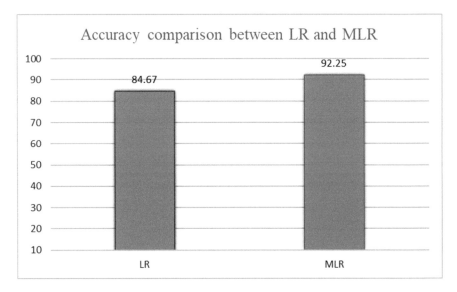

FIGURE 29.4   Comparison of accuracy between the two algorithms.

28.5   FUTURE SCOPE

This is authors' pilot study which is applied on collected data from Govt. of Odisha for Jeypore region. In this included these two algorithms LR and MLR, the authors are extending this study by applying different ML approaches like SVM, decision tree, and many more and DL approaches like CNN, RNNs, and LSTMs on same dataset for better results.

KEYWORDS

- **annual average temperature**
- **artificial intelligence**
- **Caribbean agro-meteorological initiative**
- **climate change**

- **internet of things**
- **machine learning**
- **multiple linear regression**

REFERENCES

1. Yohannes, H., (2016). *A Review of the Relationship between Climate Change and Agriculture (No. RESEARCH), 7*, 335. doi: 10.4172/2157-7617.1000335.
2. Khajuria, A., & Ravindranath, N. H., (2012). Climate change in the context of the Indian agricultural sector. *J. Earth Sci. Clim. Change, 3*, 110. doi: 10.4172/2157-7617.1000110.
3. Roudier, P., Muller, B., D'Aquino, P., Roncoli, C., Soumaré, M. A., Batté, L., & Sultan, B., (2014). The role of climate forecasts in smallholder agriculture: Lessons from participatory research in two communities in Senegal. *Climate Risk Management, 2*, 42–55.
4. Vogel, J., Letson, D., & Herrick, C., (2017). A framework for climate services evaluation and its application to the Caribbean Agro Meteorological Initiative. *Climate Services, 6*, 65–76.
5. Soares, M. B., Alexander, M., & Dessai, S., (2018). Sectoral use of climate information in Europe: A synoptic overview. *Climate Services, 9*, 5–20.
6. Tall, A., Coulibaly, J. Y., & Diop, M., (2018). Do climate services make a difference? A review of evaluation methodologies and practices to assess the value of climate information services for farmers: Implications for Africa. *Climate Services, 11*, 1–12.
7. Patil, K. A., & Kale, N. R., (2016). A model for smart agriculture using IoT. In: *2016 International Conference on Global Trends in Signal Processing, Information Computing and Communication (ICGTSPICC)* (pp. 543–545). IEEE.
8. Mendelsohn, R., (2008). The impact of climate change on agriculture in developing countries. *Journal of Natural Resources Policy Research, 1*(1), 5–19.
9. Modise, W., & Mphale, K. M., (2018). Weather forecasting: From the early weather wizards to modern-day weather predictions. *J. Climatol. Weather Forecasting, 6*, 229. doi: 10.4172/2332-2594.1000229.
10. Dharmaraj, V., & Vijayanand, C., (2018). Artificial intelligence (AI) in agriculture. *Int. J. Curr. Microbiol. App. Sci., 7*(12), 2122–2128.
11. Melissa Dell, Benjamin F. Jones, & Benjamin A. Olken, (2012). "Temperature Shocks and Economic Growth: Evidence from the Last Half Century," *American Economic Journal: Macroeconomics, 4*(3), 66–95, http://dx.doi.org/10.1257/mac.4.3.66.

CHAPTER 29

Data Transmission in Multi-Hop Cluster Using Blockchain in a Clustered Wireless Sensor Network (CWSN)

BANDITA SAHU, RANJEET KUMAR PANIGRAHI, and JEMARANI JAYPURIA

Department of Computer Science, GIET, University, Gunupur, Odisha, India

ABSTRACT

Wireless sensor network (WSN) performs a major role in many of the application areas. Such networks are implemented with a cluster concept and perform the data transfer operation efficiently. In such clustered WSN, data move from one node to another node freely. It is prone to different types of attacks. In this chapter, we have proposed a novel approach of making the data secured by applying the concept of blockchain. As the data flow through several nodes, blocks are created at each level to make the data secured. We have applied this concept to secure the data from man in middle attack and non-repudiation. Theoretical analysis of the proposed approach is performed with respect to time. The experimental analysis of the proposed method with respect to several parameters proves that this proposed approach performs better.

29.1 INTRODUCTION

Nowadays wireless sensor network (WSN) is extensively used for many applications. A group of sensors are deployed in an area for sensing the data from the environment is called the WSN [1]. For transmission of data, these sensors use wireless communication media. Such sensors are called nodes. Depending upon certain criteria, these nodes are classified into different group. Each group is named as cluster [2]. Each cluster is headed by a head node named as cluster head (CH). The nodes under this head node in that cluster are called followers or the non-cluster head (NCH). Data sensed by the NCHs are transmitted to the base station (BS) through this head node or CH. The cluster in which data transmission takes place using more than one link is called a multi-hop cluster [3]. When data transmitted from one end to the sink node or the BS is transmitted over in-secured channel. Hence, data protection plays an important role in multi-hop clustering. For this security, various encryption algorithms [4] have designed. However, blockchain [5, 6] is the emerging concept for providing security to data. It is used to keep track of the transaction made between the two parties. As its name, it is a chain of block or the so-called security code. Each block consists of three fields such as the transaction data, hashed value of the previous block and a timestamp. In order to provide security to the data to be transmitted in a multi-hop cluster, blockchain can be used. Once blockchain is implemented in clustering, the transmission of data is recorded. No intermediate node will be allowed to alter the data. Even if a node tries to alter the data, the alteration is to be reflected at each node. Blockchain is used in various fields such as, data sharing, digital voting, title transfer, money transfer, etc. In multi-hop clustering, we have to perform data sharing. As the data is to be shared from the NCH to the CH and then to the BS. There may be some other node like sub-cluster head (SCH) or any intermediate node who participates in sharing or transmitting the data to the BS. Once blockchain is used in this, no intermediate nodes can alter the data to be shared.

In this chapter, we have implemented the concept of blockchain in transmission of data in a multi-hop cluster. Data security is done using hashed block. The transaction of data is recorded with the time stamp. As compared to other cryptographic algorithm, our approach is more secured. Brute force attacks, man in middle attacks are quite impossible for such approach.

29.2   RELATED WORKS

The authors in Mallikarjuna et al. [8] have described the privacy preservation of data while transmitting in a clustered WSN. They have used midpoint clustering algorithm for formation of cluster. For authentication purpose identity, a based polynomial signature is used. Encryption and decryption of data is done by an integrity protocol based on bit sequencing message.

For the transmission of spatiotemporal data, a novel approach is designed [8]. For cluster generation, the authors have used the concept of LEACH [9]. It is a two-step process. The first step is the setup phase and the second one is formation phase. In terms of cluster formation we can't say it is less time taking, rather it is an energy-efficient algorithm with respect to other parameters. Blowfish algorithm is used for data encryption and decryption.

LEACH [9] is an algorithm used for clustering. For routing of data in WSN, this algorithm is used. It is a two-step process. The overhead of the network is more due to dynamic cluster selection. The algorithm complexity is also high because of the setup and steady phase. The authors [10] have described a new approach called efficient and secure routing protocol through SNR based dynamic clustering (ESRPSDC). Based on node energy, the CH nodes are selected. The energy consumption is more due to the dynamic selection of CH. SNR is another parameter used to identify the head node in the cluster. Thereby, the algorithm complexity increases.

Data transmission process is based on demand as presented by the authors in 2018 [11]. Sensed data are transmitted from the NCH based on demand generated at CH end. It reduces the number of frequency data transmitted. However, data security is missed. Fault-tolerant multi-hop clustering (FTMH) [12] is another approach used for efficient clustering. The authors have identified the faulty node and managed other nodes to work as obvious. They have only managing the faulty node. However, the security of data is uncovered. In selective data transmission (SDT) [13], the concentration of the author is on the reduction of data transmission frequency but not on data security. Energy-efficient SNR based clustering in UWSN with data encryption (EESCDE) [14] is an algorithm that satisfies efficient utilization of residual energy. Data is secured by the implementation of Hill cipher as the encryption technology. For securing data in CWSN, an approach is proposed in 2015 [16] to identify the selfish CH. In order to prevent the selfishness behavior of CH a new node named as inspector node (IN) is created.

29.3   PROPOSED MODEL

In the proposed algorithm, at the initial step, we have performed clustering with selection of CH, SCH, and NCH. Later blockchain is implemented to secure data.

The algorithm can be divided intofour phases:

- selection of CH;

- formation of the cluster;
- selection of SCH; and
- perform blockchain.

1. Selection of CH: Signal-to-noise ratio (SNR) [14] value of each node is computed with respect to its distance from the BS. The nodes are sorted according to the decreasing value of their SNR. Number of clusters is decided as per the requirement and no of nodes deployed in the WSN. We have restricted the number of nodes on a cluster to 10. The minimum number of nodes in a cluster is 6. The number of nodes includes CH, NCH, and SCH in that cluster. Nodes with higher SNR value are initialized as the CH. In our approach, we have assumed the number of clusters in our network is 6. And total number of nodes deployed in this area is 42. The CHs are selected as follows and shown in Figure 29.1.

   - **Step 1:** For i=1 to n//n is the total number of nodes Compute SNR[i].
   - **Step 2:** Sort the list SNR according to decreasing order max = SNR[1].
     For j =1 to k For i=1 to n.
     If SNR[i]>max=SNR[i] CH[k]=SNR[i].
   - **Step 3:** Return CH[].

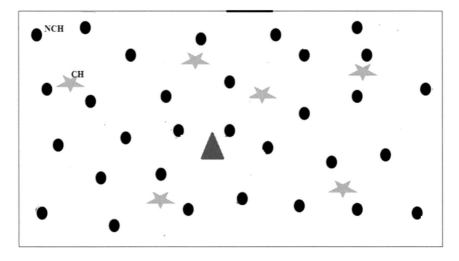

FIGURE 29.1   Selection of CH.

2. Formation of the Cluster: It is the second step of our proposed approach. Once the CHs are selected, the clusters are formed based on the distance from the CH. The Euclidean distance of each node is computed from the pre-identified CHs. The CHs which is present at minimum distance from a node is considered as its respective Head node. The number of followers node in a cluster ranges from 5 to 10. If a cluster has less number of nodes as compared to its range, the NCHs are transferred from one cluster to another cluster with some constraint. The constraint includes the number of nodes present in that cluster and distance from the respective nodes. If the nodes in the new cluster exceed its range, it stops its adoption. The remaining nodes search a new CH nearby them where they are to be accommodated. The distance between the CH and NCH is at assumed to be 10 meters in maximum. The clusters formed are as shown in Figure 29.2. The algorithm designed for this step is as follows.

- **Step 1:** for i =1 to n-k
  for j=1 to k
  Compute $D[i][k]=J(x_{CHj} - x_i)^2 + (y_{CHj} - y_i)^2$
- **Step 2:** Initialize count[k] = 0///for each cluster Step 3: extract D[i] = minimum(D[i][k])
- **Step 4:** for j = 1 to k
  for I = 1 to n-k
  if (count[k]<10 && min D[i][k]) head[i]=k;
  count[k]++;
  else (count[k]> = 10 and mind[i][k]) find next min D[i][k]
  if(count[k]<10)
  head[i] = k; count[k]++

3. Selection of Sub Cluster Head (SCH): In order to reduce the overhead of CH, a SCH is formed. The node with next highest SNR in a cluster is considered as SCH. Each cluster consists of a CH, a SCH and some follower nodes. All the followers or the NCHs send the sensed data to the SCH. It works as an assistant to the CH. It aggregates the data received by all NCH in that cluster. The aggregated data is sent to the CH. It removes the headache of data storing and data aggregation at the CH. The CH stores the data received from the SCH and transmit it to the BS. At the CH end, it receives the same data frequently from the corresponding SCH, the data is dropped. Otherwise, the data is

stored at the CH memory to send it to the BS. The SCHs selected based on the SNR values are as shown in Figure 29.3.

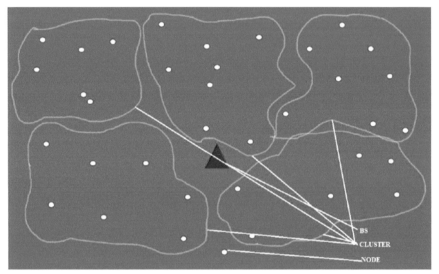

FIGURE 29.2 Cluster formation.

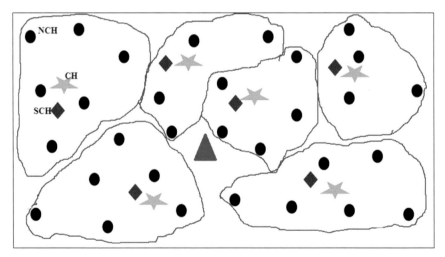

FIGURE 29.3 Selection of SCH in each cluster.

4. Implementation of Blockchain: Initially data are sensed by the NCHS. Frequently these data are transmitted to the SCH in that cluster. As the data move in the wireless medium, it is prone to

man in middle attack. The data may be changed at any time by any person. There is another possibility of repudiation. The sender may deny the sending of message. In order to avoid such attacks blockchain can be used throughout the path it is traveling. As the data is traveling through multiple hops at each hop, a block is created with the data. The timestamp of data transaction is recorded in that block. Four blocks are created as the sensed data is stored at four places. Block 1 is created at NCH. It contains the data, time at which the data is sensed and sent to the SCH, address of the next node that is SCH. The data field of previous node is Null. Block 2 is created at SCH. It contains the data, timestamp, address of previous field as address of NCH and address of next node as CH's address. We have created block 3 at CH. It holds the data, timestamp, address of SCH and address of BS. The last block is created at BS. The next address field is null here. Figure 29.4 describes how blockchain is used for secure data transmission in multi-hop clustering.

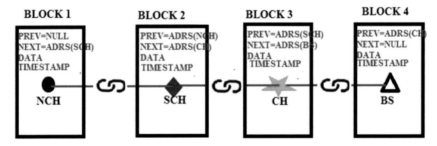

FIGURE 29.4   Hierarchy of data transmission.

29.4   THEORETICAL ANALYSIS

As the clustering is performed on the basis of SNR value, approximately all the nodes come under any of the CH. However, in other algorithms like LEACH, the nodes at the boundary regions are uncovered. The number of time s required to form clusters is less in our approach. Cluster formation is the major task in clustered WSN. The time required for this is minimized in our approach. It requires only one phase to finish the classification. CH selection is easy. The time required to compute the CH is order of n. Once the SNR values are computed, k numbers of highest SNR values are extracted from the list. It takes n times for extracting one highest value. For extracting k values, it takes k X n times. Hence the time complexity of the CH selection algorithm is O(n).

Selection of SCH made on n-k nodes in worst case. But as we are selecting the SCH in a cluster itself, no of comparison ranges from 5 to 9. The number of NCH in a cluster is restricted to 6 to 10 including one CH which is exempted from comparison. One SCH is selected for each cluster. For selecting k cluster, we require k X 9 comparison in worst case. Hence the time complexity of the SCH selection algorithm is written as O(k).

29.5   EXPERIMENTAL ANALYSIS

A random deployed WSN is created with 42 nodes. The nodes are classified into six clusters based on cluster classification constraints. The SNR value of each node computes as shown in Table 29.1. We have compared the number of iteration performed by CH. The iteration of LEACH, Efficient, and secure routing protocol for WSNs through SNR based dynamic clustering (ESRPSDC), energy-efficient SNR based clustering in underwater WSN (EESCDE). Figure 29.5 shows the comparison of number of iteration with respect to number of nodes. If the number of nodes increased in a network, the fault probability may increases. More number of nodes participates in data transferring process. Hence, data is prone to attack. The effect of increased number of on fault probability is shown in Figure 29.6. If the number of nodes increased, CH overheads will also increase. Thereby, energy utilization increases. The amount of energy required for the CH is based on transfer energy, reception energy and processing energy. Figure 29.7 represents the energy utilization with respect to number of nodes. Theoretical analysis of several algorithms is performed in previous section. The execution times are compared and shown in Figure 29.8. Data may be lost while moving from one node to another node. It is less in single hop clustering. However, the transmission loss [15] increases in multi-hop network. The transmission loss of various algorithms is compared and shown in Figure 29.9. Many algorithms exist for encryption of data to be transmitted. The message complexity varies from algorithm to algorithm. The use of blockchain concept makes the message more complex. These complexities are compared in Figure 29.10. We have compared the rate of data security with respect to increasing number of nodes in Figure 29.11.

TABLE 29.1   SNR Value of Randomly Deployed Nodes

Sl No. of Nodes	SNR Value	SL No. of Nodes	SNR Value	SL No. of Nodes	SNR Value	SL No. of Nodes	SNR Value
1.	17.770605	12	16.823666	23	16.355032	36	13.331367
2.	16.831207	13	16.548714	24	15.835801	37	12.726056
3.	16.608000	14	11.710144	25	15.304141	38	16.985802
4.	16.099493	15	10.007893	26	17.083771	39	15.443583
5.	15.374161	16	17.270308	27	16.653517	40	15.280247
6.	14.558638	17	13.888174	28	16.237606	41	14.126097
7.	13.948957	18	12.888263	29	11.646091	42	16.637484
8.	13.279752	19	11.657891	30	8.977273		
9.	15.892528	20	8.999619	33	14.518232		
10.	15.320229	21	17.152811	34	13.738029		
11.	14.551729	22	16.801319	35	12.746639		

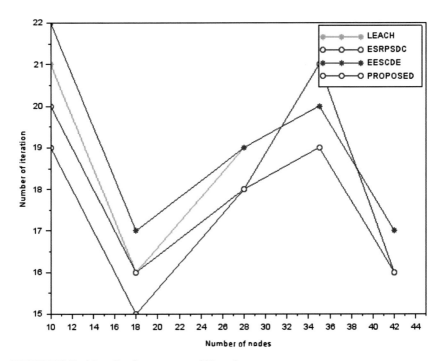

FIGURE 29.5   No. of nodes versus no. if iteration.

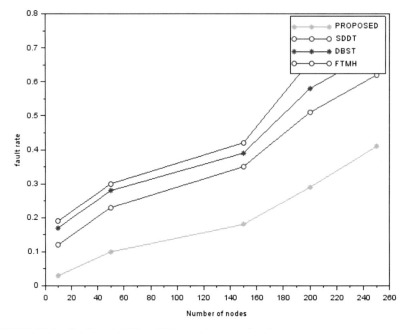

FIGURE 29.6   Fault probability with increasing no. of nodes.

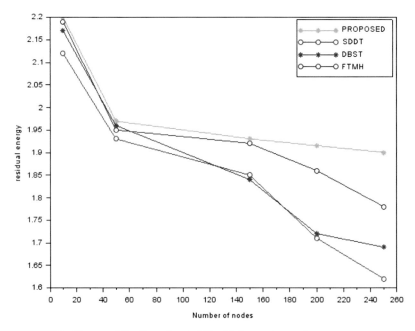

FIGURE 29.7   Energy utilization of each CH with increasing nodes.

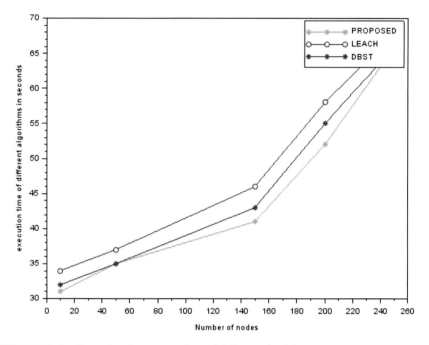

FIGURE 29.8   Execution time comparison of different algorithm.

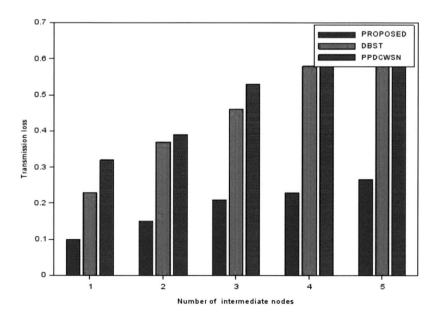

FIGURE 29.9   Transmission loss with respect to no. of intermediate nodes

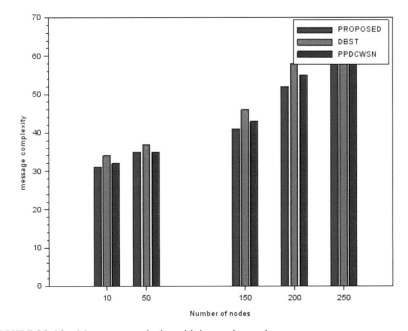

FIGURE 29.10 Message complexity with increasing nodes.

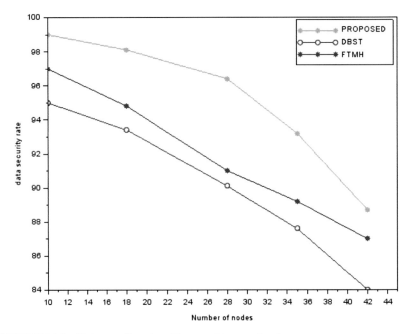

FIGURE 29.11 Data security rate with respect to no. of nodes.

29.6 CONCLUSION

We have proposed a new framework to provide security to the data. Efficient clustering is also been implemented with the concept of SNR. The cluster formation technique used in this chapter take less time as compared to other algorithms as it accomplished in a single-phase only. The execution time of the algorithm is less. The algorithm complexity as analyzed is minimized. To prevent data from man in middle attack and non-repudiation blockchain concept is used. The data cannot be manipulated at any point in the traffic as because of the blockchain. The proposed algorithm is an efficient clustering algorithm that provides high degree of data security. In future, we will use different routing protocol with block chin

KEYWORDS

- **blockchain**
- **cluster head**
- **fault-tolerant multi-hop clustering**
- **noncluster head**
- **selective data transmission**
- **signal-to-noise ratio**
- **sub-cluster head**

REFERENCES

1. Kamimura, N. W. J., & Muratai, M., (2004). Energy-efficient clustering method for data gathering in sensor networks. In: *Proc. BASENETS.*
2. Li, S. D. L., & Wenl, X., (2006). An energy efficient clustering routing algorithm for wireless sensor networks. *J. China Univ. Posts Telecommun., 13,* 7175.
3. Yi, X., Deng, L., & Liu, Y., (2010). Multi-hop clustering algorithm for wireless sensor networks. *International Conference on Measuring Technology and Mechatronics Automation* (pp. 728–731). Changsha City.
4. Sahu, P., & Sahu, B., (2019). *Demand Based Secured Data Transmission in WSN.* International Conference on Intelligent Computing and Remote Sensing.
5. Yang, J., et al., (2019). A trusted routing scheme using blockchain and reinforcement learning for wireless sensor networks. *Sensors, 19*(4), 970.
6. Ekonomou, E., & Booth, K. (2019). *Securing Data Transfer in Wireless Sensor Networks.*

7. Reddy, K. M., & Mula, S. (2015). *Privacy Preserving of Data transmission for Cluster Based Wireless Sensor Network.*

8. Adimoolam, M., Sugumaran, M., & Rajesh, R. S., (2018). A novel efficient clustering and secure data transmission model for spatiotemporal data in WSN. *International Journal of Pure and Applied Mathematics, 118*(8), 117–125.

9. Heinzelman, W., Chandrakasan, A., & Balakrishnan, H., (2000). *Energy-Efficient Communication Protocols for Wireless Micro Sensor Networks (LEACH) in HICSS* (Vol. 8, pp. 3005–3014). Maui, Hawaii.

10. Ganesh, S., & Amutha, R., (2013). Efficient and secure routing protocol for wireless sensor networks through SNR based dynamic clustering mechanisms. *Journal of Communications and Networks, 15*, 422–429.

11. Sahu, B., Pattnaik, R., Dash, P., & Panda, B., (2018). On-demand data transmission in multi-hop cluster in underwater sensor network. In: *2018 3rd International Conference for Convergence in Technology (I2CT)* (pp. 1–5). Pune.

12. Sahu, B., Prahallad, S., & Dash, P., (2018). Fault tolerant dynamic multi-hop clustering in under water sensor network. *2018 International Conference on Information Technology (ICIT).* IEEE.

13. Sahu, B., & Khilar, P. M., (2015). Selective data transmission in SNR based clustered under water wireless sensor network (CUWSN). *2015 International Conference on Man and Machine Interfacing (MAMI).* IEEE.

14. Bandita, S., & Khilar, P. M., (2016). Energy efficient SNR based clustering in underwater sensor network with data encryption. *International Conference on Distributed Computing and Internet Technology.* Springer, Cham.

15. Yaacoub, E., & Abu-Dayya, A., (2012). Multihop routing for energy efficiency in wireless sensor networks. *Wireless Sensor Networks-Technology and Protocols* (pp. 165–186). InTech Press.

16. Ishaq, Z., Seongjin, P., & Younghwan, Y., (2015). A security framework for cluster based wireless sensor networks against the selfishness problem. *2015 7th International Conference on Ubiquitous and Future Networks.* IEEE.

CHAPTER 30

An Open-Source Web-Based OWL Ontology Editing and Browsing Tool: Swoop

V. B. NARASIMHA, B. SUJATHA, and S. NAGAPRASAD

Osmania University, Amberpet, Hyderabad, Telangana – 500007, India

ABSTRACT

After the evolution of the semantic web and its supporting technologies, ontologies have recently received popularity in the area of knowledge management and knowledge sharing. They basically interlace the human understanding of symbols with their machine processability. To facilitate knowledge sharing and re-usage of information,ontologies were developed in artificial intelligence (AI). Ontology has become common in the fields of cooperative information systems and information retrieval, intelligent information integration, knowledge management, and electronic commerce. A variety of tools is available for the purpose of ontology, some are licensed tools, and the rest are open source tools. Each tool is specialized in their own way. For performing diverse tasks different set of people use different ontology building/management tools. One such used hypermedia inspired ontology browser and editor-based OWL tool is Swoop. This chapter describes the efficiency of using Swoop tool for ontology development which is indeed a main technique used in the semantic web searching process.

30.1 INTRODUCTION

The Semantic Web is also said to be a globally linked and accessed distributed database (DB) or network of linked data/information [1–3]. The ultimate aim of Semantic Web is to simply process the information by machines. From the time when the data is huge predicting both meaningful and meaning-less representation, a standardized system has been proposed by W3C [4]. Semantic Web is an important aspect used for knowledge representation on the web. There are many tools available for modeling, editing, and handling ontologies for web space. The ultimate aim of the chapter is to discuss the various tools for Web Ontology editing and to share the main purpose of using SWOOP-open source tool.

30.2   WEB MINING AND SEMANTIC WEB

The request provided by the user of web is regularly patterned using web mining. Web mining application does knowledge extraction from the web pages content. For this process web mining applies three techniques, they are: (i) web content mining; (ii) web structure mining; and (iii) web usage mining. The three techniques are designed in a way to work uniquely for perfect knowledge extraction. Semantic web converts web-based information to knowledge-based information. For a progressive work, semantic web uses web representation techniques like XML, RDF, and OWL. To follow ontology, Semantic Web utilizes several kinds of ontology building and managing tools, some of them are: OilEd, SWOOP, pOWL, Protégé, and OntoEdit.

30.3   VARIOUS ONTOLOGY BUILDING/MANAGEMENT TOOLS

Tools for ontology building or managing in web space are available in a wide range. Let us take some of them and discuss:

- **OilEd [6]:** This tool uses DAML+OIL [14] for the user to build web ontologies and it also uses OWL as a standard. It does not support a complete environment for development of Ontology, but provides adequate functionalities for building ontologies. This tool uses FaCT Reasoner for consistencies. It is available Java project under GPL License as an open-source tool. Text formats like OIL, OWL RDF/XML and DAML + OIL are the import formats and saves ontologies in the document form of DAML + OIL only.
- **Semantic Web Ontology Overview and Perusal (SWOOP) [11]:** An OWL ontology browser and editor developed using Java with hyper-media, scalable, and simple is said to be semantic web ontology overview and perusal (SWOOP). Keeping the W3C standards in mind, SWOOP has been developed with the supporting information engine-Pellet.

Features like history buttons, bookmarks, and address bar for loading ontology entities gives a feel of having another web-browser. Multiple ontology environments in which relationships and entities of a variety of ontologies to be compared, merged, edited, etc. HTML renderer is used for editing ontology. By means of various color codes plus font styles ontology changes were emphasized, for example, axioms that are added and deleted represented with diverse representation. Ontology rollback and changelog option provide a way for undo/redo actions. This tool import OWL, XML, RDF, and text formats which are used to save ontologies that are edited. The architecture of SWOOP is based on a model-view-controller (MVC) design model.

- **pOWL:** Web-based and open source PHP ontology for managing and editing is said to be pOWL and this is based on web which supports editing, RDFS/OWL viewing with random dimension. RDQL query builder and full text search for resources and literals are provided [8]. API is used for accessing the functionality with authentication. DB tables like PostgreSQL, Oracle, MySQL, etc., are used to store models which is then stored in the main memory for faster access and response. Different RDF syntax such as XML, N3, and Ntriples are used for the purpose of importing and exporting. The output is availed in HTML format. By the GNU-GPL terms, pOWl is availed free.
- **Protégé [9]:** For building knowledge-based applications and domain models with ontologies, a Java-based open-source free tool Protégé is used. Creating, manipulating, and visualizing ontologies in several formats of representation, a set of supportive knowledge-modeling structures and actions are availed in this tool. In creation of knowledge models plus for entering information in Protégé will be then customized for providing friendly domain support. Plug-in architecture can be used for extending Protégé. Through the help of Java API, facilities in varied way can be availed through the programs. With the accordance of OKBC-open knowledge base connectivity protocol, Protégé provide support in building frame-based ontologies. It also supports OWL ontologies. To produce graphical representation of the editing ontologies, the tool uses special type of plug-ins.
- **OntoEdit:** It is a portion of OntoStudio that is based on Eclipse framework developed by IBM [7]. Open plug-in structure approach is applied in OntoEdit. This is an environment for developing ontology designs and for maintaining them. Multi-lingual development support and knowledge model based on frame languages are also supported

by this tool. Importing and exporting different formats like OWL, Flogic, RDF/RDFS, OXML, plus some more design that are WebDav supported is done by OntoEdit [17]. It is a commercial product with professional versions.

30.4   ARCHITECTURE OF SWOOP

The architecture of SWOOP is based on MVC [13]. Ontology-centric information related to workspace of SWOOP such as change-logs, checkpoints, and ontologies that are currently loaded are stored in Swoop-model component. Swoop UI object's (like QNames, chosen OWL item, import's view setting, etc.), key parameters are defined using Swoop-model component. Changes to be reflected in UI that is based on Swoop Model changes are maintained and performed by Swoop-model listener class. Plugin based system is utilized for controlling which in turn loads new reasoners and renderers. Plug in framework ensures code modularity and provides hands to external developers for connecting easily to SWOOP projects. The SWOOP code is written in Java and maintenance is done using subversion repository, it also uses third party libraries and being a WonderWeb OWL API is the major advantage which uses OWL representation model. Figure 30.1 shows a pictorial view of SWOOP architecture.

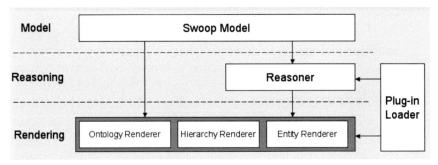

FIGURE 30.1   SWOOP architecture.

30.5   FEATURES OF SWOOP

- SWOOP basically look and feel like another web browser for the user [16]. The features that resemble the web browser are address bar for

entering ontology URL, class, property, etc. History and bookmarks are utilized for traversing and saving data for later reference.
- SWOOP ontology editing is always done inline using the HTML renderer.
- For multiple ontology, browsing, mapping, comparing supports are carried out. Ontology versioning system control is another advantageous feature in SWOOP [15].
- Partitioning modules which splits the ontology in a well organized way as class or property makes user comfortable in viewing the test ontology gaps.
- Even though the tool used in SWOOP for querying purpose known as SWOOP Query tool not advanced, it works and gives the best result as fast as possible for simple queries. Figure 30.2 gives a SWOOP overview.
- For tracking errors a reasoner debugger has been fixed at the backend.
- FOAF, NATO, AKT-Portal, OWL-S Grounding, etc., are some of the examples of built-in ontology that helps as user guide for steps to be followed.

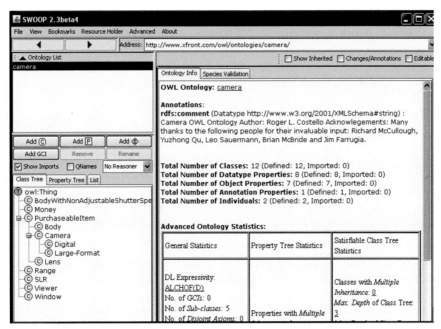

FIGURE 30.2   Overview of SWOOP.

30.6 IMPLEMENTATION RESULT OF ONTOLOGY TOOLS (TABLE 30.1)

TABLE 30.1 Ontology Tools and Their Features

Feature	OilEd	SWOOP	pOWL	Protégé	OntoEdit
Base Language	DAML+OIL	OWL	PHP+OWL	OKBC+ CLOS based meta model	F-Logic
Import/ Exports from/to Languages	RDF URI's; limited XML Schema, export: HTML.	Imp: OWL, XML, RDF, and text exp: RDF(S), OIL, and DAML	RDF syntaxes (XML, N3, Ntriples)	RDF, RDFS, DAML+OIL; XML, OWL, Clips; UML	RDFS, F-Logic, DAML+OIL; RDB, schemas
Exception Handling	No	No	Yes	No	No
Ontology storage	Files	HTML Models	MySQL, PostgreSQL, Oracle, RDQL	Files and DBMs (JDBC)	Files
Availability	Free	Free	Free	Free	Free
Ontology library	Yes	No	Yes	Yes	No

30.7 CONCLUSION

Effective tools are the major requirement for development of best ontology. Software tools are readily available to fulfill the need for Semantic Web. In this chapter, we have disclosed some kinds of ontology tools along with their features. The SWOOP tool is very much useful in creation of small and medium ontologies which can later be add-on with complex software. This is one of the main advantages in using SWOOP. This tool look and feel like a web browser which provides a comfort zone for the user to build and manage ontologies. As the tool is an open-source, user is free to access it.

KEYWORDS

- **model-view-controller**
- **ontology**
- **open knowledge base connectivity**
- **semantic web**
- **semantic web ontology overview and perusal**
- **swoop**

REFERENCES

1. Buraga, S., (2006). *XML Technologies (in Romanian).* Polirom.
2. Buraga, S., (2004). *Semantic Web (in Romanian).* Matrix Rom.
3. Daconta, M., Obrst, L., & Smith, K., (2003). *The Semantic Web.* Wiley Publishing.
4. World-Wide Web Consortium's Technical Reports, (2006). Boston. http://www.w3.org/TR/ (accessed 20 July 2020).
5. Jena. (2019). http://jena.sourceforge.net/ (accessed 20 July 2020).
6. OilEd. (2018). http://www.oiled.net/ (accessed 20 July 2020).
7. OntoStudio. (2019). http://www.ontoprise.de/ (accessed 20 July 2020).
8. pOWL. (2019). http://powl.sf.net/ (accessed 20 July 2020).
9. Protégé. (2019). http://protege.stanford.edu/ (accessed 20 July 2020).
10. *Redland RDF Application Framework.* http://librdf.org/ (accessed 20 July 2020).
11. *SWOOP (Semantic Web Ontology Overview and Perusal).* (2019). http://www.mindswap.org/2004/SWOOP/ (accessed 20 July 2020).
12. *World-Wide Web Consortium's Technical Reports*, (2006). Boston. http://www.w3.org/TR/ (accessed 20 July 2020).
13. Erich, G. R. J., Richard, H., & Vlissides, J. (1994). *Design Patterns: Elements of Reusable Object-Oriented Software.* Addison-Wesley.
14. Davies, J., Fensel, D., & Van, H. F., (2003). *Towards the Semantic Web.* John Wiley & Sons.
15. Su, X., & Ilebrekke, L., (2002). A comparative study of ontology languages and tools. *Conference on Advanced Information System Engineering (CAiSE'02).*
16. Bhaskar, K., & Savita, S., (2010). A comparative study ontology building tools for semantic web applications. *International Journal of Web and Semantic Technology (IJWesT), 1*(3).
17. Thabet, S., (2015). Ontology development: A comparing study on tools, languages, and formalisms. *Indian Journal of Science and Technology, 8*(24), doi: 10.17485/ijst/2015/v8i34/54249.

CHAPTER 31

A Real-World Application Based on Blockchain

AUROPREMI ASPRUHA

Integrated Test Range, Defence Research Development Organization(DRDO), Chandipur, Odisha 756025

ABSTRACT

Blockchain is surely a creative creation and the decentralized process of blockchain technology provides an important way to secure data. The growth of digitalization has been improved day by day. This chapter mainly focuses about explaining various applications of blockchain to provide a secure way of communication for a progressive digitalized government, i.e., food supply chain, education, and health care system. It also describes about the challenges which mainly occur during implementing blockchain concept in the above-mentioned applications.

31.1 INTRODUCTION

For making India digitalized, the government has long been adapting information technology for communication and information security for securing the information. Fortunately, Satoshi Nakamoto developed the concept of blockchain; which is an opportunity for information security to forward it to the next level. So that the next generation of digital India have an open operational platform not only for security but also for improve mutual trust between government, endeavor, and general public. This technology had been used in different field for accuracy and transparency of information. The general definition of blockchain is: it is a decentralized mechanism having distributed ledger to provide security for commonly accessible data from tampering and re-modification. Our government has long been promoting for development of digitalized government. One of the important points for digitalization is information security and we can use blockchain in security as an opportunity.

The benefit of blockchain applications are: (i) accuracy and transparency of data; (ii) accessibility of data without time delay; (iii) sharing of information; (iv) minimize the participation of the third party, etc. In the food supply chain, the data differ from person to person for personal benefit. Through blockchain, we can manage the data gap between the users. This chapter provides an application for food chain so that all the end-user gets the benefit. In education, system, and health care system if we add some applications related to this technology then we get quality product and store the information in a secure way.

This chapter first describes about basic blockchain then divided in three major parts, in first section we describe security on food supply chain followed by education section then we describe how blockchain is used in healthcare environment.

31.2 BLOCKCHAIN

Blockchain [1] is a combination of cryptographic technology, mathematical procedure, algorithm, and economical model. It connected with node to node through a distributed algorithm to solve various database (DB) synchronization problems. The main objective of blockchain is as follows:

1. **Decentralized:** Not based on any centralized mechanism.
2. **Transparent:** Every node is transparent to each other; because of its transparency, we trust blockchain.
3. **Open Source:** This technology is open to everyone. Each one can check the data publicly.
4. **Autonomy:** All the data are save, every node trust each other as well as the whole system.
5. **Immutable:** No changes take place on the system. Every node stores the value unless some attacker can take the control more than 51% [3].
6. **Anonymity:** No need to know anyone's personal address, only need to know the blockchain address.

31.3 REAL WORLD APPLICATIONS OF BLOCKCHAIN

1. **Food Supply Chain:** Food is a vital part for any society. According to several surveys, the consumers always demand good quality of food. Generally, when we buy food, we always search for the production details of products (Figure 31.1).

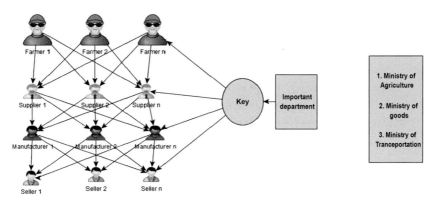

FIGURE 31.1   Blockchain application in food supply chain

But not all the information is always true. Nowadays foodborne diseases increases because of the addition of [2] low quality ingredients. But we cannot find out which part of the supply chain in involved for this work. So if all the information is shared among all the member of the chain then it is possible to counter this problem. Figure 31.1 shows the decentralized process of food supply chain. The farmer sends their product to the supplier and all the information regarding product as well as price. Hence, all the farmer and suppler have same information. Manufacture companies are received all information of farmer through supplier and all information from supplier itself. If we consider all as anode then all node are managed by information department. Information department provides public key and private key to the entire node. Ministry of agriculture, ministry of food, ministry of transportation, etc., can handle all the information [6].

2.   **Education:** The basic problem in educational departments is issuing wrong certificate, leakage of question paper and the scholarship issues. Sometime some people produce some fake certificate [4] and get the job (Figure 31.2).

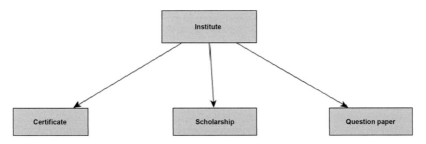

FIGURE 31.2   Blockchain application in education system.

If we add blockchain in a certificate, issuing application then the production of fake certificate can be minimize. Figure 31.3 shows how the information is communicated. The institute provides the data to the root node. The entire root node store same data of student from the institute. Whenever a student or a company wants to access those data then they retrieve the data from root server. In this way, we can stop the production of fake certificate [5].

3. **Healthcare System:** When we think about up gradation of a smart health care management system, then the efficient approach is blockchain (Figure 31.4).

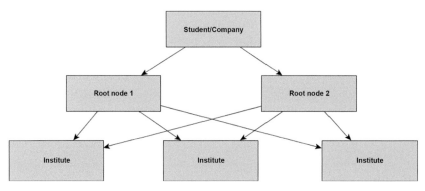

FIGURE 31.3   Retrieve data through blockchain.

Nowadays, patients are familiar with the health-related application [6]. The main advantage of this applications are, it store all health-related information including blood group, blood pressure value, blood sugar value, etc., so that it is easy to monitor the progress and also used for research-oriented observation. If we add blockchain technology in this application then we upgrade it to another level. Another issue about health sector is the proper information of drugs. So mane company give the wrong information about the expiry date of medicine, the composition of medicine, etc., if we implement a blockchain-based application in the health sector then we get all the correct data and all the details about composition from the company directly. No production of fault medicine is possible. In Figure 31.4, we describe the block diagram of blockchain for healthcare system. All the company provides the information to the root node so that when a user or doctor wants any data they can access directly from the root server.

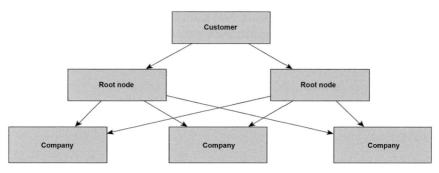

FIGURE 31.4 Blockchain in healthcare system.

31.4 PROBLEMS FOR ESTABLISHING THE BLOCKCHAIN

1. **Cost of Establishment a New Blockchain:** When we think about a new development, then we have to consider about new organization and different new system. So establishing new infrastructure required a lot of money. For develop city it is practical but for some rural area it is so difficult to establish the blockchain [1].

2. **The Long-Term Preservation of Blockchain Platform Records:** We need to preserve all the record carefully, because all the data are related to government. So the important problem for blockchain is how to store the data. According to the concept, each and every data should be stored for all the time. One effective solution is cloud-based storage. But it is also challenging. We have applied proper security on cloud.

3. **Information Security of the system:** Each individual can update data to the blockchain technology. So correctness of information is another important issue. If the provider sends the wrong information then we cannot trust the system.

4. **Maintenance of the System:** As we know, everyone can update the data, so some maintenance team should be required to verify the working procedure of the entire node [5].

31.5 CONCLUSION

Although blockchain is difficult to implement but it makes all communication transparent and trustworthy. If all the schemes of government are implemented correctly then we can think about a progressive country. This chapter suggests implementing the blockchain application in food sector, education sector and healthcare sector. If the information is secure and free from attack then there are no trust issues for any communication and this play a vital role in information security.

KEYWORDS

- **blockchain**
- **digitalization**
- **food supply chain**
- **healthcare sector**
- **healthcare system**
- **information security**

REFERENCES

1. Guo, Y., & Liang, C., (2016). *Blockchain Application and Outlook in the Banking Industry*.
2. Zhu, H., & Zhou, Z. Z., (2016). *Analysis and Outlook of Applications of Blockchain Technology to Equity Crowd Funding in China*.
3. Guo, Y., & Liang, C., (2016). *Blockchain Application and Outlook in the Banking Industry*.
4. Zhu, D., (2016). *Fraud Detections for Online Businesses: A Perspective from Blockchain Technology*.
5. Zhao, J. L., Fan, S., & Yan, J., (2016). *Overview of Business Innovations and Research Opportunities in Blockchain and Introduction to the Special Issue*.
6. Mettler, M., (2016). *Blockchain Technology in Healthcare: The Revolution Starts Here*.
7. Tse, D., Zhang, B., Yang, Y., & Cheng, C., (2017). *Blockchain Application in Food Supply Information Security*.
8. Hou, H., (2017). *The Application of Blockchain Technology in E-Government in China*.
9. Seebacher, S., & Schüritz, R., (2017). *Blockchain Technology as an Enabler of Service Systems: A Structured Literature Review*.
10. Efanov, D., & Roschin, P., (2018). *The All-Pervasiveness of the Blockchain Technology*.

CHAPTER 32

Modified Design of Conformal Circular Patch for 15GHz Application

RIBHU ABHUSAN PANDA and DEBASIS MISHRA

Department of Electronics and Telecommunication, Veer Surendra Sai University of Technology, Burla, Odisha – 768018, India,
E-mail: ribhupanda@gmail.com (R. A. Panda)

ABSTRACT

The traditional circular patch has been modified leading to a butterfly structured patch. The proposed patch has been implemented on the substrate of 30 mm × 42 mm of the area with a height of 1.6 mm. The dielectric material FR4-epoxy has been used for the substrate. This antenna designed to be operated at a frequency of 15 GHz which is used for many applications including 5G communication. The maximum distance between the two arcs has been calculated from the design frequency. Emphasis has been given to the S_{11} (<10 GHz) plot from which the return loss and the resonant frequency have been determined giving a vibrant idea about the frequency of operation of proposed design. Design and simulation have been carried out by HFSS software and parameters like antenna gain, surface current distribution, antenna gain, directivity, etc., have been observed from the corresponding simulated results. Copper material is used for patch and the same material has been used for the ground plane.

32.1 INTRODUCTION

Microstrip antennas are having ascendancy like small size, robust nature and different shaped patches have been used in recent years [1]. These antennas are used to operate in microwave frequencies. Some circular patches have been designed and modified for various applications [2–4]. A perturbed elliptical patch has been presented in the year of 2016 for 50 GHz applications [5]. Biconvex and Biconcave patch have been implemented with circular slots to enhance gain and bandwidth in the year of 2017 and 2018, respectively [6, 7]. Some patch antenna designs evolved for various 5G applications in year of 2014 [8]. In this tabloid, the conformal circular microstrip antenna has been modified to enhance the gain and to increase the performance in terms of loss due to reflection of the wave and standing wave ratio. Two perturbed semi-circular structures with arc-shaped sides have been combined which forms the butterfly-shaped patch. The 5G uses spectrum LTE frequency range (i.e., 600 MHz to 6 GHz) as well as the millimeter wave bands (i.e., 15 GHz and higher frequencies [9] [10]. The reshaped structure has been designed to be operated at frequency of 15 GHz that has been tested for 5G network applications. 5G is an advanced emerging mobile technology providing high speed data of 20 gigabits per second. It has a wide range of applications in the field of mobile broadband communication, automobile vehicles, industries [11], sensors, IoT [12], etc. This technology challenges us to develop an antenna having low cost, minimal weight, compact but capable of maintaining high performance. The proposed butterfly structured patch is designed to provide a high gain and high directivity. As it is lightweight antenna and consumes less space so it is easy to fabricate.

32.2 DESIGN SPECIFIC PARAMETERS FOR THE ANTENNA

The modified design has been premeditated at a frequency of 15 GHz which is applicable for many applications including 5G communication. The arc to arc distance has been taken 20 mm. Two inverted semicircles with arc-shaped sides have been connected to resemble a butterfly structured patch. The patch has been designed with a height of 0.01 mm on top of the substrate. The patch has been assigned with copper material. The ground plane of dimension 30 mm × 42 mm × 0.01 mm and the conducting material copper is assigned to it. The dimensions of the substrate are taken as 30 mm × 42 mm with a height of 1.6 mm. The FR4-epoxy dielectric material having relative permittivity 4.4 and relative permeability 1. And Figure 32.1 illustrates the final design of the antenna using Ansys HFSS.

32.3   STRUCTURAL ANALYSIS AND SIMULATION

Finite element method is the method used in Ansys HFSS (high-frequency structure simulator) software which is used to simulate the novel shaped patch antenna. The perturbed patch is designed on the substrate having dielectric constant of 4.4. By using the HFSS software, S11-parameter, antenna peak gain, Standing Wave Ratio measured in terms of voltage, radiation efficiency, etc., have been determined. From the simulated plots of all these antenna parameters, it has been observed that the proposed structure can be operated at 15 GHz. The calculation of the maximum distance between the two extreme arcs has been done taking the boundary condition into consideration [1]. Considering the restructured circular patch with dielectric substrate and the ground as the model representing a transmission line, feed width is considered 3 mm for impedance matching (Figures 32.2 and 32.3).

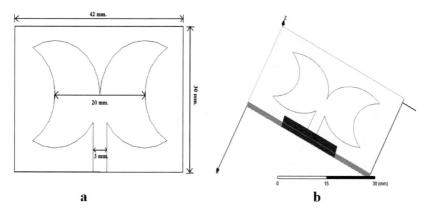

a   **b**

FIGURE 32.1   (a) Shape of the modified circular patch,(b) design using HFSS.

The impedance matching between source and load is defined by VSWR (voltage standing wave ratio). The desired value of VSWR is 1, anything nearer to the unity is considered as good for antenna design. For the proposed patch antenna, VSWR is found to be 1.06. The VSWR curve has been revealed in Figure 32.3 the ratio of radiated power to the accepted power from the transmitter is known as efficiency associated with radiation. This radiation efficiency of proposed structure has been found 0.47 as shown in Figures 32.4. Figures 32.5 and 32.6 illustrate the maximum gain and maximum directivity respectively. Radiation patterns at different angles (90, 5 and 0) are shown in Figure 32.7.

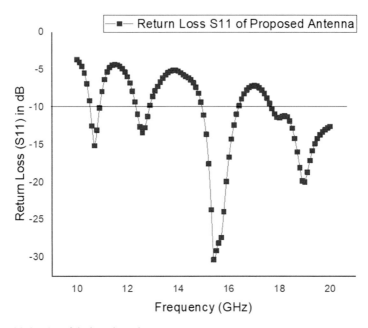

FIGURE 32.2 S_{11} of designed patch antenna.

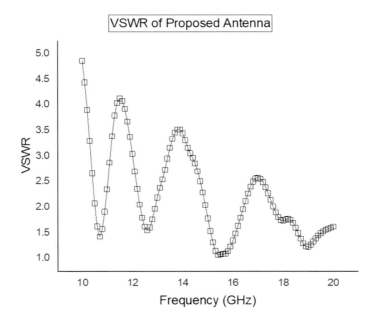

FIGURE 32.3   VSWR designed patch antenna.

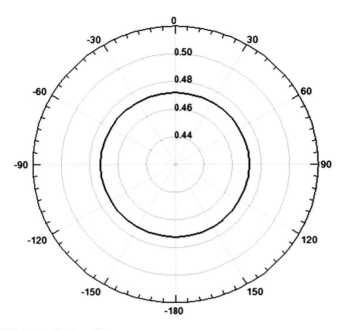

FIGURE 32.4   Radiation efficiency.

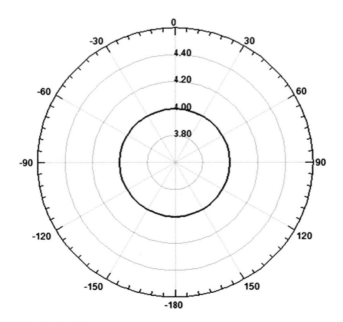

FIGURE 32.5   Peak gain.

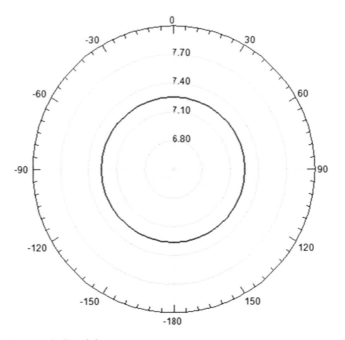

FIGURE 32.6 Peak directivity.

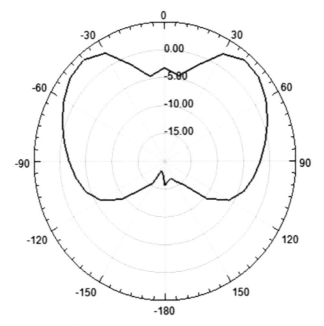

FIGURE 32.7 2D-radiation at ø = 90°.

Surface current distribution and 3D radiation pattern give an idea about the radiation mechanism of the antenna. The red color indicates that the current is properly distributed and the radiation characteristic is good for the designed structure (Table 32.1). Surface current distribution and the 3D antenna gain pattern are shown in Figures 32.8 and 32.9 respectively.

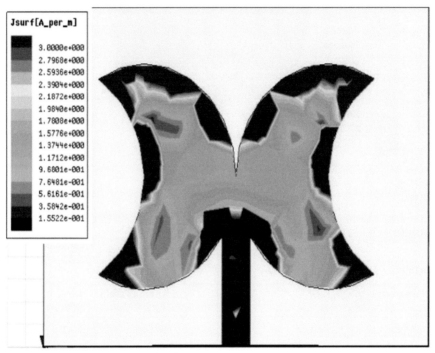

FIGURE 32.8   Surface current density.

32.4   CONCLUSION

The compact, efficient, low cost, butterfly patch has a return loss of –30.28 dB at resonance frequency 15.24 GHz. The proposed modified structured patch has a high gain of 4.01 dB and high directivity of 7.28 dB. The radiation efficiency has been found to be 0.47 dB. The butterfly-shaped patch antenna is useful for 5G communications as it has good S_{11}, gain, directivity, and radiation efficiency.

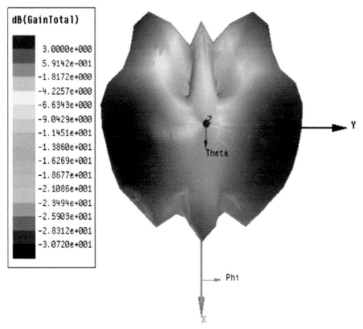

FIGURE 32.9   3D gain pattern.

TABLE 32.1   Simulated Parameters

Parameters	S_{11} [dB]	VSWR	Antenna Gain [dB]	Directivity [dB]
Values	−30.28	1.06	4.01	7.28

KEYWORDS

- 5G communication
- antenna gain
- conformal circular patch
- directivity
- high-frequency structure simulator
- radiation efficiency
- S-parameter
- voltage standing wave ratio

REFERENCES

1. Balanis, C. A., (1997). *Antenna Theory-Analysis and Design*. John Wiley & Sons, Inc.
2. Sung, Y. J., & Kim, Y. S., (2004). Circular polarized microstrip patch antennas for broadband and dual-band operation. *Electronics Letters, 40*(9), 520–521.
3. Guo, Y. J., Paez, A., Sadeghzadeh, R. A., & Barton, S. K., (1997). A circular patch antenna for radio LAN's. *Antennas and Propagation, IEEE Transactions on, 45*(1), 177–178.
4. Iwasaki, H., (1996). A circularly polarized small-size micro strip antenna with a cross slot. *IEEE Trans. Antennas Propagat., 44*, 1399–1401.
5. Panda, R. A., Mishra, S. N., & Mishra, D., (2016). *Perturbed Elliptical Patch Antenna Design for 50 GHz Application LNEE* (Vol. 372, pp. 507–518) Springer India.
6. Panda, R. A., Mishra, D., & Panda, H., (2017). *Biconvex Patch Antenna with Circular Slot for 10 GHz Application* (pp. 1927–1930). IEEE SCOPES-2016.
7. Panda, R. A., Mishra, D., & Panda, H., (2018). *Biconcave Lens Structured Patch Antenna with Circular Slot for Ku-Band Application* (Vol. 434, pp. 73–83). LNEE, Springer.
8. Beale, C. D., Jose, L. G., Antonio, D. D., Dimitri, K., & Laurent, D., (2014). Millimeter-wave access and backhauling: The solution to the exponential data traffic increase in 5G mobile communications systems? *IEEE Communications Magazine*, 88–94.
9. Pi, Z., & Khan, F., (2011). An introduction to millimeter-wave mobile broadband systems. *Commun. Mag. IEEE*, 101–107.
10. https://en.m.wikipedia.org/wiki/5G (accessed 20 July 2020).
11. António, M., Kazi, M. S. H., Shahid, M., & Jonathan, R., (2018). A survey of 5G technologies: Regulatory, standardization, and industrial perspectives. *Digital Communications and Networks, 4*(2), 87–97.
12. Farris, Orsino, A., Militano, L., Iera, A., & Araniti, G., (2018). Federated IoT services leveraging 5G technologies at the edge. *Ad Hoc Networks, 68*, 58–69.

CHAPTER 33

Cloud-Based Data Analytics: Applications, Security Issues, and Challenges

MURALI KRISHNA SENAPATY,[1] GITANJALI MISHRA,[1] and ABHISHEK RAY[2]

[1]Assistant Professor, GIET University, Gunupur, Odisha, India

[2]Professor, KIIT University, Bhubaneswar, India

ABSTRACT

Big data is rising quickly nowadays and it is very difficult to maintain the data by having individual system architecture. So in the present scenario, it is approachable to go for a cloud service to maintain data easily by hiring the required configurations. But there we observe many security issues and challenges. In this chapter, we tried to observe the need of cloud to maintain big data and simultaneously represented different security measures need to focus on handling data. Also, there are different challenges observed on storing data, processing huge data in the cloud using its techniques, transportation of data in a multi-cloud. As the Big data is growing exponentially, the cloud can be the best service if we use it with proper authentication and security features.

33.1 INTRODUCTION

Nowadays, the real-time processing and analytics of big data obtain acknowledged a noteworthy volume for consideration. With the initiation of the Internet and different social media the data has been increased in every day. The word 'BIG DATA' refers to handling different data sets, analyzing them, and capturing where the data volume, its complexity, and rate of growth varies between them. Different tools are available to handle big data. They are Hadoop, HPCC, Storm, Qubole, Statwing, Cassandra, CouchDB, PentahoFlink, Cloudera, Openrefine, Kaggle, RapidMiner, Data Cleaner, and Hive. Big data analytics (BDA) means it will analyze large volumes of data or big data to get patterns and some useful information by which decision can be taken. BDA is a process consisting of three steps:

- collecting;
- organizing; and
- analyzing.

BDA will help to organize the information in such a way that one can easily understand and will also help identify the data which is most important for business and future business decisions. After analysis, someone should get some knowledge. The cloud computing (CC) provides services to customers based on weights like allocation of system resources, data stores and its computing power. The CC service is used to elaborate on the availability of data centers to many users using internet. It provides mainly three types of services such as SaaS, PaaS, and IaaS. This chapter deals with big data in clouds with its pros, cons, and security issues with challenges [1–4, 8, 14].

BDA means the software or applications which are used to process large-scaled data to process parallelly to discover hidden values [5].

If IoT (internet of things) will be added with CC then we are getting heterogeneous data whose size is big. This data need to be stored in cloud and then processing and communicating and analyzing in a secured manner is required, so that it will achieve the target levels. So, for its different security issues and challenges have been introduced [6].

Data analytics-as-a-service (DAaaS): It uses the software-as-a-service (SaaS) delivery model. Here we tried to focus on securing data which is stored by a cloud service provider (CSP's) infrastructure and which is used to limit the access by multiple organizations. These organizations are collaborating with each other and for the data analytics is crucial collectively [7].

To minimize the cost, time, and effort in the cloud environment are done by means of resource sharing. If clustering is, being done then the complexity can be minimized [9].

33.2   IMPORTANCE OF CLOUD TO HANDLE BIG DATA OF DIFFERENT APPLICATION AREAS

Since last, few years' big data is playing very key role in many different applications and the big data leads to sometimes to take crucial decision making, do diagnose critical cases, and predict the future expectations [19, 20].

In almost all areas, the role of big data becomes a common need in present and future scenarios. In the future, it will increase exponentially as today everything is improving technologically with sensory and IoT for automation.

Some major areas where the data is huge in size and which can be maintained in the cloud database (DB) in a cost-effective manner. But the data security becoming a challenging issue. They are:

- In presents days, every company need to take smart decisions in the business competitive world. So, they need to maintain their historic and survey data and to use it for analyzing for keeping updates of their present progress or pitfalls, also about the steps to take in near futures for business improvements. In this, the big data helps the companies to take up business decisions with a statistical approach.
- It helps the doctors to analyze and identify the depth of seriousness based on different testing data.
- It helps to globalize and maintain the patient histories of health problems similar patients and the treatments have undergone, their success rates, so that the data helps to analyze the historical data for updating the approach of treatment, to identify a new direction for treatment, to take measuring steps well in advance to treat better.
- In the stock market, a real-time analysis is required on the Big data to suggest the customer for taking better decisions.
- Nowadays in smart cities of almost all are based on sensory systems, IoT devices, CC footages. So, the role of big data and as it is high in size so its storage in cloud becomes essential. So, there enough security is required to maintain the data with proper authenticity so that intruders will not be able to access, also many times a real time analysis is required on data take smart decisions in many cases such as: fire accidents, disasters, thefts, entry of new intruders to city, crimes, monitoring sensitive areas, better traffic control.
- Nowadays, the E-governance is implemented in many government offices for digitalization. So, there is a need for maintaining the

huge data of people to deal with many activities. A dedicated super computer is to handle the data of each and every citizen and they need to be connected to every citizen to know their updates, legal documents and transactions, taxes, etc.

• Many product manufacturing companies need to maintain their product types, quality, manufacturing cost, transport cost, advertisement cost, growth rate of sales, areas of sales, etc. With these data, they need to analysis about their product demand, quality of product based of feedbacks and by customer requirements to improve the product performance.

33.3 OVERVIEW OF CLOUD-BASED DATA ANALYTICS

There are many data mining techniques are exist used for mining and extracting knowledge from a large data set. But the effective response within a short span of time is challenging. When we combine the data analytics and mining together with CC, then it needs an effective solution for producing useful information in a stipulated time.

For implementing the BDA, services there are many models exist such as:

• **Data Analytics Software as a Service (DASaaS):** It offers definite data mining procedures as a service to end-users.
• **Data Analytics Platform as a Service (DAPaaS):** It offers an appropriate platform for the developers to develop their own applications without worrying about the infrastructure.
• **Data Analytics Infrastructure as a Service (DAIaaS):** It offers a group of resources to execute the data mining applications [1].

Data science is the secret insolence for an organization to influence Big Data for gaining profit out of it. When data mining and machine learning (ML) tools are applied on big data, then they represent the BDA. The intersection of big data, cloud, and data science represent cloud-based BDA [2].

The latest data mining techniques and its related tools can be useful to extract information from a large and complex dataset. These techniques are helpful in making informed decisions in many scientific and business application areas such as: a collection of tax, Sales Reports, high-energy physics, and social media. So, by combining the BDA and knowledge discovery techniques, it will produce new insights within a short span of time. In Figure 33.1 it presents about the big data along with cloud architectures to focus on different datamining techniques to interact with Big data.

Some research work is going on different BDA applications and optimization. The G-Hadoop is used for large-scale distributed data processing on many data centers to attain high throughput and fault tolerance.

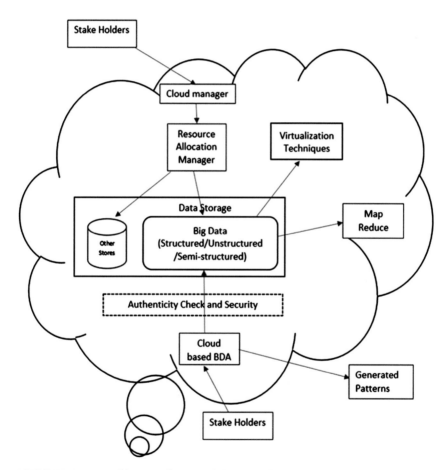

FIGURE 33.1 An architecture diagram of cloud-based big data.

BDA is also used for high performance by reducing the cost by resource allocation. To achieve this AROMA is a system which is used to automate Hadoop configuration.

Cloud environment dynamically manages the virtual machines (VMs) by deployment as per requirements. OpenNebula provides resource provisioning on local infrastructure as well as an Amazon EC2 cloud layer between service and physical infrastructure widen the classical benefits of VM to distributed infrastructure to make the system fault-tolerant.

Big data and analytics help in the education sector also. An enormous quantity of data is generated from learner's interaction with other parties. That data can be stored and processed to get some useful intonations to take the decision. Like by the activities of the learner we can predict his or her level of performance and take some necessary actions to improve the quality of learners. Here we get the knowledge development by interaction between the learners and by which it improves the learning capabilities.

There is a linked tool called social networks adapting pedagogical practice (SNAPP) is already being used by many educators to monitor and examine learners' activities in a group [13].

A wide variety of BDA frameworks are delivered to different companies to serve better services. They are Google cloud platform (GCP), Microsoft Azure, Amazon Web Services (AWS), etc. [15].

33.4   CONNECTION AMONG BIG DATA, CLOUD, AND DATA SCIENCE

1. **Big Data and Cloud:** Cloud provides different services by isolating the complexity and challenges concerning building a scalable flexible self-service application. Big data also required is the same as a cloud. Hadoop covers the complexity of distributed processing from the end-user. The distributed technology of CC allows effective management of big data. So for analyzing the data and acquiring required pattern from Cloud-based big data the parallel computing can be added advantage [12, 19].

2. **Big Data with Data Science:** For the organization of large data, analysis and storage are the two essential factors. The first large volume of data is generated. To store these data is not economically feasible. This is the major concern and security is the biggest issue in CC which comes into the picture. For this hiring is required for skilled data analysts, data engineers, and above all data scientists. The data scientist is a or group of persons who must possess (analysis, statistics, and programming) and he/she is also expected to work on new platforms which is the requirement of the organization [14].

3. **Data Science and Cloud:** For an organization of large data, the analysis part and storing the data are the two important challenges. At first, a large volume of data is generated. Then to store the data is not economically feasible. So, it is a major concern and also the security is the biggest issue in CC. It needs the employment of skilled data analysts, data engineers and data scientists. The data scientist is a group of persons who must possess (analysis, statistics, and programming) and he/she is also expected to work on new platforms which are the requirement of an organization. Figure 33.2 represents the cloud, big data, and their roles.

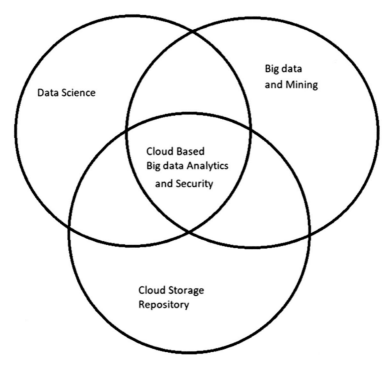

FIGURE 33.2   Cloud-based big data analytics.

4. **Cloud-based Big Data Analytics (BDA):** The connection of big data, cloud, and data science give cloud-based BDA and it helps for data-driven decision (DDD) making. DDD refers to the practice of taking decisions on the analysis of data rather than purely on intuition. Businesses can produce a giant amount of insights, values, and efficiency using DDD. In Figure 33.2, it presented about the

cloud-based BDA and its relation with cloud storage, data science, and big data mining.

Cloud-Based Big Data Analytics = Big Data + Data Science + Cloud

33.5 BIG DATA IN CLOUD

Big data business drivers: A number of business drivers are at the core of this success and explain why big data has quickly risen to become one of the most coveted topics in the industry. Seven main business drivers can be identified [18, 19]:

- The digitization of society;
- The plummeting of technology costs;
- Connectivity through cloud computing;
- Increased knowledge about data science;
- Social media applications;
- The upcoming internet-of-things (IoT); and
- Cloud computing.

CC is the technology which delivers different types of services or resources through the Internet. These resources contain many tools and applications like data storage, servers, DBs, networking, and software, etc. Different deployment models are as follows [12]:

- **Private Clouds:** They are reserved for the specific client usually one business or organization.
- **Public Clouds:** The services are available to all via the internet. Some of the service providers provide services free and some are taking charges based on their use.
- **Hybrid Clouds:** It is a combination of the private and public cloud makes hybrid clouds. This type of model gives more flexibility and optimizes the user's infrastructure and security.

The different CC techniques implementation using file systems such as Google file system and HDFS to handle big data is represented in Figure 33.3.

The traditional platform could not handle such big data. So CC has been used to accommodate these DBs. Many architectural decisions are considered before handling big data are: performance, scalability, reliability, availability, location, and placement, sensitive data, disaster recovery.

1. **Cloud Analytic Providers:** They help the organizations to generate reports from many platforms in the cloud. The provider can facilitate many features such as grouping, clustering, and forecasting using its different analytical tools. The cloud analytic providers generally have advanced tools to find complex reports.

The popular providers are:

- **Microsoft power BI;**
- **Host analytics;**
- **Zoho reports;**
- **Domo;**
- **IBM Cognos analytics.**

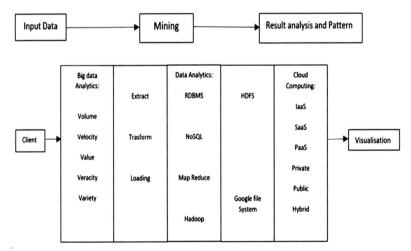

FIGURE 33.3 Patterns by implementing cloud computing techniques on big data.

And the cloud analytics tools available are:

- AWS analytics products;
- Google cloud analytics products;
- HD insight;
- Data lake analytics;
- Machine learning studio.

2. **Data Analytics Tools:** The importance of data analytics in the market has generated many openings worldwide. Many data analytics tools are the open-source tools and they are popular, user-friendly, and performance-oriented.

They are R Programming, Tableau Public, Python, SAS, Apache Spark, Excel, RapidMiner, KNIME, QlikView, Splunk, etc.

33.6   CLOUD-BASED BIG DATA ANALYTICS (BDA) FRAMEWORKS

The Apache Pig, Hive are used for processing big data whereas Hadoop, Spark is also used for storing the data. Hbase and HadoopDB are used for storing any structured data [16] (Figure 33.4).

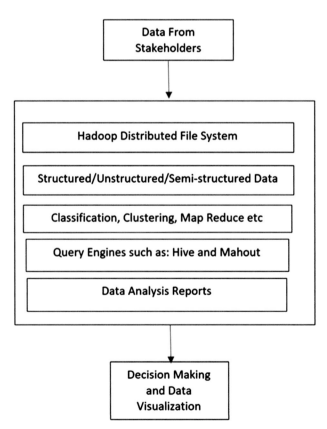

FIGURE 33.4   Flowchart: Data from stackholders to data visualization.

Figure 33.4 represents the flow of data from the different stockholders to data visualization through different stages such as distributed file system, classification of data, etc.

Different data mining tools are used for mining knowledge from a large dataset but here effective response within a time frame is the criteria. So BDA needs processers with high performance for giving better results. So BDA with cloud provides an effective solution in a small span of time [11]. There are many models for this such as:

- **Analytics Infrastructure:** The analytics infrastructure applications, utilities, services are available for performing data preparation, esti-mation, and validation.
- **Analytics Platform:** It clouds it provides an environment which provides service to the developer to build applications.
- **Analytics Software:** There are many analytics software such as Sisense, Looker, Periscope, Domo, etc., available as service to the users. However, a right analytic tool to be selected for a particular business.

1. **Hadoop:** The Hadoop distributed file system (HDFS) is developed in java. The Hadoop framework can process a large set of data which resides in different clusters of nodes.

   - **Hadoop Distributed File System (HDFS):** When a large amount of data available in different machines across different clusters the HDFS supports to hold data and provides efficient access and process them. Hadoop having a command-line interface which allows interacting with the HDFS. It also provides sufficient authentication to access different nodes [10].

2. **Map Reduce:** The tasks it mainly contains are Map and reduce. The map takes a data set and then converts it into another data set such as independent chunks so that it can be sorted out. Then the output of the Map can be used to reduce into a smaller set of data. Both input and output data are stored in the file system.
   The main advantage of Hadoop is it scalable, robust, and fault-tolerant. But is not suitable for smaller data sets [10].

3. **Spark:** It is an open-source cluster computing engine used for processing and analyzing large data sets. The spark runs on the top of the Hadoop to improve computational speed. In spark, it introduced memory caching abstraction which allows performing multiple operations on

the same data in parallel. The main components are spark core; spark SQL, cluster manager, MLlibrary, GraphX. The spark supports multiple languages and supports advanced analytical framework.

The components of the spark are:

- **Spark Core:** The spark core provides much functionality such as dispatching, scheduling, and I/O operations.
- **Spark SQL:** The important part of spark is the spark SQL which supports the structured and semi-structured data.
- **Spark Streaming:** The RDD (resilient distributed datasets) is the basic data structure of spark. Each data set of RDD can be grouped into no. of logical partitions.
- **Library of Machine Learning (ML):** Spark MLLibrary works with Python NumPy and R language. So spark supports ML in the data science area.
- **GraphX:** It is used for graph analysis and computation. We can view data in different graphs, join graphs. It is a part of the Apache Spark project which contains a variety of graph algorithms.

33.7   ISSUES IN BIG DATA WHILE USING CLOUD STORAGE

Issues in big data with cloud: The CC having many security issues. As it contains many technologies such as DBs, big data handling, and maintenance, operating systems (OSs), resource scheduling, network, load balancing, memory management, etc.

While handling the big data in a cloud many CC techniques are required. A set of analytical tools are used in the cloud to help the clients extracting information from the big data.

The challenges in CC while handling big data are:

- Protection and security of data.
- The cost of managing big data in cloud.
- Experts required to handle data.
- Government policies and procedures to control the assets maintained in the cloud.
- When data is moved from internal memory to cloud then it leads to the compliance problem.
- Managing data present in multiple clouds.

- Performance affected by the CSP.
- Data migration problem.

To maintain big data in the cloud is an easy way facility as per resources are concerned but also it brings challenges. So suitable solutions must be applied to handle the challenges.

As because many types of data are produced from different sources like social media data, sensory systems, advertisements, news, pictures. Those data of different categories of semi-structured, structured, unstructured, audio, and image-based and of in different formats.

To verify them and representing them in a common form is a challenging task. When such data is growing more and more so, to maintain the data and mine the knowledge from them is also challenging as because for every storage have its own limitations.

Also, to transport data from one cloud to another will take a longer time. Even though we have faster-accessing communication speed with a high capacity memory of present days then also it will be challenging for the growth of data day by day.

In big data, its management involves a lot of disciplines such as warehousing, integration, data quality, data governance, administration of data, data processing, etc. Here issues of metadata, accessing data, updating, and reference is a problem.

It needs am extensive parallel processing for processing of huge data in exabytes and also new analytic algorithms are required for getting information within a time frame. So, here necessary care to be taken for processing query in a stipulated time [17, 18].

When the analysis of big data is given to a third party due to the unavailability of experts or tools for processing then it will increase the risk of safety. It is because many times the data is analyzed in a third-party data center.

Nowadays the data volume is increasing at an exponential rate and the hardware requirement for using data is also quite large. So, the analytical algorithms which are used must support the present data and also to the complex dataset of the future. To process such data a scalable architecture and a high-level concurrency is required.

The different service models used in the cloud are Saas, Paas, and Iaas and deployment models are public-private community and hybrid. When the data are organized in cloud, service provider faces many security issues on providing software and infrastructure, whereas the customer also faces the security issues.

33.8 CHALLENGES IN BIG DATA ALONG WITH CLOUD

Nowadays cloud storage is used for almost many sectors, so the data security is very important. The security issues are of different levels:

- Issues related to network protocol and security in distributed nodes;
- Issues in DB management, operating systems;
- Issues during authenticating a user;
- Issued on giving access rights to the nodes;
- Protecting data;
- Resource allocation and scheduling;
- Data transaction and management.

33.8.1 SECURITY CHALLENGES

While handling the big data in the cloud and even the experts face many challenges such as:

- Possibility of generating fake data, it is because the fake data may be generated by cybercriminals and uploaded somewhere. So, here these are to be clarified using fraud detection techniques.
- The outside may access the mapper codes which may lead to a faulty process. It is because big data maintained may not have enough security layers to stop intruders.
- When we use encryption and decryption on sensitive data then it slows down the performance. But if we don't apply these security measures on sensitive data then it loses data security.
- Sometimes the IT specialists may be corrupted who may mine the sensitive information and may sale it for personal benefits.
- The security audit may not be present during handling the data.
- As the NoSQL is popular nowadays so due to maintaining the NoSQL DBs the internal security will be reduced [1, 19].

Many IoT devices generate a huge amount of data which needs the cloud for storage. We know nowadays in most of the sectors we find automation using the sensory systems such as smart grid, e-health, hospital, agriculture, etc. Most of the IoT devices generate the data which is stored in the cloud in general and analyzed using the analytical tools [6, 20].

33.9 CONCLUSION

The brief observation on handling big data in the cloud and then focused on the security issues have done. Here when we outsource the cloud and data analytics then analyzing the data security will be a point of discussion. So, we have identified the benefits of the cloud to handle big data and tried to represent the challenges when we maintain data and use it for analysis.

We can plan to identify the efficient security mechanisms in the cloud and choose the suitable one for handling the variety of big data. So, data security and high functionality together suitable for cloud-based BDA.

It is important to implement sometimes the cryptography so that various tested practical data sets can be analyzed [7].

In the cloud when big data maintained it will have the challenges of security in different types such as: when there is distributed node and communication between them is exist then there is a need for proper authentication. It can be at different levels such as cryptic techniques, communication in different nodes, authentication required in the node and apps.

There is a need for proper data storage so that it will not tamper during the communication. So we can find many challenges and between big data and cloud and them summarize them.

Storing the big data using a traditional storage media is very difficult as because the hard disk drive will be failed. Also, as there is a quick increase in the size of data so the old-style approach will not be suitable. There are a variety of data to maintain in the cloud is structured, unstructured, semi-structured, and heterogeneous data. Sometimes the data compression techniques can be used to compress data so that volume can be reduced [8, 17, 18].

KEYWORDS

- **Amazon web services**
- **big data**
- **cloud analytics**
- **data science**
- **data-driven decision**
- **GraphX**
- **Hadoop**
- **map reduce**

REFERENCES

1. Bhanu, S. M., & Koushik, A., (2017). A study of big data analytics in clouds with a security perspective. *International Journal of Engineering Research and Technology (IJERT)* (Vol. 6, No. 01). ISSN: 2278-0181. http://www.ijert.org (accessed 20 July 2020).
2. Neha, M., & Rajesh, P., (2017). Issues and challenges in convergence of big data, cloud, and data science. *International Journal of Computer Applications (0975–8887), 160*(9).
3. Manju, S. (2017). Big data analytics challenges and solutions in cloud. *American Journal of Engineering Research (AJER)* (Vol. 6, No. 4, pp. 46–51). e-ISSN: 2320-0847 p-ISSN: 2320-0936.
4. Domenico, T., (2013). *Clouds for Scalable Big Data Analytics*. Published by the IEEE Computer Society, 0018-9162/13/$31.00 © IEEE.
5. Qinghua, L., Zheng, L., Maria, K., Liming, Z., & Weishan, Z., (2015). *CF4BDA: A Conceptual Framework for Big Data Analytics Applications in the Cloud* (Vol. 3, pp. 2169–3536). IEEE. Translations.
6. Heshan, K., Ibrahim, K., Abdulatif, A., Zahir, T., & Xun, Y., (2016). Secure data analytics for cloud-integrated internet of things applications. *IEEE Cloud Computing*. Published by the IEEE Computer society, 2325-6095/16/$33.00 © IEEE.
7. Somayeh, S. M., & Amjad, F., (2019). Toward securing cloud-based data analytics: A discussion on current solutions and open issues. *IEEE Access*. Digital object identifier 10.1109/ACCESS.2019.2908761.
8. Nabeel, Z., Al-Haj, A., & Sufian, M. K., (2017). Cloud computing and big data is there a relation between the two: A study. *International Journal of Applied Engineering Research,* (Vol. 12, No. 17, pp. 6970–6982). ISSN: 0973-4562.
9. Al-Shawakfa, E., & Hiba, A., (2018). An empirical study of cloud computing and big data analytics. *Int. J. Innovative Computing and Applications, 9*(3).
10. Saneh, L. Y., & Asha, S., (2017). Review paper on big data analytics in cloud computing. *International Journal of Computer Trends and Technology (IJCTT), 49*(3).
11. Sasikala, M., (2017). An investigation of cloud computing in big data analytics. *International Journal of Computer Science and Mobile Computing* (Vol. 6, No. 4, pp. 407–412). ISSN: 2320-088X.
12. German, T., Nicolas, F., & Svetan, R., (2019). A cloud-based framework for shop floor big data management and elastic computing analytics. *Computers in Industry, 109*, 204–214.
13. Mohammad, S., Shamim, H. M., Amril, N., Ghulam, M., & Atif, A., (2019). Harnessing the power of big data analytics in the cloud to support learning analytics in mobile learning environment. *Computers in Human Behavior, 92*, 578–588.
14. Bala, M. B., & Shivika, P., (2017). Challenges and benefits of deploying big data analytics in the cloud for business intelligence. *International Conference on Knowledge Based and Intelligent Information and Engineering Systems, KES2017.* Marseille, France.
15. Subia, S., & Samar, W., (2018). Performance analysis of big data and cloud computing techniques: A survey. *International Conference on Computational Intelligence and Data Science (ICCIDS 2018).* 1877-0509 © The Authors. Published by Elsevier Ltd.
16. Samiya, K., Kashish, A. S., & Mansaf, A. (2015). *Cloud-Based Big Data Analytics: A Survey of Current Research And Future Directions*. Department of Computer Science, Jamia Millia Islamia, New Delhi.

17. Dimpal, T., & Pradeep, T., (2018). Integration of cloud computing and big data technology for smart generation. *8th International Conference on Cloud Computing, Data Science and Engineering (Confluence)*. Noida, India. Publisher: IEEE.
18. Solanki, V. K., Kumar, R., & Khari, M., (2019). In: Balas, V. E., (ed.), *Internet of Things and Big Data Analytics for Smart Generation*. Springer.
19. Khari, M., (2018). A comprehensive study of cloud computing and related security issues. In: *Big Data Analytics* (pp. 699–707). Springer, Singapore.
20. Lekha, R. N., & Sujala, D. S., (2014). Research in big data and analytics: An overview. *International Journal of Computer Applications (0975-8887), 108*(14).

CHAPTER 34

ROBIARM: Vision-Based Object Detection System

VIKRAM PURI,[1] SANDEEP SINGH JAGDEV,[2] CHUNG VAN LE,[1] and BHUVAN PURI[3]

[1]Duy Tan University, Da Nang, Vietnam

[2]Ellen Technology (P), LTD, Jalandhar, Punjab, India

[3]D. A. V Institute of Engineering and Technology, Jalandhar, Punjab, India, E-mail bhuvanpuri239@gmail.com

ABSTRACT

Research in the field of robotics is to attract industry and academia to tackle real-world hurdles. The synergy of robotics with other technologies such as the internet of things (IoT), artificial intelligence (AI), augmented reality (AR), and computer vision (CV) is a key to unlock the potential of robotics. In this study, we propose a robotic arm system enabled with computer vision and sensor to detect, grasp, and move objects from one location to another location. Two different sensors ultrasonic and infrared (IR) sensors are used to increase object detection accuracy. OpenCV libraries installed in the main controller raspberry pi for image processing. In addition, object detection and location extraction techniques are executed with the aid of image processing methods. The result outcomes from our proposed system show that the robotic arm is more accurate and stable compared to other studies.

34.1 INTRODUCTION

Nowadays, Robotics has made human life very convenient, not limited to industrial applications, but also in the research area of education, medical science, and entertainment [1]. Many companies namely Boston dynamics and researchers have worked and developed a bunch of robots to fulfill various requirements according to their research area and also to create a fruitful synergy of human-robot interaction. In addition, robots can easily deal with complex tasks that are difficult to deal for humans. Presently, numerous robots have captured the market but still, the robotic arm is one of the most favorable robots [2]. Usage of Robotic ARM is the most important tool in the factories for the assembling process especially car assembly, and big manufacturing machines. In order to control coordination and movement of the robotic arm, accuracy, stability, and precision play an important role. Figure 34.1 represents robotics connected with other related technologies.

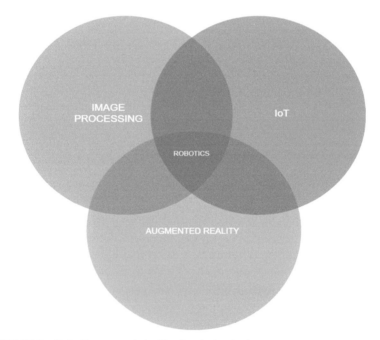

FIGURE 34.1   Robotics connected with other technologies.

Object recognition technology has rapidly increased in many domains such as object detection in 3D and 2D images, movement detection and with the handshaking of AI, these robots behaving like humans are called humanoid robots (HR). Moreover, HR also learns from its mistakes and improvises the skills with the use of reinforcement learning (RL). These days' robots provide a helping hand to humans in their day to day life.

Vision is one of the intelligent core technologies in robotics. The research area in the vision called "Computer Vision" is now considered from a scientific point of view for investigating how artificial computer vision can make a robot as human and what algorithm underlies it [3].

Open-source computer vision (OpenCV) is a real-time library program developed by Gary Bradsky in 1999 (OpenCV Introduction). It is an open-source library for both educational and commercial purposes. It supports C, C++ and Python interfaces and optimizes nearly 2500 algorithms [4]. OpenCV plays a supportive role in the development of computer vision into a new futuristic world and enables millions of people to enhance their limits in productive work.

Many researchers have proposed and developed the robotic arm and Visual system over the last few decades. Furuta [5] proposed a method to control trajectory tracking using a sensor-based feedback system. With the use of a laser beam, the proposed algorithm is used to achieve the desired coordinates for the robotic ARM. Manasinghe [35] proposed an algorithm for the industrial robotic arm to contour problem Cartesian velocity and joint torque. In this work, simulation is established to compute coordinates of each joint in the robotic arm. Koga [6] developed a virtual model for the robotic arm to calculate the joint coordinates when it picks up and places the object. Efe [7] presented a scheme to adjust the robotic arm fuzzy sliding controller with the use of the adaptive neuro-fuzzy inference system (ANFIS). Wang [8] proposed a robotic arm that's fixed on the mobile robot to detect the signs or numbers. Image Processing and detection are done through the use of microcamera fitted on the robotic arm. Juang [1] developed a robotic arm system to grab the objects with the visual recognition this system is equipped with two webcams: one webcam is employed to catch commands on screen and the other one is used for word recognition.

The main concept of this chapter is to give vision capability to the robotic arm through the use of image processing. An ultrasonic sensor is also integrated with the camera on the robotic arm to check and calculate the distance between the object and the robotic arm.

34.2 METHODOLOGY

34.2.1. ARCHITECTURE

Our proposed work is categorized into different sections namely image capturing, processing, sensing the distance of the object, flow of algorithm and robotic arm functions (Figure 34.2).

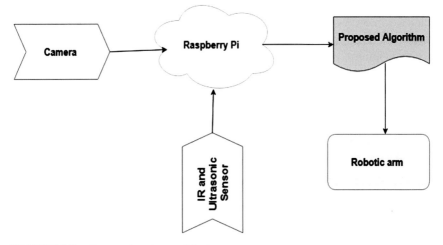

FIGURE 34.2   Proposed system architecture.

Raspberry Pi serves as the backbone of our proposed system. Image processing is done through the use of a camera and OpenCV. Raspbian operating system(OS) (Raspberian Operating System) used in the raspberry pi which based on Debian operating system provides over 35000 pre-installed packages and pre-compiled software such as python, sonic-pi, java. In addition, it's more than a pure OS. OpenCV, installed through Linux commands, provides library packages to process images taken through the camera installed with raspberry pi. These libraries are based on the python. Two different sensors namely ultrasonic and IR sensor check the distance and location of the object the sensor works on the principle to calculate the distance of the reflected wave. The formula for distance calculation of reflected wave is:

$$D = \tfrac{1}{2}\,T * SS \qquad\qquad (1)$$

where; D is distance, T is time, and SS is the speed of sound. SS varies with the humidity and temperature.

For the robotic arm, Servo motors are employed. These motors work on the pulse width modulation (PWM) principle and control through three wires: (1) power, (2) ground, and (3) signal. In PWM, there are three different pulses in which motor will work, minimum pulse, maximum pulse and the last one is repetition pulse and rotates around 180° (both sides 90°). PWM pulse decides the position of the motor shaft stamped with time duration. In the proposed work, three servo motors are employed inside the robotic arm, one for up and down and another one is to grasp the object. The last one is for rotating the arm.

34.3 CIRCUIT DIAGRAM

In the proposed study, Raspberry Pi plays a major role in controlling the robotic arm and capturing the images whilst processing it according to the requirement. Raspberry Pi is a small credit card based computer operated on Debian based OS. Table 34.1 represents the technical specifications of the raspberry pi.

TABLE 34.1 Technical Specification of Raspberry Pi

SL. No.	On-Chip	Feature
1.	CPU	Quad-Core ARM Cortex
2.	RAM	2 GB
3.	Storage	MicroSDHC card
4.	Power	5 Volt-2 ampere
5.	Graphics	Broadcom VideoCore

The camera is connected to Raspberry Pi through the USB port. IR sensor is based on three pins: (1) power, (2) ground, and (3) output and connected to raspberry pi general-purpose input-output (GPIO) pin. Three servo motors are also connected to GPIO PWM pins (see Figure 34.3).

Algorithm 1: Proposed Workflow

capturing images of Objects through the camera ;
processing these images through OpenCV ;
while *checking the location of object through sensors* **do**
    if *Detection at correct distance* **then**
        Robot Arm grasp object;
    else
        not move;
    end
end

Algorithm 1 presents our proposed workflow. In our proposed work, the Camera captures the image which is processed through the use of installed OpenCV in the raspberry pi. Sensors check the location of the object. If the object is within the range of the robotic arm, it will grasp otherwise robotic arm will not move.

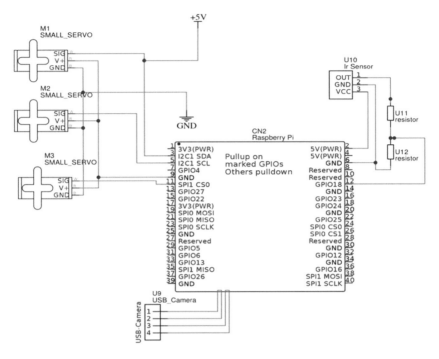

FIGURE 34.3   Circuit diagram of proposed work.

34.4   RESULTS AND DISCUSSION

In this section, Figures 34.4 and 34.5 represent the outcome of the results from our proposed work. In Figure 34.4, the terminal shows the direction of the robotic arm which moves forward or backward. Figure 34.5 shows the detection of an object with a yellow screen. The most noteworthy point of this proposed system is that it works remotely through the use of Internet. Through the use of SSH, the robotic arm captures images and grasps the crucial objects.

34.5   CONCLUSION

In this study, a vision-based robotic arm is proposed. This proposed system provides an experience to tackle real-world problems such as remote surveillance, tasks impossible for humans. The integration of day to day life problems with complex visions can improvise the image processing research. OpenCV libraries installed on the raspberry pi permit to focus on image processing with minimal labor. In this study, Robotic arm is equipped with a camera, ultrasonic sensor and IR sensor that increases the accuracy for the selection of objects from exact coordinates. Object detection and location extraction techniques are executed with the aid of image processing methods namely object extraction techniques, matching pre-installed templates.

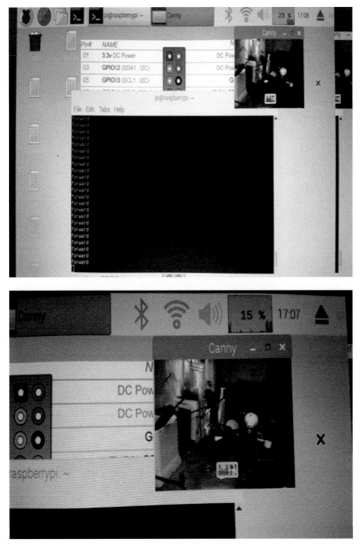

FIGURE 34.4 Left side: Direction of robotic arm; right side: detection of object.

FIGURE 34.5   Robotic arm enabled with camera.

In future work, we would employ machine learning (ML) techniques to check the quantity and quality of the object and also classify the objects belonging to a category.

KEYWORDS

- **adaptive neuro-fuzzy inference system**
- **artificial intelligence**
- **general-purpose input-output**
- **internet of things**
- **OpenCV**
- **raspberry pi**
- **robotic arm**

REFERENCES

1. Juang, J. G., Tsai, Y. J., & Fan, Y. W., (2015). Visual recognition and its application to robot arm control. *Applied Sciences*, 5(4), 851–880.
2. Manigpan, S., (2010). *A Simulation of 6R Industrial Articulated Robot Arm Using Neural Network.* Doctoral dissertation, University of the Thai Chamber of Commerce.

3. Ejiri, M., (2007). Machine vision in early days: Japan's pioneering contributions. In: *Asian Conference on Computer Vision* (pp. 35–53). Springer, Berlin, Heidelberg.

4. Bradski, G., & Kaehler, A., (2008). *Learning Open CV: Computer Vision with the Open CV Library.* O'Reilly Media, Inc.

5. Furuta, K. A. T. S. U., Kosuge, K. A. Z., & Mukai, N. O. B. U., (1988). Control of articulated robot arm with sensory feedback: Laser beam tracking system. *IEEE Transactions on Industrial Electronics*, *35*(1), 31–39.

6. Koga, M., Kosuge, K., Furuta, K., & Nosaki, K., (1992). Coordinated motion control of robot arms based on the virtual internal model. *IEEE Transactions on Robotics and Automation*, *8*(1), 77–85.

7. Efe, M. Ö., (2008). Fractional fuzzy adaptive sliding-mode control of a 2-DOF direct-drive robot arm. *IEEE Transactions on Systems, Man, and Cybernetics, Part B (Cybernetics)*, *38*(6), 1561–1570.

8. Wang, W. J., Huang, C. H., Lai, I. H., & Chen, H. C., (2010). A robot arm for pushing elevator buttons. In: *Proceedings of SICE Annual Conference 2010* (pp. 1844–1848). IEEE.

9. Munasinghe, S. R., Nakamura, M., Goto, S., & Kyura, N., (2001). Optimum contouring of industrial robot arms under assigned velocity and torque constraints. *IEEE Transactions on Systems, Man, and Cybernetics, Part C (Applications and Reviews)*, *31*(2), 159–167.

10. *OpenCV Introduction.* https://opencv.org/ (accessed 20 July 2020).

11. *Raspbian Operating System.* https://www.raspbian.org/ (accessed 20 July 2020).

CHAPTER 35

Security Analysis and Fraud Prevention in Cloud Computing

MD. IMAN ALI,[1] PRANAY JHA,[1] ASHOK SHARMA,[2] and
SUKHKIRANDEEP KAUR[3]

[1]*Research Scholar, School of Computer Application, Lovely Professional University, Phagwara, Punjab, India, E-mail: mdimanali@gmail.com (M. I. Ali), pranay1988jha@gmail.com (P. Jha)*

[2]*Associate Professor, School of Computer Application, Lovely Professional University, Phagwara, Punjab, India, E-mail: drashoksharma@hotmail.co.in*

[3]*Assistant Professor, School of Computer Science and Engineering, Lovely Professional University, Phagwara, Punjab, India, E-mail: sukhkirandeep.23328@lpu.co.in*

ABSTRACT

Cloud services nowadays became an essential requirement for the IT infrastructure to meet the demand for any organization's growth and dynamic changes. Organizations are focusing more on the business goal than managing IT infrastructure as it was there in the earlier days. They are moving the workloads to cloud services and using IAAS, SAAS, or PaaS models as per the requirement. IT infrastructures are moving from an on-premises data center to a cloud data center. By doing so, organizations are leveraging the benefits in terms of cost as they go for services, easy management, and flexibility to meet the unexpected demands. There are many other key reasons which attract customers towards cloud services.

Despite of having several benefits, cloud security is the major challenge that have been raised by most of the customers. Among all cloud security challenges, data breach is one of the topmost concerning points. There are many researchers have already been done in this area and different security frameworks have been implemented to make the cloud environment secured. Still we face cyber-attacks in every second.

To shield the cloud environment in a more secure way, the cloud service provider (CSP) and the consumer must have a concrete data security model and well-framed compliance policies in place. This chapter has discussed about the open concerns and existing frameworks of the cloud security. This chapter has demonstrated the security loopholes and the protection mechanism. Using the security logs from different cloud customers, compliance, and security incidents have been analyzed. On the basis of analyzed data and flows of traffic in existing security policies, a security framework has been proposed which can be implemented by the CSP and consumers.

35.1 INTRODUCTION

Organizations always want the data to be secured which is stored over the internet and outside the administrative control. Visibility of the security posture on real-time basis to detect vulnerability and protect the data is very critical requirement. Traffic movement between multiple clouds is also one of the most required things for cloud services but there is always lot of risk involved during such data transmission. Risk can be from the vulnerability of security compliance or breach of data by unauthorized users [8]. Important point that needs to be considered is the classification of data and defining the boundaries of the data traverse between cloud in different segment or from on-premised to cloud [9]. Few of the major issues in cloud infrastructure are listed below:

- **Visibility:** Visualization of the entire cloud environment in a single dashboard and like traffic pattern, activity monitoring, behavioral analysis, correlation of activity on cloud. Correlation of attack and threat intelligence.
- **Threat Detection and Alert:** Identify the threat and detect on time, the establishment of the threat identification with correlation of traffic and application and accessibility [7]. Suspicious activities on cloud. Depending upon the threat timely alert before to take action. Investigation of the incident happened in cloud and impact analysis.
- **Security Compliance:** This on the cloud is a major area to focus because of outside of physical security boundaries [2].

There are lot of researches already conducted and applied in the real-time environment, still cyber-attacks are happening every moment. As per the study, there is a cyber-attack in every 39 seconds [1].

35.1.1   RECENT SECURITY BREACHES

- **Uber Crypto Mining Attack:** One of the biggest attacks was Uber Crypto mining attack which has caused the loss of huge amount because of breach in cloud security controls.
- **Aadhaar India National ID Database (DB)**: UIDAI data was exposed for over a billion users. This was also a major breach which impacted huge credibility of Government of India along with the Citizens privacy.
- **Cambridge Analytica**: Unauthorized accessing the Facebook user data and using it for political benefit was a major scandal. This had exposed the security of data available on the internet.
- **Marriott Starwood Hotels:** There was a data security breach at Marriott of their guest DB and approve 500 million guest data was compromised.
- **Exactis**: Florida-based company collected data and used it for targeted advertising. This was alarmed as data breach for exposing the personal information of end-users
- **Facebook:** Cyber-attack on Facebook DB in September 2018 which compromised around 50 million user's personal information and account details.

There are plenty of such incidents happen in bits and pieces over the years. There is no such assurance of data security in cloud by cloud service provider (CSP). Even though study, analysis, and development are happening continuously but again there is a need of continuous analysis, development [3] and protection of critical infrastructure (CII). A clear framework of cloud security model is required to maintain the security compliance as per organization IT security governance policy. Cryptographic solution is also one of the important areas to focus for cloud computing (CC) infrastructure [4].

Our intent of this study is to identify the vulnerabilities, Security loopholes and open issues of cloud services in existing environment which can cause cyber-attack and data leakage. Then compare the result and solution. On the basis of analysis and comparison, proposed a framework of security policies.

35.1.2   RELATED WORK

Since every aspect of security threat cause to loss of data. Hence, data security is always an important area when we discuss about the cloud security. Multiple research and studies have already been done in this area. Existing framework related to cloud security from different papers have been listed in this section.

- Cloud security accountability, integrity, authenticity, reliability, authorization is studied [14]. In this study cloud data environment classification and integrity of the data is studied. Concept, characteristics of the cloud security was prime focused.
- A new Security model is suggested based on chain ontology in a study.
- One study has made on the basis of OWASP attack category and mapped these attacks with cloud threat with STRIDE threat model [5].
- A framework is proposed for cloud security based on Cloud Security Ontology has been framed using protégé software [13]. Which includes cloud security threats, cloud security vulnerability, cloud risk, control, and relationship.

Based on literature review of different papers, Table 35.1 shows the related work in cloud security.

35.2   METHODOLOGY

Methodology of proposed work has been divided into four classifications:

1. Identification of customer and CSP;
2. Implementation of proposed security framework;
3. Data collection related to cloud security post-implementation of proposed security framework; and
4. Testing and analysis on the basis of collected information from Step 3 (Figure 35.1).

35.3   IMPLEMENTATION

Implementation of this study has been done according to methodology defined. This has been categorized in different sections as mentioned in subsections.

TABLE 35.1 Related Work in Cloud Security

Year	Title of Paper	Proposed Solution	References
2016	New approach for ensuring cloud computing security: using data hiding methods	In this study, the vital importance of the combined use of encryption and watermarking methods in the database are described in detail.	[19]
2017	A framework for improving security in cloud computing	This paper has proposed simple and yet powerful framework using homomorphic encryption of data to store and performsecure operations in the cloud. Gentry has proposed a fully homomorphic encryption which allows us to carry computations on encrypted data (ciphertext), leads to encrypted results. When the results are decrypted will match the actions performed on plain data.	[15]
2017	Achieving flexible and self-contained data protection in cloud computing	This paper has proposed a self-contained data protection mechanism called RBAC-CPABE by integrating role-based access control (RBAC), which is widely employed in enterprise systems, with the ciphertext-policy attribute-based encryption (CP-ABE). First, we present a data-centric RBAC (DC-RBAC) model that supports the specification of fine-grained access policy for each data object to enhance RBAC's access control capabilities.	[10]
2017	A dynamic prime number based efficient security mechanism for big sensing data streams	Dynamic prime number based security verification (DPBSV) scheme for big data streams	[12]
2018	Secure framework for data security in cloud computing	This paper aims at the security model for cloud computing which ensures the data security and integrity of user's data stored in the cloud using cryptography	[11]

TABLE 35.1   *(Continued)*

Year	Title of Paper	Proposed Solution	References
2018	Formulating a security layer of cloud data storage framework based on multi-agent system architecture	Proposed a security framework and MAS architecture to facilitate security of cloud data storage. This security framework consists of two main layers as agent layer and cloud data storage layer. The propose MAS architecture includes five types of agents: UIA, UA, DERA, DRA, and DDPA.	[16]
2018	Survey on data safety and effectiveness estimation in cloud computing using cloud computing adoption framework	This paper has studied multiple frameworks by which the data will be compressed, and the data storage system will be enhanced by using optimization techniques such as Dragonfly algorithm. The data after undergoing wave let transform will be secured by generating key. The dragonfly algorithm to establish the key generation for all the compressed data. The AES and the Dragonfly algorithm are used to ensure the improved data storage. The compressed data can be retrieved by using the generated key.	[18]
2018	A framework for data storage security in cloud	This paper proposed a framework which integrates the efficient server selection, three-stage authentication, ECC, and CMA-ES to achieve the security of data in cloud.	[17]

35.3.1   IDENTIFICATION OF CUSTOMER AND CLOUD SERVICE PROVIDER (CSP)

Customer has been identified who is already running their workload on cloud since long time so that we can have data collection in a granular way. This has also been focused during the identification that customer must have verbose logging available which can be used to test and analyze the framework in later phase.

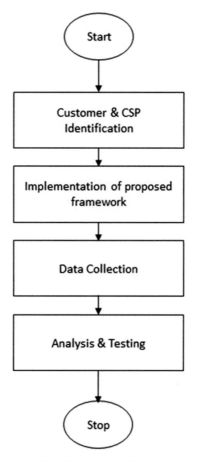

FIGURE 35.1 Stages of proposed methodology architecture.

The same customer with multiple CSPs has been chosen. These providers include Amazon, Azure, and Google. This will help in the categorization and identification of security incidents from different platforms.

35.3.2 IMPLEMENTATION OF PROPOSED SECURITY FRAMEWORK

To protect the data in the cloud which resides in the shared infrastructure model, consumer, and CSP both should give maximum attention in cloud security solution [6]. Implementation of cloud security model should be precisely designed in line with on-premises data center security solutions. Ideally, best practices should be followed and in-place before cloud before adopting any services from cloud. Security assessment should get conducted as per the classification of the data hosted in cloud. Model prototype design should look like as mentioned in Figure 35.2.

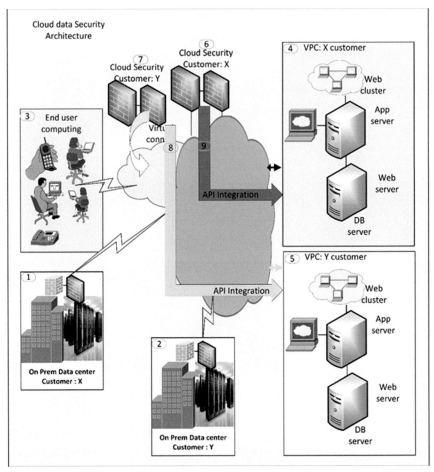

FIGURE 35.2   Cloud data security architecture.

Indexes as shown in Figure 35.2 has described in Table 35.2.

There are 2 different customer infrastructure model is shown as customer X and customer Y. Customer X has services running within the on-prem data center as well as on Cloud shared services infrastructure similarly Customer Y has same deployment model. Cloud Security solution is deployed with API integration with X and Y both customer's VPC.

TABLE 35.2 Customers from Various Datacenter Strategy

Index	Description
1.	On-Prem Data center infrastructure for Customer-X
2.	On-Prem Data center infrastructure for Customer-Y
3.	End User computing and accessing services from different cloud infrastructure
4.	Cloud VPC for Customer X
5.	Cloud VPC for Customer Y
6	Cloud Security product for Customer X
7.	Cloud Security Product for Customer Y
8.	API integration between Cloud security solution for Customer X cloud shared model with Security solution
9.	API integration between Cloud security solution for Customer Y cloud shared model with Security solution

Architecture diagram clearly indicates that whenever any organization intends for any cloud services, three important aspects need to focus as below:

- Classification of the data to be hosted in shared service model;
- Analysis of cloud security model; and
- Implement proper security solution as per organization need.

Any of the security vulnerability and flaw in the architecture can lead into major cyber-attack and can result into major loss of data any organization's reputation.

Apart from this, there should be some sort of cloud security solution implemented and available in the environment which can proactively monitor and check the compliance and report a violation of security compliance which could lead to cyber incidents. One of the Security solutions has been deployed in this research study in the consumer environment. Dome9 solution has been deployed in the customer's perimeter which will further fetch the security logs from all CSPs being used by customers for different workloads. In our scenario, we are using Amazon, Azure, and Google cloud platform (GCP). Based on the logs, Dome9 analyzes and keeps tracking of malicious behavior and issues which can lead to security breach. Below parameters have been tracked and filtered out using the Dome9 solution.

- Inventory of all workloads from different service providers.
- Compliance for all service providers and workloads.

- Incident reported on Network, IAM, and storage services.
- Identity and access management
- Policy database of workload running on AWS and Azure.
- DevOps environment
- Unauthorized access

In the next phase, we have collected the data from different sources which also includes from Dome9 which is the main source.

35.3.3   DATA COLLECTION

Using the Amazon and azure cloud infrastructure for a customer, collected the required data with the help of Dome9 cloud security product. We have tested the impact in the cloud environment from each level. Several tests have been done and found the flaws of impacted areas. The test and analysis can be done based on cloud security products like Redlock from Palo Alto network, IBM QRadar, and Dome9 from the checkpoint. These products are purely designed for cloud security. Here we have used Dome9. Cloud security analysis and vulnerability assessment are done based on this tool.

In order to test the security model of the data residing in the cloud, below parameters has been categorized:

- The pattern of the traffic from organization to cloud;
- Remote system to cloud service;
- The data transfer between/within the cloud;
- Visibility of the cloud data security risk;
- Vulnerability of the cloud-hosted data, etc.
  Multiple data have been collected from Dome9 to get the above information and it has been examined in later phase. Data collection includes:
- Inventory of different workloads in cloud;
- Compliance status of all platforms;
- Incident reported on Network, IAM, and Storage services;
- Policy database of workload running on AWS and azure.

In the next phase, we are going to analyze the data using collected information.

35.3.4   DATA ANALYSIS (DA) AND RESULT

35.3.4.1   ANALYSIS FROM DOME9

While running the Dome9 in customer's environment, architecture has been identified as showing in Figure 35.3 which will help to understand the design of infrastructure. This gives the holistic view of the architecture how cloud infrastructure looks like when DevOps resource creates cloud resources. Ideally, DevOps resource does not have any idea how the servers are connected, what are the accessibility, and what are the vulnerability. They only have limited access on cloud portal to deploy and remove workloads they don't have any view of cloud security. Any servers created either in DB, App all the servers have public and private IP address, and access to the internet, but while those are on-premise data center, this access is under manual control of security admin.

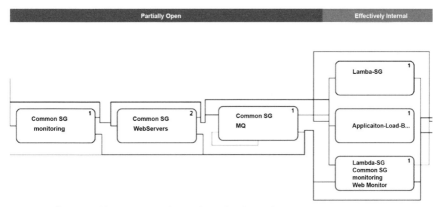

FIGURE 35.3   Architecture overview using CloudGuard Dome9.

The inventory gives the clear visibility of the integrated infrastructure. This has clearly reflected all the information of the cloud infra which includes protected assets information, instances which are running, and services which are being used. This also gives granular information about network polices. Figure 35.4 is showing inventory of workloads.

Compliance which is a major factor from organization perspective and organization security policy, data security where cyber-attack loopholes are reflecting. Different standard of security policy has been examined using Dome9 security solutions and it also shows information about non-compliant which required attention to avoid any kind of security breach. Figure 35.5 shows the compliance report of the customer's environment that is running their workloads on cloud. This has given a clear picture of the compliance gap. We need to ensure that organization cloud environment is compliant from security aspect and there are no such gaps which exist. If such security solution does not get implemented in cloud environment then it is quite difficult to see how the cloud security policies are implemented in the cloud infrastructure and how this is secured at each level.

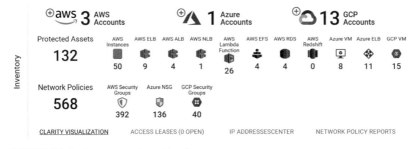

FIGURE 35.4   Inventory of workloads.

FIGURE 35.5   Compliance status of cloud-based environment.

As there is lots of security compliance gaps exist, this will lead to impact on the environment. Basis on that it generated lot of incidents as showing in Figure 35.6. We have collected the incident during the study. These incidents are related to different services and we are showing here for S3 bucket, Identity, and Access management and Network related. A cloud security administrator who is managing the environment has to fix these incidents so that attackers do not find any such gaps in the environment which may further lead to bigger issue and data leakage, cyber-attack as a result organization reputation will be in question. If these can be proactively monitored and close, such issues on time, then chances of security vulnerabilities will be very less.

Network	**108** Default Security Groups with network policies	**25** Security Groups with admin ports too exposed to the public internet	**24** Security Groups with SSH admin port too exposed to the public internet	**2** Instances are not configured within a VPC
IAM	**31** IAM Users with console password without MFA enabled	**1** Accounts without enforced Password Policy	**5** IAM Users with Inline IAM Policies applied	**19** IAM Users enabled while unused for 90 days or more
S3 Bucket	**75** S3 Buckets without server-side-encryption enabled	**75** S3 Buckets without logging enabled	**18** S3 Buckets are publicly accessible	**19** S3 Buckets without CloudTrail access logging

FIGURE 35.6   Incident reported in the cloud-based datacenter.

During this study of cloud security solutions, Importance of compliance visibility and incident on compliance polices are also shown in Figure 35.7. This is reflecting all the incident on cloud security compliance. Concern on compliance and governance of cloud data is a major point which has to address to make the environment more secured.

		AWS Dome9 Network A	AWS HIPAA	Azure Dome9 CheckUp	GCP Dome9 CheckUp	Dome9 AWS Dashboar
Dome9 Main Demo Account Accounts: 17		78.96%	66.03%	80.57%	24.34%	70.61%
Business Unit 1 Accounts: 4	⚠	79.63%	67.58%	80.57%	36.36%	71.15%
Region 1 Accounts: 1	⚠				11.11%	
AWS test (905050240067)		48.89%	37.06%			60.56%
am1987	⚠				10.34%	
am1988	⚠				11.11%	
am1990	⚠				11.11%	
am1991	⚠				11.11%	
am1996					10.34%	

FIGURE 35.7   Compliance report of a cloud-based customer.

Table 35.1 is showing compliance success and failure rate of your cloud environment.

During the study, we also fetched the policy-based incident information. This indicates the compliant status as per defined standard of security policy. During our data collection, we have generated the policy dashboard which manages the policy control of all the services being used by customer as showing in Figure 35.8.

TABLE 35.3   Compliance Success and Failure Rate of Cloud Environment

Bundle	Success Rate	# Tests	# Failed Tests	# Excluded Tests	Triggered By
AWS Inspector rules-Demo	60%	370	148	0	Policy
AWS NIST 800-53 Rev 4 (FEDRAMP)	61.33%	1792	693	0	Policy
AWS BP Matrix Update	73.70%	8126	2137	0	Policy
AWS Dome9 Best Practices – Sample	65.11%	1052	367	0	Manual
* AWS ISO 27001:2013	70.85%	4491	1309	0	Policy
AWS FedRamp Moderate Baseline	67.50%	2597	844	0	Manual
* AWS GDPR Readiness	28.98%	1194	848	0	Policy
Ticketing system demo	0%	1	1	0	Policy
AWS CSA CCM v.3.0.1	63.70%	3171	1151	0	Policy
* AWS Dome9 Best Practices	76.58%	8387	1964	0	Policy
* AWS HIPAA	65.31%	2522	875	0	Policy
* AWS NIST 800-53 Rev 4	67.64%	2599	841	0	Policy
* AWS PCI-DSS 3.2	69.59%	4440	1350	0	Policy
newTest	69.01%	2388	740	0	Policy
MyCustom-AWS PCI-DSS 3.2	69.54%	4363	1329	0	Policy
AWS entities fail	0%	305	305	0	Policy
AWS Containers Security-PREVIEW	58.26%	115	48	0	Policy
AWS NIST 800-53 Rev 4 (Pager Duty)	70.80%	4463	1303	0	Manual
My custom SOC2 (based off of ISO)	71.30%	4537	1302	0	Policy
YeshCopy of AWS Dome9 S3 Bucket Security	71.28%	1417	407	8	Policy
AWS Dome9 SOC2 based on AICPA TSC 2017	74.68%	5110	1294	0	Policy
Eran	100%	1	0	0	Policy
[SCB]* AWS CIS Foundations v. 1.1.0	70.97%	1037	301	0	Policy
**20190221 TV test AWS	4.92%	61	58	0	Policy
AWS Dome9 Best Practices SAFEBREACH	75.08%	7983	1989	0	Manual

FIGURE 35.8   Policy dashboard for cloud services.

From all above studies, this looks like the data hosted in the cloud services are not at all vulnerable and doors are open for cyber-attack until proper cloud security solution is implemented. As we have deployed security solution then its proactively monitoring the compliance status and incident reported for the cloud services. By doing this, we can avoid cyber-attack, security breaches at much extend, and customer's data will remain safe if we take timely action to fix these gaps.

35.4   DISCUSSION

This study opens a direction that there is a scope on exploring more options how cloud data can be protected while every organization are moving towards cloud services. Using this study, we deployed the security solution in customer environment and noticed a huge gap in terms of meeting the compliance which was causing several of incidents from each cloud services. Well-structured compliance policies should exist in the environment which will help in reducing the security risk and cyber threats. Apart from this study's cloud security framework, there is a need to develop more secured cryptographic algorithms for the data transmitting over the cloud.

35.5   CONCLUSION AND FUTURE WORK

The study has directed to the conclusion that any organization, private, public, or Government entity intend to adopt cloud services for any reason, maybe cost, flexibility, scalability, availability should conduct a proper security assessment. Since cloud services are a shared infrastructure model, the customer has also to take the proper responsibility for securing the data in their cloud environment.

Monitoring, visibility of the security environment, and vulnerability of the cloud data is a major concern. As stated in the implementation section, major incident, and security breaches clearly depict that data breach on the cloud has led to major organization financial and reputation loss. Cloud Security solutions are a primary area which needs to be focused on before and during the cloud service adoption and post-implementation continuous monitoring and gap analysis. Cloud services are now days become easy and profitable but maintaining the same security policies on cloud infra is a tedious task.

This study suggests placing a security solution on the cloud to protect all the workloads that are running on the cloud. This study also demonstrated that continuous cloud security analysis and gap implementation is required to keep the cloud secure. Internet accessibility by servers on cloud becomes automatic once the server is built. When any DevOps resource builds any servers, it does not have much transparency on the environment's architecture and policies. Cloud security solution helps cloud administrators to implement the right security policies and gives direction using the appropriate dashboards, compliance reports, and policies in place. Cloud security solution helps to protect the resources to follow the right policies to keep the cloud environment safe and protected from security breaches.

This study strongly demands to conduct further research to explore on cloud Security to mitigate the risk of the exponential growth of cyber threat. This study also suggests extending the study of cloud cyber forensic analysis.

KEYWORDS

- **cloud computing**
- **cloud security**
- **critical information infrastructure**
- **cybersecurity**

- **data security**
- **data-at-rest**
- **security vulnerabilities**

REFERENCES

1. Alazab, M., Sitalakshmi, V., Paul, W., Moutaz, A., & Ammar, A., (2012). Cybercrime: The case of obfuscated malware. In: Christos, K. G., Hamid, J., Elias, P., Rabih, B., & Al-Nemrat, A., (eds.), *Global Security, Safety, and Sustainability and E-Democracy* (pp. 204–211). Lecture Notes of the Institute for Computer Sciences, Social Informatics, and Telecommunications Engineering. Springer Berlin Heidelberg.

2. Brandic, I., Dustdar, S., Anstett, T., Schumm, D., Leymann, F., & Konrad, R., (2010). Compliant cloud computing (C3): Architecture and language support for user-driven compliance management in clouds. In: *2010 IEEE 3rd International Conference on Cloud Computing* (pp. 244–251). https://doi.org/10.1109/CLOUD.2010.42 (accessed 21 July 2020).

3. Dehling, T., Sebastian, L., & Ali, S., (2019). *Security of Critical Information Infrastructures*. SSRN Scholarly Paper ID 3354471. Rochester, NY: Social Science Research Network. https://papers.ssrn.com/abstract=3354471 (accessed 21 July 2020).

4. Domingo-Ferrer, J., Oriol, F., Ribes-González, J., & David, S., (2019). Privacy-preserving cloud computing on sensitive data: A survey of methods, products, and challenges. *Computer Communications, 140, 141*, 38–60. https://doi.org/10.1016/j.comcom.2019.04.011 (accessed 21 July 2020).

5. Hong, J. B., Armstrong, N., Dong, S. K., Alaa, H., Noora, F., & Khaled, M. K., (2019). Systematic identification of threats in the cloud: A survey. *Computer Networks, 150*, 46–69. https://doi.org/10.1016/j.comnet.2018.12.009 (accessed 21 July 2020).

6. Jha, P., Sartaj, S., & Ashok, S., (2019). Data control in public cloud computing: Issues and challenges. *Recent Patents on Computer Science*. http://www.eurekaselect.com/172656/article (accessed 21 July 2020).

7. Kandias, M., Nikos, V., & Dimitris, G., (2013). The insider threat in cloud computing. In: Sandro, B., Bernhard, H., Dimitris, G., & Stephen, W., (eds.), *Critical Information Infrastructure Security* (pp. 93–103). Lecture Notes in Computer Science. Springer Berlin Heidelberg.

8. Kazim, M., & Shao, Y. Z., (2015). *A Survey on Top Security Threats in Cloud Computing*. https://doi.org/10.14569/IJACSA.2015.060316#sthash.hDOJaK6L.dpuf (accessed 21 July 2020).

9. Ko, R. K. L., (2014). Data accountability in cloud systems. In: Surya, N., & Mukaddim, P., (eds.), *Security, Privacy, and Trust in Cloud Systems* (pp. 211–238). Berlin, Heidelberg: Springer Berlin Heidelberg. https://doi.org/10.1007/978-3-642-38586-5_7 (accessed 21 July 2020).

10. Lang, B., Wang, J., & Liu, Y., (2017). Achieving flexible and self-contained data protection in cloud computing. *IEEE Access,5*, 1510–1523. https://doi.org/10.1109/ACCESS.2017.2665586 (accessed 21 July 2020).

11. Mishra, N., Tarun, K. S., Varun, S., & Vrince, V., (2018). Secure framework for data security in cloud computing. In: Millie, P., Kanad, R., Tarun, K. S., Sanyog, R., & Anirban, B., (eds.), *Soft Computing: Theories and Applications* (pp. 61–71). Advances in Intelligent Systems and Computing, Springer Singapore.

12. Puthal, D., Surya, N., Rajiv, R., & Jinjun, C., (2017). A dynamic prime number based efficient security mechanism for big sensing data streams. *Journal of Computer and System Sciences, 83*(1), 22–42. https://doi.org/10.1016/j.jcss.2016.02.005 (accessed 21 July 2020).

13. Singh, V., & Pandey, S. K., (2019). Cloud security ontology (CSO). In: Himansu, D., Rabindra, K. B., Harishchandra, D., & Diptendu, S. R., (eds.), *Cloud Computing for Geospatial Big Data Analytics: Intelligent Edge, Fog, and Mist Computing* (pp. 81–109). Studies in big data. Cham: Springer International Publishing. https://doi.org/10.1007/978-3-030-03359-0_4 (accessed 21 July 2020).

14. Sun, Y., Junsheng, Z., Yongping, X., & Guangyu, Z., (2014). Data security and privacy in cloud computing. *International Journal of Distributed Sensor Networks, 10*(7), 190903. https://doi.org/10.1155/2014/190903 (accessed 21 July 2020).

15. Surbiryala, J., Li, C., & Rong, C., (2017). A framework for improving security in cloud computing. In: *2017 IEEE 2nd International Conference on Cloud Computing and Big Data Analysis (ICCCBDA)* (pp. 260–264). https://doi.org/10.1109/ICCCBDA.2017.7951921 (accessed 21 July 2020).

16. Talib, A. M., Rodziah, A., Rusli, A., & Masrah, A. A. M., (n.d.). Formulating a security layer of cloud data storage framework based on multi agent system architecture. *GSTF International Journal on Computing1*(1). https://www.academia.edu/1492593/Formulating_a_Security_Layer_of_Cloud_Data_Storage_Framework_Based_on_Multi_Agent_System_Architecture (accessed 21 July 2020).

17. Tyagi, M., Manish, M., & Bharat, M., (2018). A framework for data storage security in cloud. In: Mohan, L. K., Munesh, C. T., Shailesh, T., & Vikash, K. S., (eds.), *Advances in Data and Information Sciences* (pp. 263–272). Lecture notes in networks and systems, Springer Singapore.

18. Vaishnav, J., (2018). *Survey on Data Safety and Effectiveness Estimation in Cloud Computing Using Cloud Computing Adoption Framework*, 5.

19. Yesilyurt, M., & Yildiray, Y., (2016). New approach for ensuring cloud computing security: Using data hiding methods. *Sādhanā, 41*(11), 1289–1298. https://doi.org/10.1007/s12046-016-0558-8 (accessed 21 July 2020).

CHAPTER 36

Prediction of Diseases Through Data Mining

A. V. S. PAVAN KUMAR,[1] VENKATA NARESH MANDHALA,[2] and
DEBRUP BANERJEE[2]

[1]Department of Computer Science and Engineering, GIET University,
Gunpur, Odisha, India, E-mail: avspavankumar@giet.edu

[2]Department of Computer Science and Engineering, Koneru Lakshmaiah
Education Foundation, Vaddeswaram, Guntur, Andhra Pradesh, India,
E-mail: mvnaresh.mca@gmail.com (V. N. Mandhala)

ABSTRACT

Data mining is the computational procedure of discovering styles in huge statistics sets regarding strategies on the intersection of artificial intelligence (AI), gadget studying, facts, and database (DB) systems. It is an interdisciplinary subfield of computer technological know-how. In nowadays lifestyles, illnesses are increasing increasingly more. Data mining is one of the solutions for it; it helps us to overcome this problem by exploring old datasets. For any disease if it is identified at early stage treatment can be done easily. A wide range of data is produced in health care institutions; we will use that data to get some useful information. Data mining in the medical sector helps doctors for diagnosis and treatment of diseases, this chapter makes an effort to study and find interesting patterns from the data of patients.

36.1 INTRODUCTION

Data mining is the technique used for discovering obscure qualities from a huge quantity of information. Because the population increase as the diseases are increasing daily. The examination of this restorative info is hard while not the computer-primarily based investigation [5]. The computer-primarily based investigation, the large regions for the specialists to managing the big live of patient's knowledge sets from multiple points of read; for instance, perceive complicated indicative tests, translating past outcomes, and consolidating the disparate info along. This prediction system helps the patients and declines the meditative expenses. As per a survey, around 17.5 million deaths are due to heart attacks, which means 31% of global deaths is because of heart attacks, so, if this disease is predicted earlier, it could be very helpful to everyone [1]. Therefore, we can apply data mining on historical data and retrieve some useful information regarding heart diseases.

36.2 KNOWLEDGE DISCOVERY PROCESS

The phrases knowhow discovery in databases (KDD) and records mining are typically used interchangeably [6]. KDD is that the technique of fixing the low-level statistics into high-degree data. Therefore, KDD refers to the nontrivial elimination of implicit, antecedently unknown, and doubtlessly helpful data from knowledge in databases (DBs). Whereas statistics mining Associate in Nursing KDD are often handled as similar words, however, in real facts mining is an essential step within the KDD system. The Knowhow Discovery in DBs system comprises of a couple of steps leading from raw statistics collections to some style of recent records [7]. The repetitive methodology includes the next steps:

1. **Cleaning:** It can also be known as facts cleansing it are a phase whereby noise records and unrelated statistics are eliminated from the gathering.
2. **Integration:** At this degree, various records resources, often heterogeneous, could also be shared in a very commonplace supply.
3. **Data Selection:** At this step, the facts associated with the analysis are set on and retrieve from the records assortment.

4. **Data Transformation:** To boot said as statistics consolidation, its miles a locality whereby the chosen statistics is remodeled into forms applicable for the mining method.

5. **Data Mining:** It's far the important step whereby sensible ways are applied to extract patterns doubtless helpful.

6. **Sample Evaluation:** This step, firmly fascinating designs representing ability are recognized supported given measures.

7. **Expertise Illustration:** Is that the closing phase whereby the determined experience is visually delineated to the user. On this tempo, image techniques are accustomed to facilitate customers to perceive and interpret the info mining consequences.

In today's world health is a major issue and medical test, disease finding is taking a longer time as we all that about it [2]. Here day-by-day, we have increased in demand for medical facilities and new technology on medical also lead to increase in complexity in finding the better, most precise treatment, many hospitals are using DBs and computers to manage patient's records which made easy to manage data and in the same way we also know that artificial intelligence (AI) is also developing now, is used in many prediction and robotic application in this we are also using some of this AI technique on a DB which is formed on applying data mining techniques on a large patients DB. We are using ID3 algorithm Naïve Bayesian classifiers algorithm which can be used for prediction separately by using decision trees and tables, respectively, here we use both to overcome some limitations and to improve accuracy [3]. We can also use genetic algorithm and KNN algorithm to improve accuracy another solution for this is to use clustering based on male and female data, we can classify the data based on hierarchical clustering (Agnes).

36.3 PREDICTION OF HEART DISEASE

Mainly, Naïve Thomas Bayes algorithm, decision tree algorithm, and neural network algorithms area unit normally used.

The Naïve Bayesian classifier relies on the Thomas Bayes theorem with independence assumptions between predictors. Naïve Bayesian model is beneficial for terribly massive datasets with no difficult repetitious parameter estimation. Bayesian classifier outperforms additional subtle classification ways [4].

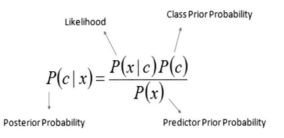

$$P(c \mid X) = P(x_1 \mid c) \times P(x_2 \mid c) \times \cdots \times P(x_n \mid c) \times P(c)$$

where, P(c|x) is that the posterior probability of sophistication given predictor, P(c) is that the previous likelihood of sophistication, P(x|c) is that the chance that is that the likelihood of predictor given category, P(x) is that the previous likelihood of predictor.

Bayes theorem provides some way to calculate the posterior likelihood, P(c|x) of from P(c), P(x), and P(x|c). Bayes classifier assumes that the impact of the worth of a predictor (x) on a given class (c) is freelance of the values of different predictors. This assumption is termed category or class conditional independence.

Every leaf node holds a category label, each internal node denotes a take a look at on associate degree attribute, each branch four denotes the end result of a take a look at and top node within the tree is that the root node. Associate degree associated call tree is incrementally developed whereas It breaks down a dataset into smaller and smaller subsets. The result is a tree with leaf and call nodes.

In 1980, a scientist J. Ross Quinlan developed a decision tree algorithm called ID3-Iterative Dichotomiser. When he developed C4.5 that was the successor of ID3. Each adopts a greedy approach. The trees area unit created in an exceedingly top-down algorithmic divide-and-conquer manner, during this formula, there's no backtracking.

Decision tree is made top-down ranging from a root node then partition the information into subsets with similar (homogenous) values. For the homogeneity of a sample ID3, formula uses entropy to calculate. If the sample is associate, degree equally divided its entropy of 1 and if the sample is totally homogenized, the entropy is zero.

Using frequency table by entropy for one attribute:

$$E(S) = \sum_{i=1}^{c} - p_i \log_2 p_i$$

C4.5 is used for classification, and C4.5 is usually remarked as a applied math classifier. C4.5 chooses the attribute of information that almost all effectively splits its set of samples into subsets enriched in one category or the opposite at the every node of the tree it's the normalized information gain. Any of the attribute with the very best normalized data gain is chosen to create the choice. Then C4.5 formula recurs on the smaller sub lists.

36.4   ARTIFICIAL NEURAL NETWORKS (ANNS)

An artificial neural network (ANN), often without a doubt referred to as a "neural community" (NN), maybe a mathematical version or procedure version supported biological neural networks, in specific phrases, is associate emulation of biological neural convenience [7]. It includes associate inter-connected establishment of artificial neurons and processes info the usage of a connectionist technique to computation. In most instances, associate ANN is associate accommodative convenience that changes its form all totally on out of doors or internal knowledge that flows via the community within the route of the educational part. In bigger smart phrases neural networks area unit non-linear applied mathematics facts modeling instrumentality. They will be wont to model difficult relationships among inputs and outputs or to find designs in info. A neural network is an associate interconnected cluster of nodes, similar to the massive network of neurons within the human brain.

A neural community got to be designed specified the software system of a difficulty and speedy of inputs produces (both 'direct' and through a relaxation approach) the favored set of outputs. Several ways to line, the strengths of the connections exist. One manner is to line the weights expressly, employing a priori ability. One another method is to 'train' the neural community with the help of feeding it education patterns and holding it amendment its weights in line with some learning rule. We're capable of categorizing the attending to apprehend things as follows:

- Supervised getting to know or associative studying whereby the network is trained by providing it with enters and matching output designs. These enter-output pairs are also provided by associate external teacher, or via the system, that contains the neural network.
- Unsupervised mastering or self-employed whereby associate (output) unit is trained to reply to clusters of patterns among the enterpopula-tion. During this paradigm, the convenience is meant to find statisti-cally salient capabilities of the enter population. In distinction with

the supervised mastering paradigm, there's no priori set of categories into that the designs area unit to be tagged rather the machine have to be compelled to develop its own illustration of the input stimuli. The subsequent diagram represents the neural network.

Outputs

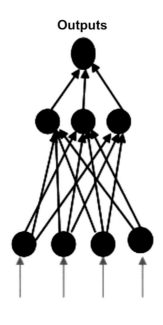

Inputs

36.5   THEORETICAL ANALYSIS

The coronary heart sickness prediction systems uses clinical dataset encompass parameters primarily based on threat elements as age, circle of relatives records, diabetes, high blood pressure, excessive cholesterol tobacco smoking, alcohol consumption, and many others. The prognosis time and enhance the diagnosis accuracy, medical diagnostic choice aid structures (MDDSS). The neural community method is used for studying the heart disease statistics. Making use of feed-forward algorithm with variable gaining knowledge of pace and momentum the heart sickness DB are skilled by way of the neural network. The input layer carries 13 neurons to represent thirteen attributes. It includes four magnificence labels particularly regular person, 1st stroke, second stroke, and stop of life. The output layer includes two neurons to represent these four coachings. The neural network is built with and without hidden layer is matless and multilayer networks area unit educated. The dataset classifies the man or woman into regular and abnormal individual based on coronary heart diseases.

Some of the causes for coronary heart disorder (CHD):

1. **Smoking:** The smoking is predominant cause of coronary heart attack, stroke, and different peripheral arterial disorder. Nearly 40% of everybody die from smoking tobacco accomplish that due of coronary heart and blood liner illnesses. A heart assault is the death of or damage to part of the coronary heart electricity because the delivery of blood to the heart strength is significantly decreased or bunged.

2. **Cholesterol:** The strange ranges of lipids (fat) in blood are risk of coronary heart illnesses. Cholesterol is a gentle, waxy substance discovered some of the lipids in bloodstream and it can also include in all the cells of the body. The excessive ranges of LDL (low-density lipoprotein) LDL cholesterol accelerate atherosclerosis increasing the hazard of heart sicknesses.

3. **Weight Problems:** That is used to explain the health condition of each person substantially above his or her I deal with healthful weight. A higher hazard or fitness trouble including heart sickness, stroke, high blood strain, diabetes.

4. **Lack of Physical Exercise:** The shortage of workout is a chance issue used for rising coronary artery sickness (CAD). Require in bodily work out increases danger of CAD increase for diabetes and excessive blood pressure. CHD or ischemic heart disease (IHD) is a broad term that can check with any condition that affects the coronary heart. This chapter additionally offers the comparison of set of rules on accuracy and data. Normal motives for abnormal heart rhythms (arrhythmias) or situations that can spark off arrhythmias consist of:

   i. Coronary heart imperfections you're conceived with (intrinsic coronary heart surrenders);
   ii. Coronary delivers path illness;
   iii. High blood pressure;
   iv. Diabetes;
   v. Smoking;
   vi. Inordinate usage of liquor or caffeine;
   vii. Drug addiction;
   viii. Valvular coronary illness.

Heart failure or congestive heart failure, this means that that the heart remains operating, however it isn't pumping blood as nicely because it have to, or getting sufficient oxygen.

Heart valve issues can cause the coronary heart not commencing enough to allow proper blood drift. Occasionally the coronary heart valves don't close and blood leaks thru, or the valve leaflets bulge into the top chamber, causing blood to float backward through them.

In addition to these, coronary heart diseases may be hereditary. That means if a person's dad and mom or grandparents has heart diseases, there may be possibility for that character to get CHD.

36.6 PROPOSED SOLUTION

With the existing algorithms the accuracy was now not up to mark and there are issues with every of the algorithm. To conquer this we will use Genetic set of rules and okay-nearest neighbor set of rules with addition to the present algorithms. Genetic algorithms attempt to contain thoughts of natural evolution. In most popular, genetic learning starts as follows. An initial populace which has a set of guidelines is created. Each rule represents a string of bits. For instance, allow us to that samples in coaching set area unit delineate by means of Boolean attributes, say B1 and B2, and directions C1 and C2. The rule is "If now not B1 and B2 then C1" is also painted as 011.

There are two units in every of sort strategies to applying GA in sample reputation:

1. Use GA as a classifier right away in computation.
2. Use a GA to optimize the results, i.e., as associate degree optimizer to line up the parameters in numerous classifiers. Most packages of gas in pattern name optimize some parameters among the category method. Hydrocarbon has been disbursed to search out associate degree most advantageous set of feature weights that enhance classification accuracy. First, a traditional feature extraction technique which has major component analysis (PCA) is applied, and so a classifier that embrace okay-NN (nearest neighbor set of rules) is employed to calculate the health characteristic for GA. Combination of classifiers is another region that fuel were wont to optimize. GA is likewise utilized in selecting the prototypes among the case-based whole sort. Second approach of genetic set of rules to optimize the result from the dataset

is additional effective to compute the correct values of observations of statistics through utilizing info mining ways.

36.6.1 *K-NEAREST-NEIGHBOR-CLASSIFIER*

The training tuples are outlined by n attributes. Every tuple represents an element in Associate in nursing n-dimensional space. In this means, all of the tuples are hold on in n-dimensional pattern house. While an unknown tuple is given, this algorithm searches the sample space for okay tuples towards this unknown tuple. For every tuple, RMS value is calculated among unknown and recognized one Euclidean or Manhattan distances may be used. KNN algorithm is one of the most effective classification set of rules. In spite of such simplicity, it is able to give surprisingly aggressive outcomes. KNN algorithm can also be used for regression troubles.

Example for KNN algorithm:

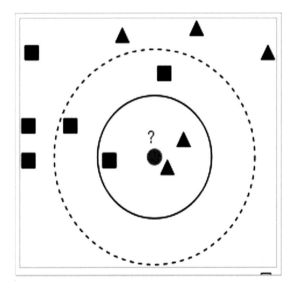

36.6.2 *CLUSTERING*

Cluster analysis or agglomeration is that the task of clustering a group of objects in such some way that objects within the same group (called a cluster) are a lot of similar (in some sense or another) to every aside from to those in alternative teams (clusters). It may be a main task of exploratory records mining, and a standard technique for applied math statistics analysis, employed in several fields; any other proposed answer for this prediction of coronary heart sicknesses is, clustering, based at the received data set. Data may be divided into clusters initially, i.e., male and female, from that clusters a hierarchy can be created. From which we are able to predict whether the sickness will come or now not. We also can enhance the accuracy the usage of each Naïve Bayes set of rules and choice tree algorithm.

36.7   EXPERIMENTAL ANALYSIS

By applying the clustering, data can be divided into two clusters initially, i.e., male, and female, from that clusters a hierarchy can be created. From which we can predict whether the disease will come or not. In the above image, data is divided into five clusters using hierarchical clustering (agglomerative).

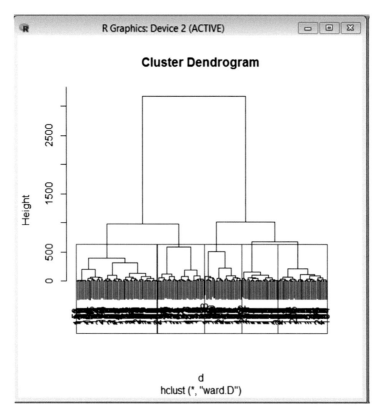

Disease	Symptom
Fever	Body pain
Cold	Headache
Skin allergy	Headache
Fever	Headache
Fever	Headache
Cold	Headache
Skin allergy	Body pain
Skin allergy	Body pain
Fever	Headache
Skin allergy	Headache
Fever	Body pain
Cold	Headache
Cold	Headache

Frequency table	Body pain	headache
Cold	0	4
Skin allergy	3	2
Fever	2	3
Grand Total	5	9

Likelihood table	Body pain	Headache		
Cold	0	4		4/14=1.29
Skin allergy	3	2		5/14=0.36
Fever	2	3		5/14=0.36
Grand Total	5	9		
	5/14=0.36	9/14=0.64		

Problem:

Patients gets a headache if he/she gets fever

P(headache | fever) = (P(fever | headache) * P(headache))/P(fever)

P(fever) = 5/14 = 0.36

P(headache) = 9/14 = 0.64

P(headache/fever) = (0.33*0.64) = 0.60

It means 60% of people who gets headache if he/she gets fever

After applying the Bayes algorithm, if we use the decision tree algorithm it can give good results.

36.8   CONCLUSION

The rate of diseases is increasing day by day. There are many people doesn't know about the diseases that even existed. Mainly the heart disease is the major death-causing disease around the world. This needs a further step of prediction. But with the inaccurate procedures, there may be fatal outcomes. So hereby with the help of developed algorithms, there is chance of increasing the accuracy. This system provides the idea of prediction of disease with the usage of different algorithms. It may also help to produce a drastic improvement in prediction technique.

KEYWORDS

- **artificial intelligence**
- **artificial neural network**
- **coronary artery disease**
- **coronary heart disorder**
- **data mining**
- **ischemic heart disease**

REFERENCES

1. Bharti, S., & Singh, S. N., (2015). Analytical study of heart disease prediction comparing with different algorithms. In: *International Conference on Computing, Communication, and Automation* (pp. 78–82). IEEE.
2. Dewan, A., & Sharma, M., (2015). Prediction of heart disease using a hybrid technique in data mining classification. In: *2015 2nd International Conference on Computing for Sustainable Global Development (INDIACom)* (pp. 704–706). IEEE.
3. Krishnaiah, V., Srinivas, M., Narsimha, G., & Chandra, N. S., (2014). Diagnosis of heart disease patients using fuzzy classification technique. In: *International Conference on Computing and Communication Technologies* (pp. 1–7). IEEE.
4. Shouman, M., Turner, T., & Stocker, R., (2012). Using data mining techniques in heart disease diagnosis and treatment. In: *2012 Japan-Egypt Conference on Electronics, Communications, and Computers* (pp. 173–177). IEEE.
5. Sudhakar, K., & Manimekalai, D. M., (2014). Study of heart disease prediction using data mining. *International Journal of Advanced Research in Computer Science and Software Engineering, 4*(1).
6. Thanigaivel, R., & Kumar, K. R., (2016). Boosted a priori: AN effective data mining association rules for heart disease prediction system. *Middle-East Journal of Scientific Research, 24*(1), 192–200.
7. Kumar, R. N., & Kumar, M. A., (2016). Medical data mining techniques for health care systems. *International Journal of Engineering Science*, 3498.

CHAPTER 37

Naïve Bayes Full-Spectral Image Analytics Using K-Nearest Based Parzen Window

R. RAJASEKHAR[1] and R. N. V. JAGAN MOHAN[2]

[1]Assistant Professor, Department of CSE, GIET University, Gunupur, Odisha, India, E-mail: rrajasekhar@giet.edu

[2]Associate Professor, Information Technology, SRKR Engineering College, Andhra Pradesh, India, E-mail: mohanrnvj@gmail.com

ABSTRACT

Blockchain and big data are a complement to each other and are an amalgamated. Blockchain platform can support any kind of digitized information. It is possible to use it in the field of Big Data is chiefly to enlarge the protection or the quality of the data. Blockchain and big data is to make sure data quality. To transform the big data like the one image data into high level build in which might be analyzed is realized. The band selection process used in pixel occurrences for full-spectral image classification is one of the challenging areas of studies for those working in the field of geo-spatial intelligence system using image analytics with big data. To use the wet center geo-spatial data sets of enormous workout full-spectral image bands using Naïve Bayes classification and massive inspection pattern is equal to the many classes with the help of K-nearest-based Parzen Window. Our main aim is to reduce the dimensionality of the data set labels. In this regard, we have probable replica of these spectral bands for each pixel value and observed that this reduction does not affect the silhouette of the request particularly. This reduction was required to defeat the expletive of dimensionality problematic.

37.1 INTRODUCTION

Blockchain is a heart of computer technology. It is cryptographically secure distributed database (DB) proficiency for storing and transmitting information. Each record in the DB is called a block and contains such as the transaction date and a link to the previous block. Blockchain and Big data are two expertise's in full swing, but they are also two balancing expertise's. We can study how the Blockchain transforms storage space and image data analysis (DA). The image segmentation and feature extraction play a crucial role for image classification and recognition. Images and videos are segmented using several procedures in each with specific purposes and digital processing methods are well known for segmentation of images. Segments are spatially pertinent regions of image or video ability to see that consists of common set of abilities. These can be color deliveries, intensity levels, texture, moving, and motionless portions of a video vision and other gauges. The feature refinement as the upcoming process. To aid in the recognition of higher-level and low-level feature extraction has features and stowed with each occurrence. The immense of study in this arena has culminated in numerous procedures. These two approaches are on spectral and spatial images. Full-spectral image is one type of spectral image. Full-spectral imaging use of imaging spectroscopy and is the replacement to hyperspectral imaging suggested by Meynart et al. [1].

Full-spectral imaging improved to develop the proficiencies of remote sensing with soil based remote sensing. In imaging spectroscopy namely hyperspectral imaging or spectral imaging contains in each pixel of an image obtains many bands of light strength data from the spectrum, in place of impartial the three bands of the RGB color model. Further, fair it is the instantaneous gaining of spatially co-registered, i.e., process of transforming various sets of data through one coordinate system. These data as like numerous photographs, data from not the same sensors, times, depths, or viewpoints. By above this area is used in computer vision, medical imaging and compiling and analyzing images and data from satellites. Image registration is essential permitted to be able to match or assimilate the data obtained from these not the same type images in voluminous spectrally nearby bands. Despite the fact, hyperspectral imaging obtains data as numerous nearby spectral bands, full spectral imaging obtains data as spectral curves. An important benefit of full-spectral imaging in excess of hyperspectral imaging is an important reduction in the data rate of recurrence and size. Full-spectral imaging bands are extracted and keep the information, i.e., in the raw data.

The information is contained in the shape of the spectral curves. From frequency on the data is produced by a full-spectral imaging-system is relative to the quantity of information in the scene or image.

37.2 FULL-SPECTRAL IMAGE ANALYTICS

Full-spectral analytics of an image is a transformation as of images or frames to logically obtained data. The main aim of image analytics is an image as the meltdown a non-moving scene and a set of frames as the meltdown scene encompassing an image still background subdivision and moving foreground segments. In this regard, the video consists of sequence of frames. More exactly, the main unbiased of image analytics is to obtain along an unstructured version of reality in the form of image and videos into a machine analyzable depiction of a set of variables. A variable is symbolized by a series of values associated to a thing. Each such value is time-stamped to make it imaginable to treat a variable as a time series. The discovery of feature objects and movements is on images has many labels with image processing or computer vision. The image analytics used for particular transformations. At this stage, analytics of an image continues to be a set of transformations on input image that add value and make a rich set of time sequence as analytically prepared the output. Initially, transformation step divisions of images into structured elements are get ready them for feature extraction that is the identification of image as low-level features. The second transformation step is the detection of relationships among these features, variables, and time. The third transformation step is the extraction of variables with time-stamped values.

37.3 ASSOCIATION OF THE VARIABLES, FEATURES, AND TIME

Classification of various associations between variables features and time is a sub-discipline of artificial intelligence (AI) termed as machine learning (ML) amalgamated with realistic statistics. The relations amid variables, features, and time in image analytics are signified as a predictive model. Previously, a predictive model that can be made, a set of instances is extracted from all the given images and/or all the given videos being analyzed. From a predictive modeling point of view there are three subsets of all extracted instances are of interest like training and test and predictive instances. A ML or statistical modeling algorithm trains a predictive model based on the set of annotated training instances. Modeling algorithms are depending on identified methods and numerous more. Test instances estimate the accuracy of a predictive model created a modeling procedure. The training process is frequently frequent with dissimilar sets of training and test illustration and algorithm parameters until the accuracy of the predictive model is at an acceptable level. After the predictive model has been guided, it is used to classify predicted instances in a process.

37.4   NAÏVE BAYES FULL-SPECTRAL IMAGE CLASSIFICATION

Naïve Bayes is one of the simplest classifiers and works surprisingly well for different applications particularly that involving characters classification. Assume X to be an image in which consists of x_1, x_2, x_3... x_n of n pixels, such that, $X=\{x_1, x_2, x_3... x_n\}$ belongs to a set of class labels $C = \{C_1, C_{2...} C_m\}$. The class of a pixel will depend upon a set of spectral bands. For the given i^{th} pixel x_i, having k spectral bands is represented as $B_j = \{b_{j1}, b_{j2} ... b_{jk}\}$. By using general approach, each x_j to be classified whose probability of occurrence $P(x_j/C_i)$ is maximum, will be the final output of C_i. To estimate the value of $P(x_j/C_i)$, this classifier Naïvely assumes that the pixels of image X are independent of each other; therefore, it is known as Naïve Bayes. Once independence is assumed, the derivation is used to compute $P(x_j/C_i)$ as follows:

$$P(x_j/C_i) = P(C_i \wedge x_j)/P(x_j) = P(x_j/C_i) P(x_j)/P(C_i)$$

$$\alpha \ P(x_j/C_i) \ P(C_i)$$

$$\alpha \ P(O_1=x_1|C_i) \ ... \ P(O_i=x_1|C_i) \ P(C_i)$$

Here, the pixels record X contains attributes or objects O_i with values x_i. The denominator P(X) is ignored because it is common for all the classes. The last line of the derivation is obtained by assuming independence between the attributes or objects.

For classification is the values of $P(O_i=x_1|C_i)$ are pre-computed and stored for all possible attribute values and classes. At the time of classification, these probability values are used to estimate the $P(C_i|X)$ as per the above derivation and the class with the high level of probability of occurrence is output.

37.5 K-NEAREST NEIGHBOR OF MULTI-PIXELS USING PARZEN WINDOW

Let X be the image, it consists of x_1, x_2, x_3... x_n of n pixels. Therefore, $X = \{x_1, x_2, x_3... x_n\}$. Let R be the Region of x_n pixels in X. To begin with, we select the pixels randomly in a region R, than we can estimate the proper classes $C_1, C_2... C_n$ by the band values of the particular pixel having band value lies in between $0 < b_i < 200$ (restricted due to convenience), The bands are having same object but maybe with different colors or orientations of the pixel. Then we find out the mean and variance of each pixel. The classes like Earth Science. On the other hand suppose that n bands b_1, b_2... b_n are independently drawn according to the probability density function P(X), and there are K out of n bands falling width the region 'R,' we have P=K/n. Therefore, we above at the following obvious estimate for P(X).

$$\text{So that, } P(X) = \frac{\dfrac{K}{n}}{V}$$

The Parzen probability density estimation formula for two-dimensional as follows:

$$P(X) = \frac{\dfrac{K}{n}}{V}$$

$$= \frac{1}{n}\sum_{i=1}^{n}\frac{1}{h^2}\emptyset\left(\frac{b_i - x}{h}\right)$$

$\emptyset\left(\dfrac{b_i - x}{h}\right)$ is called a window function.

$$P(x) = \frac{1}{n}\sum_{i=1}^{n}\frac{1}{\sqrt{2\pi\sigma}}\exp\left(\frac{-(b_i - x)^2}{2\sigma^2}\right)$$

This is the average of n Gaussian function also known as normal distribution with each object data points as a center, σ needs to be predetermined. Now, we estimate the n bands compare with 'n' classes. Therefore, that in every band can match with any of 'n' classes. Thus, we can estimate the equality or inequality of classes.

37.6 MAPREDUCE APPROACH FOR IMAGE BIG DATA WITH BLOCKCHAIN

Map-reduce is an indoctrination replica for inscription appliance that can process Big data in similar on many nodes. MapReduce gives analytical potential for analyzing huge volumes of intricate images data. The quantity of images data require to gather and managed can fall beneath the type of big data. On the other hand, big data is not only about scale and quantity. It can also engage one or more of the subsequent features like velocity, variety, volume, and intricacy. The MapReduce procedure holds two significant errands namely map and reduce. The Map job takes a set of images data and transforms it into a further set of feature extraction image pixels data, where each individual image pixels are broken down into Bandskey-Pixel value pairs. The reduce task takes the yield from the Map as an input and combines those image pixels data like Bandskey-Pixel value pairs into a lesser set of Bands. The reduce job is for all time do after the map work. In this process there are six steps are involved for image analytics MapReduce process. The following are as follows:

1. **Input Stage:** Here, we have a Record Reader that transform in each image in an input file and sends the parsed data to the mapper in the form of Band key-Pixel value pairs.
2. **Map Stage:** Map is a user-defined task in which takes a sequence of Band key-pixel value pairs and processes each one of them to make zero or more Band key-pixel value pairs.
3. **Intermediate Band Keys Stage:** They Band key-Pixel value pairs make by the mapper are well-known as intermediary keys.
4. **Combiner Stage:** A combiner is a kind of local Reducer that assemblage similar data from the map stage into particular sets. It takes the intermediate Band keys from the mapper as input and applies a user-defined code to cumulative the pixel feature extracted values in a small possibility of one mapper.
5. **Jumble-Up and Sort:** The Reducer task begins with the Shuffle and Sort step. It downloads the grouped Bands key-pixel value pairs onto the local appliance, where the Reducer is running. The individual Bands key-pixel value pairs are sorted by key into a better data catalog. The image pixels or bands data list groups the equivalent keys together so that their pixel values can be iterated simply in the Reducer task.

6. **Reducer:** It obtains the grouped Bands key-Pixel value paired images data as input and runs a Reducer function using Parzen window technique for prediction of identification purposes. Here, the pixels data can be amassed, filtered, and combined in a number of ways, and it needs a broad range of processing. Once the image bands execution is completed, it gives zero or more Bands key-Pixel value pairs to the last footstep.

7. **Output Stage:** In this, output formatters that transform the final bands key-pixel value pairs from the Reducer function to storage and transmitting the band DA information. Each Full-Spectral Image Band called a Block folder and contains such as transaction date and a linkage to the previous block using a record writer.

6. **Experimental Results:** The experimental result is a range of indicate is an explanation of the energy of a signal as a resolution of incidence. The signal function is time and frequency in which indications to the time-varying namely time-frequency spectrum depiction. The spectral Band analysis in which refers to the many methods used to determine spectra. Likewise, signals may be deterministic or Band Selection process using Blockchain and MapReduce Approach. To work out the executing times for each band selection is importance and find out the total band executing time needed to allocation of process (Table 37.1 and Figure 37.1).

TABLE 37.1   Envisage the CPU Time for Band Selection Process

SL. No.	Time(sec)	Bands
1.	40	398
2.	38	390
3.	42	410
4.	50	502
5.	60	590
6.	30	305
7.	20	210
8.	25	252
9.	40	398
10.	39	392

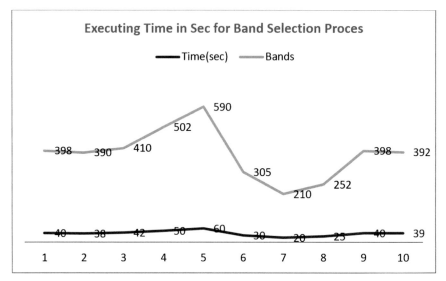

FIGURE 37.1   Envisage the CPU time for band selection process.

37.7   CONCLUSION

The process of transforming big data like image data into higher level build that can be analyzed and identified. Appliance identification of band selection process uses in pixel frequencies for full-spectral image identification is one of the demanding areas of investigate for those working in the field of geo-spatial intelligence system using image analytics with big data approach. Utilizing geo-spatial wet center data sets of enormous training full-spectral image bands using Naïve Bayes classification and immense test pattern equivalent to the numerous classes with the help of K-nearest by using Parzen Window method. Mainly focused about decrease the dimensionality of the data set labels. In this regard, we have predictable model of these spectral bands for each pixel value and observed that this reduction does not affect the silhouette of the request notably. This reduction was necessitated to defeat the expletive of dimensionality problem. It can study how the blockchain transforms storage space and image DA.

KEYWORDS

- **big data**
- **blockchain**
- **full-spectral image**
- **geo-spatial**
- **image analytics**
- **K-nearest based Parzen Window**
- **Naïve Bayes**

REFERENCES

1. Meynart, R., Bolton, J. F., Neeck, S. P., Shimoda, H., Lurie, J. B., & Aten, M. L., (2004). *Full Spectral Imaging: A Revisited Approach to Remote Sensing, 5234*, 243. doi: 10.1117/12.510485. ISSN: 0277-786X.
2. Sicong, L., Quian, D., Alim, S., & Haiyan, P. (2018). *Band Selection-Based Dimensionally Reduction for Change Detection in Multi-Temporal Hyper Spectral Images.*
3. Liangpei, Z. (2018). *Recent Advances in Hyper Spectral Image Processing.*
4. Yuri, M., Masahiro, Y., & Nagaaki, O. (2018). *Hybrid-Resolution Multispectral Imaging Using Color Filter Array.*
5. Telmo, A., Luís, P., José, B., & Jonáš, H. (2018). *Hyperspectral Imaging: A Review on UAV-Based Sensors, Data Processing and Applications for Agriculture and Forestry.*
6. Guoru, D., Qihui, W., Jinlong, W., & Yu-Dong, Y. (2018). *Big Spectrum Data: The New Resource for Cognitive Wireless Networking.*
7. Gonzalo, R. A., Richard, B., & Jon, Y. H. (2018). *Hyperspectral, Multispectral, and Multimodal (HMM) Imaging: Acquisition, Algorithms, and Applications.*
8. Ting, W., Bo, D., & Liangpei, Z. (2018). *An Automatic Robust Iteratively Reweighted Unstructured Detector for Hyper Spectral Imagery.* Senior Member, IEEE.
9. Patrick, T., Matthew, N., & Charles, G. (2018). *Hyper Spectral Imaging Sensor for Tracking Moving Targets.*
10. Nagarajan, M., & Rashmi, P. K. (2018). *Exploration of Unmixing and Classification of Hyper Spectral Imagery.*

CHAPTER 38

Mental Health Disorders and Privacy-Preserving Data Mining: A Survey

VIJAYA PINJARKAR,[1] AMIT JAIN,[2] and ANAND BHASKAR[2]

[1]PhD Scholar, Sir Padampat Singhania University, Rajasthan, India,
E-mail: vijaya.pinjarkar@spsu.ac.in

[2]Assistant Professor, Sir Padampat Singhania University, Rajasthan, India,
E-mails: amit.jain@spsu.ac.in (A. Jain), anand.bhaskar@spsu.ac.in
(A. Bhaskar)

ABSTRACT

Workplace stress for tech workforces is not uncommon. Ignorance of this may lead to mental problems like depression, anxiety, etc. The mental health records of these employees are with counselors. The digitalization era encourages counselors to store records digitally. Where privacy issues arise as, these records are freely accessible to a wide range of users including patients, psychiatrists, investigators, statisticians, and data scientists. Privacy-preserving data mining (PPDM) algorithms play a major character in the privacy preservation of records. This chapter provides a survey on mental health disorder and different PPDM techniques which are useful for mental health data.

38.1 INTRODUCTION

Mental health is an implicitly significant aspect for the class of life. It strongly touches one's performance and individual happiness. In 2015, World Health Organization (WHO) declared the sustainable development goals, which are "a universal call to action to end poverty, protect the planet and ensure that all people enjoy peace and prosperity" [1]. WHO has stated the importance of promoting mental health worldwide as well [2].

Workplace stress for tech workforces is not uncommon, but when kept unchecked it may lead to problems, such as depression and anxiety. Anxiety and depression are the widest categories seen, may be caused by office work. The most common types of mental illness are anxiety disorders, mood disorders (such as depression or bipolar disorder), personality disorders, and schizophrenia disorders.

The person suffering from mental health not only creates intolerable negative emotions, but also results in lower productivity, inability to concentrate on work and stressed relationships with colleagues. Some people do not get help for their mental illness as they are afraid of the consequences of letting their boss know about problem. However, many workplaces understand true cost of not dealing with issues like anxiety and depression, so such workplaces work to promote good mental health. The person suffering from mental health required good counselors/doctors so that they overcome the problem of mental health (anxiety, depression….).

From the several areas of health care, the last two decades have found several cases related to mental health [1]. Mental health does not always stay same all the way through human life. It may change with changes in circumstances and as move through different stages of life. There is a humiliation attached to mental health problems, means people feel uncomfortable and do not talk much about mental health problem. Many people do not even feel comfortable in communicating their feeling. But it is healthy to express and react on feelings.

The era of digitalization encourages counselors/doctors to store mental health patient data digitally. Health care is one of the fields where abundant data is generated and it is sensitive data. The digitalization of mental health data records as well as treatment notes have made individual mental health data records readily available to a wide range of users, such as patients, scholars, statisticians, psychiatrists, and data scientist's. Increased accessibility of sensitive mental health data records threatens the privacy and confidentiality.

The current era has seen data mining and machine learning (ML) techniques producing excellent outcomes for diverse applications using the data generated by applications. Data related to mental health is available with doctors or counselors but it is highly confidential data so doctors or counselors are not ready to share this type of data as it concerns about the privacy of the underlying patients. Hence, mental health care should be addressed using the upcoming variant of data mining known as privacy-preserving data mining (PPDM). In the nutshell, this chapter gives survey about mental health disorder and different PPDM techniques.

38.2   RELATED WORK

The survey is mainly divided into two parts: (a) related to mental health disorders; and (b) related to PPDM.

38.2.1   RELATED TO STUDY IN MENTAL HEALTH DISORDERS

Mental health does not always stay the same all the way through human life; it may change as circumstances change and as one move through different stages of your life.

The authors provided the survey of different data mining algorithm used for the prediction of mental health [1]. Data mining methods applied to diseases such as dementia, schizophrenia, depression, etc., can be of great help to the clinical decision, diagnosis prediction and improve the patient's quality of natural life.

Authors provided study of different cases and causes which are responsible for mental disorder [2]. Review was done to assess the capacity of mental disorders and study the various issued and challenges at community level. The WHO estimated that mental and behavioral disorders account for about 12% of global load of diseases but in India, the load is ranged from 9.5 to 102 per 1000 population.

Authors described about the necessity to treat mental health problem that prevail among children, which may lead to difficult problems if not cured at an early stage. Early diagnosis of mental health problem helps professionals to treat it, an earlier stage which improves the patient's life [3]. Research has identified that eight ML techniques (AODEsr, multi-layer perceptron (MLP), RBF Network, IB1, KStar, multi-class classifier (MCC), FT, LADTree) compared performance on various measures of accuracy in diagnosing five (Attention problem, academic problem, anxiety problem, attention deficit hyperactivity disorder (ADHD) and pervasive developmental disorder (PDD)) basic mental health problems. ML techniques used for analyzing medical data which helps in diagnosing the problem.

Authors proposed an expert system which is development by using Tools: Visual Studio 2005, NET Framework 2.0, IIS (Server), SQL Server 2000, Dreamweaver 8, Photoshop, SPSS16.0: used for data analysis (DA) and processing [4]. In this system, the undergraduates to be tested may participate in the physical fitness and mental health test by selecting a specific content. The system utilizes specific data processing program to process the data obtained from the physical fitness and mental health test, and carry out the analysis and interpretation of individual physical fitness and psychological characteristics obtained from the final data by consulting the system, psychological measurement handbook, and norm. The undergraduate physical fitness and mental health detection system possesses a great advantage.

The author designed College student's mental health assessment system based on network is a new way of modern psychological education, which makes college students communicate with each other [5]. It used Model, View, and Controller architecture technology to develop the design. The platform used MyEclipse development tools and SQL Server 2008 database (DB) to store data. Assessment system becomes the key to improve the level of education of college student's mental health.

The author introduced how to use the data mining tools IDA, based on supervised learning rules to mine the data sets on mental health of college students and help to discover useful information from three profiles: (i) how to deal with the interpersonal relationship; (ii) attitudes on life values; and (iii) how to evaluate university life [6]. This study found some common states and problem among college students. Investigated mental health of college students from different grade. The numbers of questionnaires are 484 in total with 150 questionnaires, 300 online questionnaires, and 50 on-the-spot interviews. The actual effective number of returned samples is 500, and the remaining 16 samples are invalid or valueless.

The purpose of this study is to assess work stress on mental health in male and female office [7]. This study was carried out in Government and Private Office workers in and around North Chennai. A study group consisted of 200 office workers in the clerical grade in the age group of 25–58 years of both genders. Mental health monitoring is very useful tool for early detection of mental disorder. It has been shown to be a good instrument for detecting problems of social dysfunction, psychosomatic problems, anxiety, and depression. So it can be considered more appropriate for use in primary care. For this, work the general health questionnaire (GHQ)-28 used as a tool to assess the work stress.

38.2.2   RELATED TO PRIVACY-PRESERVING DATA MINING (PPDM)

Privacy-preserving data mining is divided into two types:

- **Based on the Data Lifecycle Phase:** The data lifecycle phase on which the privacy preservation ensured is data collection, data publishing, data distribution, and the output of data mining.
- **Based on the Location of Computation:** In location-based, the classification can be broadly categorized as: central/commodity server and distributed. Figure 38.1shows different techniques and algorithm in PPDM.

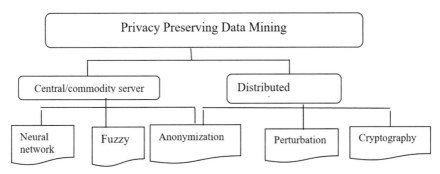

FIGURE 38.1   PPDM classification.

1. **k-Anonymity Method:** Anonymity is guaranteed by the existence of at least other k-1 undistinguishable records for each record in a DB [8]. This group of k undistinguishable records is referred to as equivalence class. It uses Generalization and suppression methods. The drawbacks of k-anonymity are: (1) Assume that each record represents a unique individual. If this is not the case, an equivalence class with k records does not necessarily link to k different individuals; (2) Sensitive attributes are not taken into consideration like… for the anonymization, which can disclose information, especially if all records in a class have the same value for the sensitive attribute.
2. **l-Diversity Method:** It expands the k-anonymity model by requiring every equivalence class to have at least 1 "well-represented" value for the sensitive attributes [9]. It uses Generalization and suppression methods. k-anonymity does not take into consideration the

distribution of the sensitive values which can lead to privacy breaches when the sensitive values are distributed in a skewed away.

3. **t-Closeness Method:** It solves the l-diversity problem of skewed sensitive values distribution by requiring that the distribution of the sensitive values in each equivalence class to be "close" to the corresponding distribution in the original table, where close means upper bounded by a threshold t (Li et al., 1985).

4. Personalized privacy achieved by creating a taxonomy tree using generalization, and by allowing the record owners to define a guarding node [10]. Owners' privacy is breached if an attacker is allowed to infer any sensitive value from the sub-tree of the guarding node with a probability (breach probability) greater than a certain threshold.

5. **Adding Noise Perturbation:** Data randomization done by adding noise with a known statistical distribution [11]. It preserves statistical properties after reconstruction of the original distribution. But the drawbacks of this method are: (a) It limits data utility to the use of aggregate distribution. (b) Masking extreme values require great quantity of noise, severely degrading data utility. (c) Noise reduction techniques can be used to accurately estimate the original individual values is more difficult.

6. **Randomization:** The randomization is a technique for PPDM, in which noise is added to the data in order to mask the attribute values of records [11]. Randomization gives advantages like: (a) offering the process of knowledge discovery along with the balance in between the utility and privacy. (b) Use of data distribution method in order to create private representation of records. (c) Not required knowledge of the distribution of other records in the data. (d) Cheapest and effective technique to secure privacy of each user.

7. **Resampling:** Comprise of drawing repeated samples from original from the original data samples [12].
Resampling generates a unique sampling distribution on the basis of the actual data. It does not involve the utilization of generic distribution tables in order to compute approximate probability values.

8. **Data Reduction:** Transformation of numerical or alphabetical digital information derived empirically or experimentally into a corrected, order, and simplified from data reduction can be applied to obtain the reduced representation of the datasets that is much smaller in volume [12].

9. **Global Recording:** An act of generalization or suppression of an attribute in original data set, through all the instances, to the same level in the respective generalization hierarchy of that attribute [8]. Global recording technique is more appropriate for categorical microdata, where it helps disguise records with strange combinations of categorical attributes. Global recording is used heavily by statistical offices.

10. **Micro Aggregation:** Protect individual records in such a way that data can be mined, distributed, and published without providing any personal information associate with specific individual [13] and [14]. In other words, instead of releasing the original values of the individual records, the system releases the mean of the group.

38.2.3   SOME AUTHORS WORKED ON PRIVACY-PRESERVING DATA MINING (PPDM) FOR MENTAL HEALTH

The authors highlighted major concerns in mental health research [15]. They developed a privacy-preserving DA approach which allows data sets to be analyzed while retaining the confidentiality of patient records.

Wang and Zhang [16] addressed the accuracy issues in PPDM through matrix-factorization. Work indicates that the matrix factorization-based data distortion scheme perturb only confidential attribute to meet privacy requirement indicates that the matrix factorization-based data distortion schemes perturb only confidential attributes to meet privacy requirements while preserving general data pattern for knowledge extraction (Table 38.1).

TABLE 38.1   Comparison of Privacy-Preserving Data Mining Methods

Paper	Method	Issue Addressed	Advantages	Open Issues
Li and Li [15]	proposed a privacy-preserving data analysis approach	inappropriate access, use, and disclosure of confidential information	Approach is designed to incorporate the following privacy elements: robust data, anonymization, data encryption, security technologies, and transparency.	big data
Wang and Zhang [16]	The matrix factorization based data distortion.	the accuracy issues	Matrix factorization-based approach provides possibility of simultaneously achieving privacy, accuracy, and efficiency.	Perturb only confidential attributes.

The digitalization era encourage counselors to store patient records digitally. Where a privacy issue arises as, these records are freely available to a wide range of users including psychiatrists, researchers, patients, and scholars, statisticians, and data scientists. To secure these records privacy and maintain records availability to authenticated users, PPDM algorithms play a major role, for privacy preservation of records.

38.3 CONCLUSION

Paper affords a survey on mental health disorders in children, college students, and adults. Mental health does not always stay the same all the way through human life. It can change as circumstances change and as one move through different stages of your life. WHO estimated that mental and behavioral disorders account for about 12% of the global burden of diseases. In India, the burden of mental and behavioral disorders ranged from 9.5 to 1.02 per 1000 population [2]. The second part of the chapter is to study about different PPDM techniques. PPDM techniques are useful to maintain the privacy of mental health data.

KEYWORDS

- **attention deficit hyperactivity disorder**
- **general health questionnaire**
- **multi-class classifier**
- **multi-layer perceptron**
- **pervasive developmental disorder**
- **privacy-preserving data mining**

REFERENCES

1. Alonso, S., De La Torre, D. I., Hamrioui, S., Lopez-Coronado, M., Calvo, B. D., Morón, L., & Franco, M., (2018). Data mining algorithms and techniques in mental health: A systematic review. *Journal of Medical Systems* (Vol. 42, pp. 1–15). Springer.
2. Reddy, B. V., Gupta, A., Lohiya, A., & Kharya, P., (2013). Mental health issues and challenges in India: A review. *International Journal of Scientific and Research Publications, 3*, 1–3.
3. Sumathi, M. R., & Poorna, B., (2016). Prediction of mental health problems among children using machine learning techniques. *International Journal of Advanced Computer Science and Applications, 7*, 552–557.

4. Haitao, H., & Xinyan, D., (2010). Development and application of undergraduate physical fitness and mental health detection system. *2010 2ⁿᵈ International Conference on Computer Modeling and Simulation*, pp. 486–488.

5. Yanhua, S., (2015). College student's mental health assessment system design based on J2EE. *2015 International Conference on Intelligent Transportation* (pp. 216–219). Big Data and Smart City, Halong Bay.

6. Yuan, C., (2014). Data mining techniques with its application to the dataset of mental health of college students. *2014 IEEE Workshop on Advanced Research and Technology in Industry Applications (WARTIA)* (pp. 391–393). Ottawa.

7. Thenmozhi, R., Kabali, B., & Kapali, S. C., (2014). Effect of administrative stress on mental health in a group of office workers-across sectional study. *2014 International Conference on Science Engineering and Management Research (ICSEMR)* (pp. 1–3). Chennai.

8. Samarati, P., & Sweeney, L., (1998). Generalizing data to provide anonymity when disclosing information. *PODS '98 Proceedings of the 17ᵗʰ ACM SIGACT-SIGMOD-SIGART Symposium on Principles of Database Systems* (pp. 809–812).

9. Machanavajjhala, A., Kifer, D., Gehrke, J., & Venkita, S. M., (2007). ℓ-diversity: Privacy beyond k-anonymity. *ACM Transactions on Knowledge Discovery from Data, 1*, 809–812.

10. Xiao, X., & Tao, Y., (2006). Personalized privacy preservation. In: *Proceedings of ACM Conference on Management of Data (SIGMOD'06)* (pp. 229–240).

11. Agrawal, R., & Srikant, R., (2000). Privacy-preserving data mining. *ACM SIGMOD International Conference on Management of Data, 29*, 439–450.

12. Liew, C. K., Choi, U. J., & Liew, C. J., (1985). A data distortion by probability distribution. *ACM Transactions on Database Systems, 10*, 395–411.

13. Li, X. B., & Sarkar, S., (2006). A tree-based data perturbation approach for privacy-preserving data mining. *IEEE Transactions on Knowledge and Data Engineering, 18*, 1278–1283.

14. Jane, V. S. V., & Arumugam, G., (2018). A survey on micro aggregation based privacy preserving data mining techniques. *International Journal of Scientific Research Engineering and Technology, 7*, 268–279.

15. Li, J., & Li, X., (2015). Privacy preserving data analysis in mental health research. *2015IEEE International Congress on Big Data* (pp. 95–101). New York, NY.

16. Wang, J., & Zhang, J., (2007). Addressing accuracy issues in privacy preserving data mining through matrix factorization. *2007 IEEE Intelligence and Security Informatics* (pp. 217–220). New Brunswick, NJ.

17. https://www1.undp.org/content/seoul_policy_center/en/home/sustainable-development-goals.html (accessed 21 July 2020).

18. http://www.who.int/mental_health/SDGs/en/ (accessed 21 July 2020).

19. Feige, E. L., & Watts, H. W., (1970). In: Bisco, R. L., (ed.), *Protection of Privacy Through Micro Aggregation in Databases, Computers, and the Social Sciences* (pp. 261–272). Wiley-Inter-science, New York.

20. Li, N., Li, T., & Venkatasubramanian, S., (2007). T-closeness: Privacy beyond k-anonymity and l-diversity. *2007IEEE 23ʳᵈ International Conference on Data Engineering* (pp. 106–115) Istanbul.

21. Shah, A., & Gulati, R., (2016). Privacy preserving data mining: Techniques, classification, and implications-a survey. *International Journal of Computer Applications, 137*, 40–46.

CHAPTER 39

Industry 4.0 with Big Data Analysis: A Review

K. MURALI GOPAL

Department of Computer Science and Engineering, School of Engineering, GIET University, Gunupur – 765022, Odisha, India, E-mail: kmgopal@giet.edu

ABSTRACT

The data produced through the industrial application of the internet of everything (IIoE) is manifest due to the enormous utilization of internet of things (IoT) devices and sensors. But the handling of generated data in the IoT devices is limited due to its limited resources like Storage, networks, and computation. By performing big data analysis (BDA) on the data produced by the IIoE, customer-level, employee-level, and operational intelligence can be enhanced. Many studies have been performed on the above paradigm; here the advanced BDA algorithm and methods are investigated that improve the smart IIoE system. Here we formulate a categorizing and classifying the works on the source of import constraints (e.g., data of sources, analytic techniques and implements, requirements, industrial analytics, analytics types, and its applications). We also discuss future research areas and challenges.

39.1 INTRODUCTION

Industry internet of everything (IIoE) is an industry which uses actuators and smart sensors for industrial process and boosts their manufacturing at the same time use sensor to control the human resources and find the behavior of the employee. Cyber-physical systems (CPS) [1], cloud computing (CC), internet of things (IoT) [2–4], Internet of services [6], automation [5], wireless technologies, concentric computing [8], and augmented reality [7] are few names supporting the ecosystem of IIoT. To deliver accuracy, efficiency, and extraordinary flexibility to the manufacturing industry-IoT, big data analytics (BDA), CC and CPS are used with IIoT activities [9, 10]. By Cross-platform combination, IIoT systems essential to safeguard real-time capability, interoperability, virtualization, decentralization, service orientation, modularity, and security all domains [11].

IIoT systems are supposed to have potentials, such as self-prediction, self-awareness, self-comparison, self-maintenance, self-configuration, and self-organization [12]. Data analysis (DA) is a process of gathering, managing, computing, analyzing, and visualizing sequential growing data in terms of 5Vs (velocity, volume, value, veracity, and variety) [13]. In the IIoT system, big data rise to unrestricted inside and outside events applicable to the employee, clients, production, business operation, and machines [14]. Across IIoT system different storage are provided like on-board, in-networks, in-memory, and huge distributed storage facilities are collected and proceed by BDA [15, 16]. In the IIoT System, the data processing facilities vary from resourceful distributed CC to resource-limited IoT Devices [17]. Analytic operations vary in terms of prescriptive, descriptive, preventive, and predictive procedures [14]. Real-time knowledge visualization over multiple IIoT systems must ensure by BDA processes. The IIoT system should be supported by BDA processes to maximize value construction to progress in business models for maximization in profit [14, 18].

IIoT and BDA have been studied separately. Industrial networks produce a large amount of data due to the large-scale utilization of sensing devices and systems. This study provides details of the IIoT and BDA basis of key operations of BDA. In this study, it builds a BDA for IIoT systems classification and categorization. The study provides a hypothetic to recognize modern automated data processes for elevating intelligence in IIoT systems.

39.2   IIOE AND BDA

It provides a detailed study of various features of big data implementation in IIoT environment. Design principles to be used before designing and installing of IIoT systems, are emphasized. BDA life cycle is discussed for suppling intelligence in IIoT systems (Figure 39.1).

In a real-world IIoT organization of smart robotics, a power and robotics firm is connected sensors to observer the maintenance needs of robots to speedy repairs before parts break down.

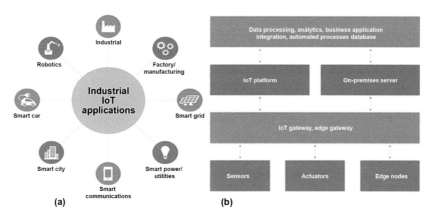

FIGURE 39.1   (a) Industrial application; (b) IIoT infrastructure.

For example, Robotics manufacturer (Fanuc) is implemented sensors within its robotics, with cloud-based data analytics, to predict forthcoming failure of components in its robots. By that, the plant manager can schedule maintenance at suitable times, reducing costs, and averting potential downtime. Figure 39.1(a) shows a few applications of IIoT. Where Figure 39.1(b) gives the IIoT infrastructure to collect information.

Seven principles should be involved in the design of IIoT systems as depicted [40]. Firstly, compatibility must be ensured amongst various technologies and communication among the technologies should be realized. Secondly, virtualization techniques must be measured for efficient service distribution through IIoT systems. Virtualization is of different types based on its data, networks, operating system (OS), platform, and applications. Thirdly, distributed storage and processing of data should be implemented for IIoT system. Fourthly, real-time feedback must be provided to the entire stakeholder. Fifthly, Service-oriented architecture must be implemented in all systems. Sixthly, system implementation should be a modular approach. Finally, security must be a concern. The above principle must be considered for designing BDA processes in IIoT systems (Figure 39.2).

Integration of processing units with onboard computations and networking facilities is referred to as CPS [22, 23]. IIoT systems work with the support of IoT devices and CPS and the massive amount of data generate [24]. Analysis of these data leads to defect-free product manufacturing and improve machine health [1, 20, 25]. IoT devices remotely sense and actuate in the IIoT system, they can be stand-alone or CPS to perform predefined actions. CPS and IoT devices produce massive data for the cloud service to perform BDA processes [27]. In-service selection, real-time service provisioning, and service orchestration can be facilitated by BDA if the number of cloud services can grow vastly [28].

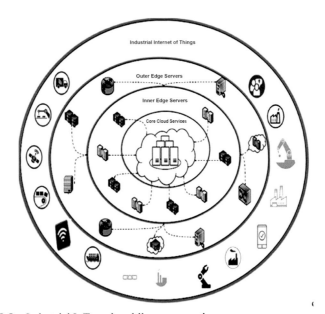

FIGURE 39.2   Industrial IoTs and multilayer computing resources.
(*Source*: Reprinted with permission from Ref. [40]. © 2019 Elsevier.)

New advancement in computing and sensing technologies has giving opportunities for big data processing. A wide variety of computing facilities and devices can able to perform processing at the sensor level, endpoint in IIoT level, edge server level as well as decentralized and centralized CC level as shown in Figure 39.2 [8, 14, 21, 29]. The IoT device and sensor has limited computational power and small in size but by using onboard smart data reduction strategies capable to reduce data streams [30]. For the load distributed of big data application, centralized computing and edge servers are helpful [31, 32]. Automating, multistage execution, and organization of BDA processes are essential in concentric computing situations [33].

Multistage interdependent application and device component (Figure 39.3) is executed the BDA process. Categories of components are as follows:

Data engineers build storage infrastructure and computing to consume, clean, adapt, shape, and transform data. IIoT environment generates and consumes data from outbound customer activities and inbound enterprise operations. To improve the worth of row data at the initial stage need extra processing as per relevance to IIoT systems. Data cleaning and wrangling methodologies select relevant datasets from the streaming data. To produce the correctness and relevance of big data, data conformity is used. Improve the data quality in terms of dropping the number of attributes and uniform data processing data transformation and data shaping is implemented.

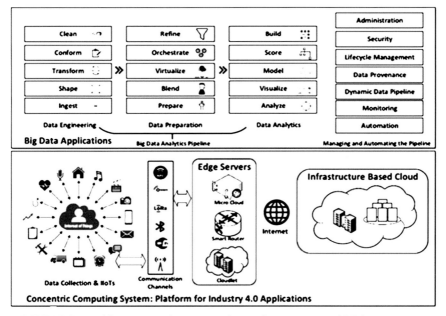

FIGURE 39.3   Multistage execution, automation, and management of BDA processes. (*Source*: Reprinted with permission from Ref. [40]. © 2019 Elsevier.)

Data preparation events take 70% to 80% of the time as big data is a collection of the large dataset so statistical methods are used to refine, handle unbalanced, unstructured data efficiently and to reduce the complexity of the data. For reducing the network traffic and latency, data locality is needed. Anomalies and outliers need to be detected for further analysis.

The training dataset is prepared by the data scientists and the models are trained and tested in a supervised environment. Then the model is deployed to find the information patterns from the industrial environment.

39.3   CATEGORIZATION AND CLASSIFICATION (FIGURE 39.4)

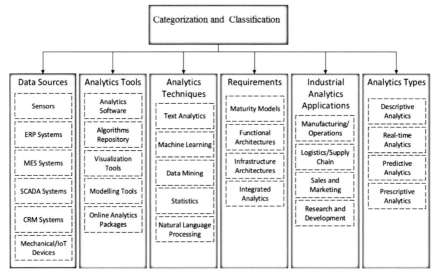

FIGURE 39.4   Classification and categorize of BDA processes [40].
(*Source*: Reprinted with permission from Ref. [40]. © 2019 Elsevier.)

39.4   ALGORITHM AND TECHNOLOGIES IN BDA FOR INDUSTRY 4.0

In this section review of the industry 4.0 with respect to the smart factory system and IoT systems with BDA is considered with respect to security, processing and automation.

Maximize production is the main objective of any industrial automation but customization in accord with customer requirement is lacking through predictive analytics. Optimization of big data through a self-organizing map (SOM) leads to mass adapted product manufacturing [34]. The design process can be improved by using the clustering-based big data optimization use k-means clustering algorithm [35].

It is a challenge to achieve zero-defect in SFS. Zero defect means certify high-quality product development during the implementation of manufacturing processes. Time series model ensures monitoring during operation. The neo-fuzzy neuro time series model method is used in the IIoT system.

Through radio-frequency identification, object tracking is carried out for smart manufacturing through wireless and networking technology [26]. The products are tracked during production and ensure that analytics implementation delivers intelligent shop monitoring. The exception diagnosis and performance analysis have been suggested and verified using Petri nets and decision tree algorithm [26]. Identifies and forecast of machine faults during machine operations and facilitate machine data gathering and organization can be done using prognostic health monitoring (PHM) [36–38].

Predictive and defensive maintenance are the key necessities of large-scale IIoT systems. BDA helps in performing prediction on historical data and online upkeep of the machinery. Integration of Hadoop, storm technology, neural network methods makes prediction more efficient [39]. Adopting big DA for intelligent predictive maintenance is unique.

39.5   CONCLUSION

The industry 4.0 aims to connect manufacturing systems to its product sellers. This chapter deliberates grow of BDA in industry 4.0 and a study of associated algorithms and technologies. A study on classifying basis of the key concepts is presented. It is evident that BDA in industry 4.0 is in its initial stage. Future research can be conducted on CPS, augmented reality, and IoT devices for improving on security, network, and data analytics capabilities.

39.6   FUTURE SCOPE

The first industrial revolution replaces the manual method by tools in the early 19th century leads to mass production. In the 21st century, the fourth industrial revolution-Industry 4.0 join with digital technologies and supply chain processes leads to optimize ad automated production. Industry 4.0 implements and connects sensors with machines and components to generate real-time data for the analysis. By applying artificial intelligence (AI) and machine learning (ML) algorithms on the generated big data and produce valuable information like the health of the machines, working environment of the employee behavior and consumer behavior. Based on the information the corrective action can be planed for optimized production.

Lack of technologies, valuable data generated by IIoT system is lost. So collecting and generating data is no more concern, efficient extraction of value from it is a concern.

Few challenges in industry 4.0 are as follows:

- Threshold specification for machine design and/or product data;

- Control systems on machine operational data;
- Quality of product and process data;
- Manual operations records for staff;
- Manufacturing performance systems;
- Manufacturing and operational costs information;
- system-monitoring deployments and fault-detection;
- Logistics information;
- Product usage, feedback of the customer information.

Industry 4.0 also open a challenge in the areas like cybersecurity of the sensor networks;privacy of the data generated by the IIoT system; processing of the Big data in CPS or cloud;connectivity, and communication of sensors and components with the machinery and environment; and system management and efficiency of production and processes [40].

Industry 4.0 is a wide-open for big DA but the implementation of cutting edge technologies like CC and fog computing is questionable. The changing market requires enhanced quality with a faster response. The across multiple sectors consumer require a customized product with best in quality and faster response, so industry 4.0 need classification on gathering, processing, and produce data from all the sources and real-time analysis by automated rules and adaptive ML.

KEYWORDS

- **big data analysis**
- **cyber-physical systems**
- **industry 4.0**
- **internet of thing**
- **prognostic health monitoring**
- **self-organizing map**

REFERENCES

1. Ilge, A., & Edward, A. L., (2016). Systems engineering for industrial cyber-physical systems using aspects. *Proceedings of the IEEE, 104*, 997–1012.
2. Martin, W., Thilo, S., & Juergen, J., (2017). The future of industrial communication: Automation networks in the era of the internet of things and industry 4.0. *IEEE Industrial Electronics Magazine, 11*(1), 17–27.

3. Charith, P., Susan, Y. L. W., Tim, B., Hamed, H., Arosha, K. B., Richard, M., Andy, C., et al., (2017). Valorizing the IoT data box: Creating value for everyone. *Transactions on Emerging Telecommunications Technologies, 28*(1), 1–17.

4. Charith, P., Chi, H. L., Srimal, J., & Min, C., (2014). A survey on internet of things from industrial market perspective. *IEEE Access, 2*, 1660–1679.

5. Mikkel, R. P., Lazaros, N., Rasmus, S. A., Casper, S., Simon, B., Volker, K., & Ole, M., (2016). Robot skills for manufacturing: From concept to industrial deployment. *Robotics and Computer-Integrated Manufacturing, 37*, 282–291.

6. Ling, L., (2016). Services computing: From cloud services, mobile services to internet of Services. *IEEE Transactions on Services Computing, 9*(5), 661–663.

7. Sunwook, K., Maury, A. N., & Joseph, L. G., (2016). Augmented reality 'smart glasses' in the workplace: Industry perspectives and challenges for worker safety and health. *IIE Trans. Occup. Ergon. Hum. Factors, 4*(4), 253–258.

8. Philip, H., (2017). *Concentric Analytics for IoT.*

9. Ibrar, Y., Ejaz, A., Ibrahim, A. T. H., Abdelmuttlib, Ibrahim, A. A., Abdullah, G., Muhammad, I., & Mohsen, G., (2017). Internet of things architecture: Recent advances, taxonomy, requirements, and open challenges. *IEEE Wirel. Commun., 24*(3), 10–16.

10. Ejaz, A., Ibrar, Y., Abdullah, G., Muhammad, I., & Mohsen, G., (2016). Internet-of-things based smart environments: State of the art, taxonomy, and open research challenges. *IEEE Wirel. Commun., 23*(5), 10–16.

11. Wang, K., (2016). Intelligent predictive maintenance (IPDM) system-industry 4.0 scenario. *WIT Trans. Eng. Sci., 113*(1), 259–268.

12. Jay, L., (2015). Smart factory systems. *Inform.-Spektrum, 38*(3), 230–235.

13. Mohsen, M., Fariza, N., Abdullah, G., Ahmad, K., Ibrahim, A. T. H., Aisha, S., & Ibrar, Y., (2017). Big IoT data analytics: Architecture, opportunities, and open research challenges. *IEEE Access*, 5247–5261.

14. Rehman, M. H., Batool, A., et al., (2016). Big data analytics in mobile and cloud computing environments. In: *Innovative Research and Applications in Next-Generation High Performance Computing* (pp. 349–367). IGI Global.

15. Geng, K., & Liu, L., (2016). Research of construction and application of cloud storage in the environment of industry 4.0. *International Conference on Industrial IoT Technologies and Applications* (pp. 104–113). Springer.

16. Gaber, M. M., Aneiba, A., Basurra, S., Batty, O., Elmisery, A. M., Kovalchuk, Y., & Rehman, M. H., (2019). *Internet of Things and Data Mining: From Applications to Techniques and Systems* (p. 1292). Wiley Interdiscip. Rev: Data Min. Knowledge Discov.

17. Rehman, M. H., Sun, L. C., Wah, T. Y., & Khan, M. K., (2016). Towards next-generation heterogeneous mobile data stream mining applications: Opportunities, challenges, and future research directions. *J. Netw. Comput. Appl.*

18. Ehret, M., & Wirtz, J., (2017). Unlocking value from machines: Business models and the industrial internet of things. *J. Mark. Manage, 33*(1/2), 111–130.

19. Li, B., Kisacikoglu, M. C., Liu, C., Singh, N., & Erol-Kantarci, M., (2017). Big data analytics for electric vehicle integration in green smart cities. *IEEE Commun. Mag., 55*(11), 19–25.

20. Dimitrios, G., & Prem, P. J., (2016). Internet of things: From internet scale sensing, to smart services. *Computing, 98*(10), 1041–1058.

21. In Lee, & Kyoochun, L., (2015). The internet of things (IoT): Applications, investments, and challenges for enterprises. *Bus. Horiz., 58*(4), 431–440.

22. Wang, L., Torngren, M., & Onori, M., (2015). Current status and advancement of cyber-physical systems in manufacturing. *J. Manuf. Syst., 37*(Part 2), 517–527.

23. Zhou, K., Liu, T., & Liang, L., (2016). From cyber-physical systems to industry 4.0: Make future manufacturing become possible. *Int. J. Manuf. Res., 11*(2), 167–188.

24. Lee, J., Ardakani, H. D., Yang, S., & Bagheri, B., (2015). Industrial big data analytics and cyber-physical systems for future maintenance and service innovation. *Procedia CIRP, 38*, 3–7.

25. Cheng, F. T., Tieng, H., Yang, H. C., Hung, M. H., Lin, Y. C., Wei, C. F., & Shieh, Z. Y., (2016). Industry 4.1 for wheel machining automation. *IEEE Robot. Automat. Lett., 1*(1), 332–339.

26. Zhong, R. Y., Xu, C., Chen, C., & Huang, G. Q., (2017). Big data analytics for physical internet-based intelligent manufacturing shop floors. *Int. J. Prod. Res., 55*(9), 2610–2621.

27. Tao, F., Cheng, Y., Da, X. L., Zhang, L., & Li, B. H., (2014). CCIoT-CMfg: Cloud computing and internet of things-based cloud manufacturing service system. *IEEE Trans. Ind. Inf., 10*(2), 1435–1442.

28. Hossain, M., & Muhammad, G., (2016). Cloud-assisted industrial internet of things (IIoT)-enabled framework for health monitoring. *Comput. Netw., 101*, 192–202.

29. Harshit, G., Amir, V. D., Soumya, K. G., & Rajkumar, B., (2017). iFogSim: A toolkit for modeling and simulation of resource management techniques in the internet of things, edge, and fog computing environments. *Software: Practice and Experience, 47*(9), 1275–1296.

30. Rehman, M. H., Jayaraman, P. P., & Perera, C., (2017). The emergence of edge-centric distributed IoT analytics platforms. In: *Internet of Things* (pp. 213–228). Chapman and Hall/CRC.

31. Lin, K., Xia, F., & Fortino, G., (2019). Data-driven clustering for multimedia communication in internet of vehicles. *Future Gener. Comput. Syst., 94*, 610–619.

32. Pace, P., Aloi, G., Gravina, R., Caliciuri, G., Fortino, G., & Liotta, A., (2019). An edge-based architecture to support efficient applications for healthcare industry 4.0. *IEEE Trans. Ind. Inf., 15*(1), 481–489.

33. Hassan, N., Gillani, S., Ahmed, E., Yaqoob, I., & Imran, M., (2018). The role of edge computing in internet of things. *IEEE Commun. Mag.,* 1–6.

34. Saldivar, A. A. F., Goh, C., Chen, W. N., & Li, Y., (2016). Self-organizing tool for smart design with predictive customer needs and wants to realize industry 4.0. *Evolutionary Computation (CEC), 2016 IEEE Congress on, IEEE* (pp. 5317–5324).

35. Saldivar, A. A. F., Goh, C., Li, Y., Chen, Y., & Yu, H., (2016). Identifying smart design attributes for industry 4.0 customization using a clustering genetic algorithm. *Automation and Computing (ICAC), 22nd International Conference on, IEEE, 2016* (pp. 408–414).

36. Fleischmann, H., Kohl, J., Franke, J., Reidt, A., Duchon, M., & Krcmar, H., (2016). Improving maintenance processes with distributed monitoring systems. *Industrial Informatics (INDIN), 2016 IEEE 14th International Conference on, IEEE* (pp. 377–382).

37. Nunez, D. L., & Milton, B., (2017). An ontology-based model for prognostics and health management of machines. *Journal of Industrial Information Integration, 6*(2017), 33–46.

38. Nabati, E. G., & Thoben, K. D., (2017). Data-driven decision making in planning the maintenance activities of off-shore wind energy. *Procedia CIRP, 59*, 160–165.

39. Jiafu, W., Shenglong, T., Di, L., Shiyong, W., Chengliang, L., Haider, A., & Athanasios, V. V., (2017). A manufacturing big data solution for active preventive maintenance. *IEEE Transactions on Industrial Informatics, 13*(4), 2039–2047.

40. Muhammad, H. R., Ibrar, Y., Khaled, S., Muhammad, I., Prem, P. J., & Charith, P., (2019). The role of big data analytics in industrial internet of things. *Future Generation Computer Systems, 99*, 247–259.

Index

Z